Analysis of Protein Post-Translational
Modifications by Mass Spectrometry

WILEY SERIES ON MASS SPECTROMETRY

Analysis of Protein Post-Translational Modifications by Mass Spectrometry

Edited by John R. Griffiths and Richard D. Unwin

Published by John Wiley & Sons, Inc., Hoboken, New Jersey

Published simultaneously in Canada

For general information on our other products and services or for technical support, please contact our Customer Care Department within the United States at (800) 762-2974, outside the United States at (317) 572-3993 or fax (317) 572-4002.

Wiley also publishes its books in a variety of electronic formats. Some content that appears in print may not be available in electronic formats. For more information about Wiley products, visit our web site at www.wiley.com.

Library of Congress Cataloging-in-Publication Data:

Names: Griffiths, John R., 1964- editor. | Unwin, Richard D., editor.
Title: Analysis of Protein Post-Translational Modifications by Mass Spectrometry / edited by John R. Griffiths, Richard D. Unwin.
Description: Hoboken, New Jersey : John Wiley & Sons, 2016. | Includes bibliographical references and index.
Identifiers: LCCN 2016024928 | ISBN 9781119045854 (cloth) | ISBN 9781119250890 (epub)
Subjects: LCSH: Post-translational modification. | Mass spectrometry.
Classification: LCC QH450.6.A53 2016 | DDC 572/.645--dc23
LC record available at https://lccn.loc.gov/2016024928

Set in 10/12pt Warnock by SPi Global, Chennai, India

10 9 8 7 6 5 4 3 2 1

Contents

List of Contributors

Roland S. Annan
Proteomics and Biological Mass
Spectrometry Laboratory
GlaxoSmithKline
Collegeville, PA
USA

Perdita E. Barran
Manchester Institute of
Biotechnology
The University of Manchester
Manchester
UK

Navin Chicooree
Cancer Research UK Manchester
Institute
The University of Manchester
Manchester, UK

Helen J. Cooper
School of Biosciences
University of Birmingham
Birmingham, UK

Andrew J. Creese
School of Biosciences
University of Birmingham
Birmingham, UK

Dominic M. Desiderio
The Charles B. Stout Neuroscience
Mass Spectrometry Laboratory
Department of Neurology
University of Tennessee Health
Science Center
Memphis, TN
USA

Sian Estdale
Covance Laboratories
Harrogate
UK

Caroline A. Evans
Department of Chemical and
Biological Engineering
University of Sheffield
Sheffield, UK

David Firth
Covance Laboratories Ltd.
Harrogate, UK

Florian Gnad
Proteomics and Biological Resources
Genentech Inc
South San Francisco, CA
USA

John Griffiths
Cancer Research UK Manchester
Institute
The University of Manchester
Manchester
UK

David J. Harvey
Department of Biochemistry
University of Oxford
Oxford
UK

Éva Hunyadi-Gulyás
Institute of Biochemistry
Biological Research Centre of the
Hungarian Academy of Sciences
Szeged
Hungary

Éva Klement
Institute of Biochemistry
Biological Research Centre of the
Hungarian Academy of Sciences
Szeged
Hungary

Jennie R Lill
Proteomics and Biological Resources
Genentech Inc
South San Francisco, CA
USA

Ying Long
Key Laboratory of Cancer
Proteomics of Chinese Ministry of
Health, Xiangya Hospital
Central South University
Changsha, Hunan
P. R. China

Dean E. McNulty
Proteomics and Biological Mass
Spectrometry Laboratory
GlaxoSmithKline
Collegeville, PA
USA

Katalin F. Medzihradszky
Department of Pharmaceutical
Chemistry
University of California San
Francisco
San Francisco, CA
USA

Kamila J. Pacholarz
Manchester Institute of
Biotechnology
The University of Manchester
Manchester
UK

Rebecca Pferdehirt
Proteomics and Biological Resources
Genentech Inc
South San Francisco, CA
USA

Naila Rabbani
Warwick Systems Biology Centre
University of Warwick
Coventry
UK

Timothy W. Sikorski
Proteomics and Biological Mass
Spectrometry Laboratory
GlaxoSmithKline
Collegeville, PA
USA

Duncan L. Smith
Cancer Research UK Manchester
Institute
The University of Manchester
Manchester
UK

Paul J. Thornalley
Warwick Medical School, Clinical
Sciences Research Laboratories
University of Warwick
Coventry
UK

Richard Unwin
Centre for Advanced Discovery and
Experimental Therapeutics (CADET)
Central Manchester University
Hospitals NHS Foundation Trust
Manchester
UK

Rosie Upton
Manchester Institute of
Biotechnology
The University of Manchester
Manchester
UK

Xianquan Zhan
Key Laboratory of Cancer
Proteomics of Chinese Ministry of
Health, Xiangya Hospital
Central South University
Changsha, Hunan
P. R. China

Preface

While preparing a recent review article in *Mass Spectrometry Reviews* on the analysis of post-translational modifications (PTMs) by mass spectrometry, we realized that, although there is much excellent work and many new tools being developed in this area, the field was lacking a coherent resource where these advances could be easily and readily accessed both by experts and those wishing to begin such studies. We subsequently decided that there was a need for a more comprehensive description of some of these modifications, and their analysis by mass spectrometry, in the form of a textbook. Since a detailed discussion of multiple modifications was required, it rapidly became apparent that this would require the support of experts in their own specialized fields. We are, therefore, grateful that a number of mass spectrometrists from around the world whom we, and others involved in proteomics, consider to be experts in the analysis of specific PTMs, agreed to contribute to this effort.

The aim of the book is to provide the reader with an understanding of the importance of the protein modifications under discussion in a biological context, and to yield insights into the analytical strategies, both in terms of sample preparation, chemistry, and analytical considerations required for the mass spectrometric determination of the presence, location, and function of selected important PTMs.

The scene is ably set with a concise introduction to the general strategies employed in PTM analysis by mass spectrometry, covering some of the key technologies which are referred to in more detail in subsequent chapters. Of course, well-known and more thoroughly investigated modifications such as phosphorylation, glycosylation, and acetylation are described in this work in great detail. However, other PTMs are garnering interest within the field and play major roles in protein function both in normal cellular regulation and in the disease setting. These PTMs are generally less well studied to date, and include, for example, tyrosine sulfation, glycation, nitration, and citrullination – the conversion of arginine to citrulline. The analysis of ubiquitination and SUMOylation, both of which involve the addition of a second, small protein to the target in a complex regulation of protein localization, activity, and stability completes the array of modifications included in this book. In addition, the book rounds off with a description of one of the current "hot topics" in mass spectrometry: that of top-down studies of intact protein structure and modification, using the example of the characterization of monoclonal antibodies.

As editors, it has been our joint pleasure and privilege to have been given the opportunity to read at first hand these works and to compile them into a book of which we are very proud. On behalf of both of us we would like to express our sincere thanks and appreciation for the hard work and generosity given by all of the contributors.

Finally, to you the reader, we hope that you are able to use this book in your research, either as a reference book to dip into from time to time, to introduce you to new methodologies or new ideas to help support your work, or as a means of gaining a greater understanding of the analysis of PTMs by mass spectrometry from some expert scientists.

April 2016

John R. Griffiths
Richard D. Unwin
Manchester, UK

1

Introduction

Rebecca Pferdehirt, Florian Gnad and Jennie R. Lill

Proteomics and Biological Resources, Genentech Inc., South San Francisco, CA, USA

1.1 Post-translational Modification of Proteins

While the human proteome is encoded by approximately 20,000 genes [1, 2], the functional diversity of the proteome is orders of magnitude larger because of added complexities such as genomic recombination, alternative transcript splicing, or post-translational modifications (PTMs) [3, 4]. PTMs include the proteolytic processing of a protein or the covalent attachment of a chemical or proteinaceous moiety to a protein allowing greater structural and regulatory diversity. Importantly, PTMs allow for rapid modification of a protein in response to a stimulus, resulting in functional flexibility on a timescale that traditional transcription and translation responses could never accommodate. PTMs range from global modifications such as phosphorylation, methylation, ubiquitination, and glycosylation, which are found in all eukaryotic species in all organs, to more specific modifications such as crotonylation (thought to be spermatozoa specific) and hypusinylation (specific for EIF5a), which govern more tight regulation of associated proteins. Taken together, over 200 different types of PTMs have been described [5], resulting in an incredibly complex repertoire of modified proteins throughout the cell.

The addition and subtraction of PTMs are controlled by tight enzymatic regulation. For example, many proteins are covalently modified by the addition of a phosphate group onto tyrosine, serine, or threonine residues in a process called phosphorylation [6]. Phosphorylation is catalyzed by a diverse class of enzymes called kinases [7], whereas these phosphomoieties are removed by a second class of enzymes referred to as phosphatases. The tight regulation of kinases and phosphatases often creates "on/off" switches essential for regulation of sensitive signaling cascades. There are some exceptions to this rule however, and the hunt is still underway for the ever-elusive hypusine [8]

Analysis of Protein Post-Translational Modifications by Mass Spectrometry,
First Edition. Edited by John R. Griffiths and Richard D. Unwin.
© 2017 John Wiley & Sons, Inc. Published 2017 by John Wiley & Sons, Inc.

removing enzyme or putative enzymes responsible for the removal of protein arginine methylation. However, it is also possible that proteins bearing these PTMs are modulated or removed from the cell by other mechanisms of action. For example, proteolysis is rarely (if ever) reversible, and many proteins (e.g., blood clotting factors and digestive enzymes) are tightly governed by irreversible cleavage events where the active form is created after proteolysis of a proenzyme.

While PTMs such as phosphorylation and lysine acetylation exist in a binary "on/off" state, many other PTMs exhibit much more complex possible modification patterns. For example, lysine residues can be modified by covalent attachment of the small protein ubiquitin, either by addition of a single ubiquitin or by addition of ubiquitin polymers. In the latter case ubiquitin itself is used as the point of attachment for addition of subsequent ubiquitin monomers [9]. To add another layer of complexity, ubiquitin has seven lysines (K6, K11, K27, K29, K33, K48, and K63), each of which may be used as the point of polyubiquitin chain linkage, and each of which has a different functional consequence. For example, K63-linked chains are associated with lysosomal targeting, whereas K48-linked chains trigger substrate degradation by the proteasome. Thus, even within one type of PTM, multiple subtypes exist, further expanding the functional possibilities of protein modification.

In addition, many proteins are modified on multiple residues by different types of PTMs. A classic example is the PTM of histones. Histones are nuclear proteins that package and compact eukaryotic DNA into structural units called nucleosomes, which are the basic building blocks of chromatin and essential for regulation of gene expression. The C-termini of histones are composed of unstructured tails that protrude from nucleosomes and are heavily modified by methylation, acetylation, ubiquitylation, phosphorylation, SUMOylation, and other PTMs [10]. Overall, 26 modified residues on a single-core histone have been identified, and many of these residues can harbor multiple PTM types. In a generally accepted theory referred to as the "histone code," the combination of PTMs on all histones comprising a single nucleosome or group of nucleosomes regulates fine-tuned expression of nearby genes.

As we begin to uncover the modified proteome, the importance of the interplay between multiple different PTMs has become increasingly apparent. One classic example is the involvement of both protein phosphorylation and ubiquitylation in the regulation of signaling networks [11]. Protein phosphorylation commonly promotes subsequent ubiquitylation, and the activities of ubiquitin ligases are also frequently regulated through phosphorylation. In a recent study by Ordureau et al., quantitative proteomic studies were employed to describe the PINK1 kinase–PARKIN UB ligase pathway and its disruption in Parkinson's disease [12]. The authors describe a feedforward mechanism where phosphorylation of PARKIN by PINK1 occurs upon mitochondrial damage, leading to ubiquitylation of mitochondria and mitochondrial proteins by PARKIN. These newly formed ubiquitin chains are then themselves phosphorylated by PINK1, which promotes association of phosphorylated PARKIN with polyubiquitin chains on the mitochondria, and ultimately results in signal amplification. This model exemplifies how intricate interactions between multiple different PTMs regulate protein localization, interactions, activity, and ultimately essential cellular processes.

Recent advances in mass spectrometry methods, instrumentation, and bioinformatics analyses have enabled the identification and quantification of proteome-wide PTMs. For example, it is now a common practice to identify ten thousand phosphorylation sites in a single phosphoproteome enrichment experiment [13]. In addition, precise quantitation allows a deeper understanding of the combinations and occupancy of PTMs within a given protein. Such MS-based PTM analyses have led to previously impossible discoveries, advancing our understanding of the role of PTMs in diverse biological processes.

1.2 Global versus Targeted Analysis Strategies

Detection of PTMs by mass spectrometry can be achieved via global or targeted methods. The biological pathway of interest usually determines the type of PTM to be analyzed and associated methods. In a more targeted approach, researchers decide to investigate PTMs, because a protein of interest shows a higher than expected molecular weight or multiple bands by western blot after application of a stimulus, thus prompting speculation as to whether this could be due to PTM. Either way, the first step in PTM mapping is to determine the type of PTM of interest. In some cases the observed mass shift in a mass spectrometer indicates a certain PTM type. Many PTMs, however, result in the same mass addition (e.g., +42 Da for both acetylation and trimethylation). One powerful strategy in determining PTM identity involves the employment of the enzymes responsible for PTM removal. For example, after antibody enrichment of a modified protein, the antibody-bound protein can be incubated with general phosphatases, deubiquitinating enzymes (DUBs), or deSUMOylating enzymes (SENPs), and PTM removal can be assayed by western blot. Another method for PTM identification is western blotting with PTM specific "pan-antibodies." Many commercially available antibodies exist for this purpose, recognizing common PTMs such as acetylation, methylation, ubiquitylation, and phosphorylation or even more rare PTMs such as crotonyl-, malonyl- or glutaryl-lysine modification. Once the type of PTM that is decorating a protein has been identified, the next step is to attempt to map the amino acid residue(s) that bear this modification.

One of the first applications of mass spectrometry in protein research was the mapping of a PTM on a single protein [14]. A commonly used approach involves protein-level immunoprecipitation followed by separating the captured proteins by SDS-PAGE, excising the higher molecular weight band, and performing in-gel tryptic digestion followed by LC-MS/MS. By searching for mass shifts indicative of the suspected modification(s), PTM-containing peptides can be identified and the PTM site mapped back to the protein. The strategy of identifying proteins in complex mixtures by digesting them into peptides, sequencing the resulting peptides by tandem mass spectrometry (MS/MS), and determining peptide and protein identity through automated database searching is referred to as shotgun proteomics and is one of the most popular analysis strategies in proteomics [15]. This protein-level enrichment approach, however, is dependent on sufficient levels of the modified protein compared to unmodified and the availability of protein-specific antibodies for immunoprecipitation. It is also possible that modifications may occur within the antibody epitope, blocking enrichment of the modified form altogether.

Researchers are commonly interested in analyzing PTMs from a complex mixture of proteins rather than on only one substrate. This can be a challenge, since modified peptides often occur in substoichiometric levels compared to unmodified versions and also may ionize less efficiently by electrospray ionization (ESI). However, several enrichment strategies exist, allowing for reduction of sample complexity and easier detection of the modified peptide species. Peptide-level immunoprecipitation using antibodies specific to a given PTM is an increasingly popular method of enrichment prior to MS. While this strategy can be employed for any PTM enrichment, it has been most commonly used for mapping ubiquitination sites. Tryptic digestion of ubiquitinated proteins generates a diglycine remnant attached to the ubiquitinated lysine residue (K-GG) that can be recognized by antibodies. The resulting mass shift of +114.0429 Da can be detected by MS/MS. Not only has K-GG peptide immunoaffinity enrichment enabled the identification of hundreds of ubiquitination sites on a global level but it has also been shown to enhance identification of ubiquitination sites on individual proteins, when compared to protein-level IP coupled with MS/MS [16].

To understand the biological significance of a specific PTM, it is also important to determine the PTM site occupancy or percentage of a protein's total population that is modified. Quantification of site occupancy can be accomplished by combining antibody peptide enrichment with stable isotope-labeled internal standards of the same sequence, a method termed stable isotope standards and capture by anti-peptide antibodies (SISCAPA) [17]. By coupling immunoprecipitation with stable isotope dilution multiple reaction monitoring (SID-MRM), absolute quantitation of both modified and unmodified protein populations can be determined in a high-throughput, multiplexing-compatible fashion [18].

In addition to antibody-based enrichment approaches, several strategies for chemical enrichment of PTM-containing subproteomes have been developed.

These approaches can also be coupled with the use of stable isotope standard peptides and SRM/MRM for accurate quantification of PTM dynamics. The most widely studied PTM, with the most variety of enrichment methods available, is phosphorylation. Global analysis of serine, threonine, and tyrosine phosphorylation can be achieved by a combination of peptide fractionation using strong cation exchange (SCX) followed by further enrichment with immobilized metal affinity chromatography (IMAC). The SCX/IMAC approach allows for enrichment of phosphorylated peptides to over 75% purity and ultimately identification of over 10,000 phosphorylation sites from 5 mg of starting protein [13, 19]. Another common approach for selective enrichment of the phosphoproteome is using metal oxide affinity chromatography (MOAC) such as titanium dioxide (TiO_2) [20] or aluminum hydroxide ($Al(OH_3)$) [21]. MOAC methods have been reported to achieve higher sensitivity than IMAC (at the cost of lower specificity though). The combination of multiple enrichment approaches may ultimately be the best approach.

Phosphopeptide enrichment strategies can also be applied on crude protein extract to enrich for entire phosphoproteins. Enriched fractions are typically separated by two-dimensional gel electrophoresis (2D-GE) or sodium dodecyl sulfate–polyacrylamide gel electrophoresis (SDS-PAGE). In either case, each observed protein spot/band is quantified by its staining intensity, and selected spots/bands are excised, digested, and analyzed by MS. The advantage of phosphoprotein enrichment is that intact proteins are separated, and the molecular weight and isoelectric point of proteins can be determined. This greatly aids in protein identification by MS. However, protein-level enrichment has several disadvantages, including loss of small or hydrophobic proteins during precipitation steps, less specific enrichment when compared to phosphopeptides, and difficulty in identifying low-abundance proteins or modifications [22].

In summary, both targeted and global methods for PTM identification have been significantly tuned in recent years but are still facing challenges. The choice of method is usually dictated by the biological question. However, global strategies are becoming increasingly popular due to their versatility, sensitivity, and ability to collect a wealth of data, triggering new hypotheses that ask for validation by targeted experiments.

1.3 Mass Spectrometric Analysis Methods for the Detection of PTMs

Mass spectrometers are powerful, analytical tools that have evolved rapidly over the past few decades to become the instrument of choice for protein and peptide characterization. Mass spectrometry is often used in parallel to other techniques such as western blot analysis or protein microarrays for detecting and quantifying PTMs. One of the main advantages of mass spectrometry is the ability to rapidly analyze many samples in a high-throughput manner. Mass spectrometric analyses can be divided into three main strategies: "bottom-up," "middle-down," and "top-down" proteomic approaches [23]. Laboratories typically employ bottom-up proteomic methodologies to characterize PTMs. Proteins of interest are purified and proteolytically digested with an enzyme such as trypsin, with resultant peptides being separated by reversed-phase chromatography or another analytical method compatible with mass spectrometric analysis. One of several fragmentation methods and ion detection methodologies can then be employed (see Sections 1.3.1–1.3.4 for description of the various types of bottom-up proteomic analyses). It is common to associate "data-dependent" MS/MS analysis with bottom-up approaches, where resulting peptide spectra are then pieced back together *in silico* to give an overview of the protein and its PTMs.

In top-down proteomics, intact protein ions or large protein fragments are subjected to gas-phase fragmentation for MS analysis. Here, a variety of fragmentation mechanisms can be employed to induce dissociation and mass spectrometric analysis of the protein including collision-induced dissociation (CID), electron transfer dissociation (ETD), and electron capture dissociation (ECD) [24–26]. High-resolution mass detectors such as the quadrupole–time of flight (Q-TOF), Fourier transform ion cyclotron resonance (FT-ICR), or orbitrap mass spectrometers are typically employed as the spectra generated from top-down fragmentation tend to be highly charged and therefore difficult to resolve without high-resolution power. Top-down proteomics to date has been a less popular tool for characterizing PTMs than bottom-up analysis. However, it is an invaluable tool in cases where a bottom-up approach would lose contextual information about combinatorial PTM distribution (e.g., in the case of histone PTM analysis [27]). The middle-down approach has more commonly been employed as a strategy whereby a proteolytic enzyme can be used to generate longer polypeptides from a protein of interest and has shown utility in analyzing complex PTMs such as the histone code [28, 29]. Compared to middle-down and top-down methods, the bottom-up approach often offers better front-end separation of peptides, typically equating to higher sensitivity and selectivity. There are however some limitations to the bottom-up approach including the risk of low sequence coverage, particularly when employing a single proteolytic enzyme such as trypsin where cleavage may result in peptides yielding chemophysical properties with poor analytical attributes, such as size or substandard hydrophobicity.

1.3.1 Data-Dependent and Data-Independent Analyses

The type of mass spectrometric analysis performed for PTM detection depends on whether a single protein with a single PTM is being analyzed or if it is a global approach, such as a global phosphoproteomic analysis. When targeting a single protein or a subset of proteins for a PTM of interest, a straightforward strategy is to perform an enzymatic digestion followed by data-dependent MS/ MS analysis of peptides. In this approach, the intact molecular weight of each peptide in the full MS scan is analyzed, and then a selection of the most abundant peptides in the full MS scan are sequentially selected for fragmentation using one of several fragmentation methods. The resulting spectra are then analyzed either through *de novo* sequencing or more commonly using a search algorithm such as SEQUEST [30], Mascot [31], or Andromeda [32]. Peptides are then scored using an algorithm to calculate the false discovery rate or validated through manual spectral interpretation or by incorporation of a synthetic standard.

In traditional data-dependent acquisition (DDA), a proteomic sample is digested into peptides, separated often by reversed-phase chromatography, and ionized and analyzed by mass spectrometry. Typically instruments are programmed to select any ions that fall above a certain intensity threshold in full MS for subsequent MS/MS fragmentation. Although a powerful and highly utilized technique, the method is indeed biased to peptides that are of higher abundance, and lower level moieties such as post-translationally modified peptides may go undetected using DDA. Several years ago an alternative methodology called data-independent acquisition (DIA) was introduced which has slowly been gaining momentum [33]. In DIA analysis, all peptides within a defined mass-to-charge (*m/z*) window are subjected to fragmentation; the analysis is repeated as the mass spectrometer walks along the full *m/z* range. This results in the identification of lower level peptides, for example, post-translationally modified species present at substoichiometric levels compared to their nonmodified counterparts. It also allows accurate peptide quantification without being limited to profiling predefined peptides of interest and has proved useful in the biomarker community where quantitation on complex samples is routinely employed. The DIA method has matured in terms of utility over the past few years with the introduction of more user friendly and accurate search algorithms and spectral library search capabilities [34, 35]. Its utility as a tool to identify complex, low level, and isobaric amino acids has also recently been reported [36, 37].

1.3.2 Targeted Analyses

In addition to data-dependent approaches, targeted methods also exist whereby specific ion transitions can be monitored. These various targeted approaches are summarized in Figure 1.1, each of which has been employed to characterize post-translationally modified peptides.

Precursor ion scanning (PIS) is a sensitive mode of mass spectrometric operation primarily performed on triple quadrupole instruments, which has been

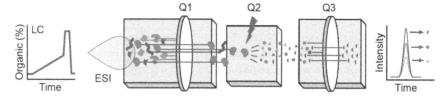

Figure 1.1 The selected reaction monitoring technique. Molecular ions of a specific analyte are selected in Q1 and fragmented in Q2. electrospray ionization (ESI). Molecular ions of one or several contaminants are isolated and fragmented together. A specific fragment ion from the target analyte (transition) is selected in Q3 and guided to the detector. The number of target fragment ions is counted over time, resulting in an SRM trace for each transition. On the far right, cycles through three transitions, corresponding to three different fragments of the target analyte, and the corresponding three SRM traces are shown. Source: Picotti and Aebersold 2012 [38]. Reproduced with permission from Macmillan Publishers Ltd.

employed for the analysis of predefined PTMs. In PIS, the third quadrupole of a triple quadruple mass spectrometer is fixed on a selected *m/z*, typically that being a neutral loss ion, for example, 79 Da for a phosphate anion observed in negative ion mode of detection, whereby the (PO_3) species is derived from the CID of phosphorylated moieties [39]. This method is highly selective and sensitive and has been applied to other PTMs beyond the analysis of phosphopeptides. Another targeted method traditionally employed for identification of post-translationally modified peptides is neutral loss scanning (NLS) [40]. NLS experiments monitor all pairs of precursor ions and product ions that differ by a constant neutral loss consistent with the PTM of interest. However, with the exponential improvements in speed and sensitivity for instruments such as the orbitrap and Q-TOF, these methods are less commonly employed than several years ago.

1.3.3 Multiple Reaction Monitoring

Multiple reaction monitoring (MRM), also known as selected reaction monitoring (SRM), is a targeted mass spectrometric methodology that is not limited to the analysis of PTM modified peptides but has been used extensively as a sensitive method to analyze various types of peptides. In MRM analyses MS/MS is applied to detect and quantify selected peptides of interest, such as those previously identified in differential discovery studies or specific post-translationally modified forms of a known peptide. Here, the specificity of precursor to product transitions is harnessed for quantitative analysis of multiple proteins in a single sample. Software tools such as MRMaid [41] or Skyline [42] allow rapid MRM transition generation and method construction for targeted analyses.

1.3.4 Multiple Reaction Monitoring Initiated Detection and Sequencing

Multiple reaction monitoring-initiated detection and sequencing (MIDAS™) [43] has been a well-utilized method for the analysis of PTM modified peptides with application to acetylated [44], phosphorylated [45], and ubiquitinated [46] species. MIDAS is a hypothesis-driven approach that requires the primary sequence of the target protein to be known and a proteolytic digest of this protein to be performed. MIDAS allows one to perform a targeted search for the presence of post-translationally modified peptides with detection based on the combination of the predicted molecular weight (measured as mass–charge ratio) of the PTM modified proteolytic peptide and a diagnostic fragment which is generated by specific fragmentation of modified peptides during CID performed in MS/MS analysis. Sequence information is subsequently obtained which enables PTM site assignment.

1.4 The Importance of Bioinformatics

The ultimate goal of proteomics is to obtain a picture of the entire complement of proteins without gaps. Genomics has already achieved this goal at the level of DNA and RNA by mapping complete genotypes. Proteomics, however, aims to describe phenotypes that display a significantly more complex functional diversity in a dynamic environment. Historically, proteomics tried to approach this challenge by establishing comparably primitive approaches such as two-dimensional gels, which gave the genomics field a competitive edge. In the last decade, however, mass spectrometry has become the method of choice, and recent advances allow the measurement of expression and modification states of thousands of proteins in a single experiment. In the last few years, the number of identified PTM sites, in particular, phosphorylation sites, has increased up to 100-fold [47]. Furthermore, mass spectrometry enables the reconstruction of protein interactions in networks and complexes. Shotgun proteomics is the most widely used approach generating thousands of spectra per hour. Therefore computational methods have to face a huge amount of generated data and a combinatorial explosion in the number of potential molecular states of proteins. In the early era of mass spectrometry as a high-throughput technology, computational analysis was commonly considered the "Achilles heels of proteomics" [48] because of the alarmingly high false discovery rates accompanied with the absence of adequate statistical methods. Fortunately, the establishment of stringent standards by the community [49] and the development of robust computational methods dragged the false discovery rates down to one percent and reduced the fraction of unassigned spectra to 10% [50].

The primary problem that all computational approaches try to solve is to assign a given MS/MS spectrum to a peptide sequence within the shortest amount of time. The most common approach is to generate theoretical fragment masses for candidate peptides from a specified protein sequence database and map these against experimental spectra. The pool of possible peptides is mainly defined by the proteolytic enzyme, mass tolerance, and specified PTM. Numerous software tools have been developed to this end [51], and they mainly differ in scoring the similarities between calculated and experimental spectra and in the statistical validation of results. SEQUEST [30] is one of the first and most commonly used tools for MS/MS-based proteomics. Its scoring scheme is based on spectral correlation functions that basically count "matched peaks," defined as the number of fragment ions common between the computed and experimental spectra. Mascot [31] extends this approach by estimating the probability of observing the shared peak count by chance. Because Mascot is a commercial software, the underlying algorithms are not provided. The search engine Andromeda [32], which is integrated into the freely available MaxQuant platform [52], also employs probabilistic scores. Notably, because selection of precursor ion for fragmentation is performed with low resolution to ensure high sensitivity, coeluting peptides with similar masses are frequently cofragmented. While the resulting "chimerical" MS/MS spectra [53] usually distort the detection and quantitation of peptides, Andromeda includes an algorithm that detects the "second" peptide and uses this information to increase the identification rate.

Other computational tools such as Protein Prospector [54] employ empirical scoring schemes that incorporate the number of matched peaks as well as the fraction of total peak intensities that can be explained by them. But when it comes to the identification of PTM sites, all methods face the same issue of the combinatorial explosion of theoretical peptides in cases where too many variable modification types are allowed. Consequently, spectra-to-peptide searches are usually restricted to up to three modifications. However, Byonic [55], which is also based on the principle of matching experimental to theoretical spectra, allows a larger number of modification types by setting an upper limit on the total occurrence of each modification. Furthermore, Byonic provides "wildcard" searches that allow the detection of unanticipated modifications by searching within specified mass delta windows.

In addition to the combinatorial explosion of theoretical peptides, another challenge in the analysis of PTMs is the precise localization of PTMs within peptides. Since PTM sites of the same protein commonly display distinct behaviors [56], it is imperative to determine their exact localizations. To this end, Ascore [57] assesses the probability of correct site localization based on the presence and intensity of site-determining ions. The corresponding algorithm essentially reflects the cumulative binomial probability of identifying site-determining ions. The same concept is used by the "localization probability score," [56] which is integrated into MaxQuant.

After the identification of peptides and associated PTMs, output scores of database search tools are translated into estimated false discovery rates. To this end, "target-decoy searching" [58] is commonly applied. The main idea of this approach is to search MS/MS spectra against a target database that contains protein sequences and reversed counterparts. Under the assumption that false matches to sequences from the original database and matches to decoy peptide sequences follow the same distribution, peptide identifications are filtered using score cutoffs corresponding to certain FDRs.

Taken together, technological advances and accompanied developments of computational methods now allow the routine identification of thousands of proteins, including PTM sites, giving a global and hopefully soon a complete picture of the proteome. Bioinformatics approaches have mastered many problems in the analysis of proteomics data but are still facing several challenges including the decryption of unmatched spectra. The accumulation of detected PTM sites across studies has been managed by various databases, including UniProt (www.uniprot.org) [2], PhosphoSite (www.phosphosite.org) [59], and PHOSIDA (www.phosida.com) [60].

Acknowledgements

We thank Allison Bruce from Genentech for help with the illustration.

References

1 Clamp M, Fry B, Kamal M, Xie X, Cuff J, Lin MF, Kellis M, Lindblad-Toh K, Lander ES. Distinguishing protein-coding and noncoding genes in the human genome. *Proc Natl Acad Sci U S A* 2007;**104**:19428–19433.

2 Consortium TU. UniProt: a hub for protein information. *Nucleic Acids Res* 2015;**43**:D204–D212.

3 Ayoubi TA, Ven WJVD. Regulation of gene expression by alternative promoters. *FASEB J* 1996;**10**:453–460.

4 Jensen ON. Modification-specific proteomics: characterization of post-translational modifications by mass spectrometry. *Curr Opin Chem Biol* 2004;**8**:33–41.

5 Walsh C. *Post-translational Modification of Proteins: Expanding Nature's Inventory*. Roberts and Company Publishers; 2006.

6 Johnson LN. The regulation of protein phosphorylation. *Biochem Soc Trans* 2009;**37**:627–641.

7 Manning G, Whyte DB, Martinez R, Hunter T, Sudarsanam S. The protein kinase complement of the human genome. *Science* 2002;**298**:1912–1934.

8 Dever TE, Gutierrez E, Shin B-S. The hypusine-containing translation factor eIF5A. *Crit Rev Biochem Mol Biol* 2014;**49**:413–425.

9 Komander D, Rape M. The ubiquitin code. *Annu Rev Biochem* 2012;**81**:203–229.

10 Bannister AJ, Kouzarides T. Regulation of chromatin by histone modifications. *Cell Res* 2011;**21**:381–395.

11 Hunter T. The age of crosstalk: phosphorylation, ubiquitination, and beyond. *Mol Cell* 2007;**28**:730–738.

12 Ordureau A, Sarraf SA, Duda DM, Heo J-M, Jedrychowski MP, Sviderskiy VO, Olszewski JL, Koerber JT, Xie T, Beausoleil SA, Wells JA, Gygi SP, Schulman BA, Harper JW. Quantitative proteomics reveal a feedforward mechanism for mitochondrial PARKIN translocation and ubiquitin chain synthesis. *Mol Cell* 2014;**56**:360–375.

13 Villén J, Gygi SP. The SCX/IMAC enrichment approach for global phosphorylation analysis by mass spectrometry. *Nat Protoc* 2008;**3**:1630–1638.

14 Stenflo J, Fernlund P, Egan W, Roepstorff P. Vitamin K dependent modifications of glutamic acid residues in prothrombin. *Proc Natl Acad Sci U S A* 1974;**71**:2730–2733.

15 Nesvizhskii AI. Protein identification by tandem mass spectrometry and sequence database searching. In: *Mass Spectrometry Data Analysis in Proteomics*. New Jersey: Humana Press; 2006. p 87–120.

16 Anania VG, Pham VC, Huang X, Masselot A, Lill JR, Kirkpatrick DS. Peptide level immunoaffinity enrichment enhances ubiquitination site identification on individual proteins. *Mol Cell Proteomics* 2014;**13**:145–156.

17 Anderson NL, Anderson NG, Haines LR, Hardie DB, Olafson RW, Pearson TW. Mass spectrometric quantitation of peptides and proteins using Stable Isotope Standards and Capture by Anti-Peptide Antibodies (SISCAPA). *J Proteome Res* 2004;**3**:235–244.

18 Kuhn E, Addona T, Keshishian H, Burgess M, Mani DR, Lee RT, Sabatine MS, Gerszten RE, Carr SA. Developing multiplexed assays for troponin I and interleukin-33 in plasma by peptide immunoaffinity enrichment and targeted mass spectrometry. *Clin Chem* 2009;**55**:1108–1117.

19 Gruhler A, Olsen JV, Mohammed S, Mortensen P, Færgeman NJ, Mann M, Jensen ON. Quantitative phosphoproteomics applied to the yeast pheromone signaling pathway. *Mol Cell Proteomics* 2005;**4**:310–327.

20 Larsen MR, Thingholm TE, Jensen ON, Roepstorff P, Jørgensen TJD. Highly selective enrichment of phosphorylated peptides from peptide mixtures using titanium dioxide microcolumns. *Mol Cell Proteomics* 2005;**4**:873–886.

21 Wolschin F, Wienkoop S, Weckwerth W. Enrichment of phosphorylated proteins and peptides from complex mixtures using metal oxide/hydroxide affinity chromatography (MOAC). *Proteomics* 2005;**5**:4389–4397.

22 Fíla J, Honys D. Enrichment techniques employed in phosphoproteomics. *Amino Acids* 2012;**43**:1025–1047.

23 Aebersold R, Mann M. Mass spectrometry-based proteomics. *Nature* 2003;**422**:198–207.

24 McLafferty FW, Horn DM, Breuker K, Ge Y, Lewis MA, Cerda B, Zubarev RA, Carpenter BK. Electron capture dissociation of gaseous multiply charged ions by Fourier-transform ion cyclotron resonance. *J Am Soc Mass Spectrom* 2001;**12**:245–249.

25 Syka JEP, Coon JJ, Schroeder MJ, Shabanowitz J, Hunt DF. Peptide and protein sequence analysis by electron transfer dissociation mass spectrometry. *Proc Natl Acad Sci U S A* 2004;**101**:9528–9533.

26 Wells JM, McLuckey SA. Collision-induced dissociation (CID) of peptides and proteins. *Methods Enzymol* 2005;**402**:148–185.

27 Han J, Borchers CH. Top-down analysis of recombinant histone H3 and its methylated analogs by ESI/FT-ICR mass spectrometry. *Proteomics* 2010;**10**:3621–3630.

28 Moradian A, Kalli A, Sweredoski MJ, Hess S. The top-down, middle-down, and bottom-up mass spectrometry approaches for characterization of histone variants and their post-translational modifications. *Proteomics* 2014;**14**:489–497.

29 Sidoli S, Lin S, Karch KR, Garcia BA. Bottom-up and middle-down proteomics have comparable accuracies in defining histone post-translational modification relative abundance and stoichiometry. *Anal Chem* 2015;**87**:3129–3133.

30 Eng JK, McCormack AL, Yates JR. An approach to correlate tandem mass spectral data of peptides with amino acid sequences in a protein database. *J Am Soc Mass Spectrom* 1994;**5**:976–989.

31 Perkins DN, Pappin DJC, Creasy DM, Cottrell JS. Probability-based protein identification by searching sequence databases using mass spectrometry data. *Electrophoresis* 1999;**20**:3551–3567.

32 Cox J, Neuhauser N, Michalski A, Scheltema RA, Olsen JV, Mann M. Andromeda: a peptide search engine integrated into the MaxQuant environment. *J Proteome Res* 2011;**10**:1794–1805.

33 Egertson JD, Kuehn A, Merrihew GE, Bateman NW, MacLean BX, Ting YS, Canterbury JD, Marsh DM, Kellmann M, Zabrouskov V, Wu CC, MacCoss MJ. Multiplexed MS/MS for improved data independent acquisition. *Nat Methods* 2013;**10**:744–746.

34 Distler U, Kuharev J, Navarro P, Levin Y, Schild H, Tenzer S. Drift time-specific collision energies enable deep-coverage data-independent acquisition proteomics. *Nat Methods* 2014;**11**:167–170.

35 Röst HL, Rosenberger G, Navarro P, Gillet L, Miladinović SM, Schubert OT, Wolski W, Collins BC, Malmström J, Malmström L, Aebersold R. OpenSWATH enables automated, targeted analysis of data-independent acquisition MS data. *Nat Biotechnol* 2014;**32**:219–223.

36 Parker BL, Yang G, Humphrey SJ, Chaudhuri R, Ma X, Peterman S, James DE. Targeted phosphoproteomics of insulin signaling using data-independent acquisition mass spectrometry. *Sci Signal* 2015;8:rs6–rs6.

37 Sidoli S, Lin S, Xiong L, Bhanu NV, Karch KR, Johansen E, Hunter C, Mollah S, Garcia BA. Sequential window acquisition of all theoretical mass spectra (SWATH) analysis for characterization and quantification of histone post-translational modifications. *Mol Cell Proteomics* 2015;14:2420–2428.

38 Picotti P, Aebersold R. Selected reaction monitoring-based proteomics: workflows, potential pitfalls and future direction. *Nat Methods* 2012;9(6):555–566.

39 Annan RS, Carr SA. The essential role of mass spectrometry in characterizing protein structure: mapping post-translational modifications. *J Protein Chem* 1997;16:391–402.

40 Casado-Vela J, Ruiz EJ, Nebreda AR, Casal JI. A combination of neutral loss and targeted product ion scanning with two enzymatic digestions facilitates the comprehensive mapping of phosphorylation sites. *Proteomics* 2007;7:2522–2529.

41 Mead JA, Bianco L, Ottone V, Barton C, Kay RG, Lilley KS, Bond NJ, Bessant C. MRMaid, the web-based tool for designing multiple reaction monitoring (MRM) transitions. *Mol Cell Proteomics* 2009;8:696–705.

42 MacLean B, Tomazela DM, Shulman N, Chambers M, Finney GL, Frewen B, Kern R, Tabb DL, Liebler DC, MacCoss MJ. Skyline: an open source document editor for creating and analyzing targeted proteomics experiments. *Bioinformatics* 2010;26:966–968.

43 Unwin RD, Griffiths JR, Leverentz MK, Grallert A, Hagan IM, Whetton AD. Multiple reaction monitoring to identify sites of protein phosphorylation with high sensitivity. *Mol Cell Proteomics* 2005;4:1134–1144.

44 Evans CA, Griffiths JR, Unwin RD, Whetton AD, Corfe BM. Application of the MIDAS approach for analysis of lysine acetylation sites. In: Hake SB, Janzen CJ, editors. *Protein Acetylation.* Totowa, NJ: Humana Press; 2013. p 25–36.

45 Unwin RD, Griffiths JR, Whetton AD. A sensitive mass spectrometric method for hypothesis-driven detection of peptide post-translational modifications: multiple reaction monitoring-initiated detection and sequencing (MIDAS). *Nat Protoc* 2009;4:870–877.

46 Mollah S, Wertz IE, Phung Q, Arnott D, Dixit VM, Lill JR. Targeted mass spectrometric strategy for global mapping of ubiquitination on proteins. *Rapid Commun Mass Spectrom RCM* 2007;21:3357–3364.

47 Choudhary C, Mann M. Decoding signalling networks by mass spectrometry-based proteomics. *Nat Rev Mol Cell Biol* 2010;11:427–439.

48 Patterson SD. Data analysis–the Achilles heel of proteomics. *Nat Biotechnol* 2003;21:221–222.

49 Carr S, Aebersold R, Baldwin M, Burlingame A, Clauser K, Nesvizhskii A. The need for guidelines in publication of peptide and protein identification data

working group on publication guidelines for peptide and protein identification data. *Mol Cell Proteomics* 2004;**3**:531–533.

50 Cox J, Mann M. Quantitative, high-resolution proteomics for data-driven systems biology. *Annu Rev Biochem* 2011;**80**:273–299.

51 Nesvizhskii AI, Vitek O, Aebersold R. Analysis and validation of proteomic data generated by tandem mass spectrometry. *Nat Methods* 2007;**4**:787–797.

52 Cox J, Mann M. MaxQuant enables high peptide identification rates, individualized p.p.b.-range mass accuracies and proteome-wide protein quantification. *Nat Biotechnol* 2008;**26**:1367–1372.

53 Houel S, Abernathy R, Renganathan K, Meyer-Arendt K, Ahn NG, Old WM. Quantifying the impact of chimera MS/MS spectra on peptide identification in large-scale proteomics studies. *J Proteome Res* 2010;**9**:4152–4160.

54 Clauser KR, Baker P, Burlingame AL. Protein prospector role of accurate mass measurement (±10 ppm) in protein identification strategies employing MS or MS/MS and database searching. *Anal Chem* 1999;**71**:2871–2882.

55 Bern M, Kil YJ, Becker C. Byonic: advanced peptide and protein identification software. *Curr Protoc Bioinforma* 2012;Chapter 13:Unit13.20.

56 Olsen JV, Blagoev B, Gnad F, Macek B, Kumar C, Mortensen P, Mann M. Global, in vivo, and site-specific phosphorylation dynamics in signaling networks. *Cell* 2006;**127**:635–648.

57 Beausoleil SA, Villén J, Gerber SA, Rush J, Gygi SP. A probability-based approach for high-throughput protein phosphorylation analysis and site localization. *Nat Biotechnol* 2006;**24**:1285–1292.

58 Elias JE, Gygi SP. Target-decoy search strategy for increased confidence in large-scale protein identifications by mass spectrometry. *Nat Methods* 2007;**4**:207–214.

59 Hornbeck PV, Chabra I, Kornhauser JM, Skrzypek E, Zhang B. PhosphoSite: a bioinformatics resource dedicated to physiological protein phosphorylation. *Proteomics* 2004;**4**:1551–1561.

60 Gnad F, Ren S, Cox J, Olsen JV, Macek B, Oroshi M, Mann M. PHOSIDA (phosphorylation site database): management, structural and evolutionary investigation, and prediction of phosphosites. *Genome Biol* 2007;**8**:R250.

2

Identification and Analysis of Protein Phosphorylation by Mass Spectrometry

Dean E. McNulty, Timothy W. Sikorski and Roland S. Annan

Proteomics and Biological Mass Spectrometry Laboratory, GlaxoSmithKline, Collegeville, PA, USA

2.1 Introduction to Protein Phosphorylation

Much of the activity in the cellular proteome is under the control of reversible protein phosphorylation. Phosphorylation-dependent signaling regulates differentiation of cells, triggers progression of the cell cycle, and controls metabolism, transcription, apoptosis, and cytoskeletal rearrangements. Signaling via reversible protein phosphorylation also plays a critical role in intracellular communication and immune response. Phosphorylation can function as a positive or negative switch, activating or inactivating enzymes. It can serve as a docking site to recruit other proteins into multiprotein complexes or serve as a recognition element to recruit other enzymes that add other post-translational modifications (PTMs) or additional phosphorylation sites. Phosphorylation can trigger a change in the three-dimensional structure of a protein or initiate translocation of the protein to another compartment of the cell. Disruption of normal cellular phosphorylation events is responsible for a large number of human diseases [1–3]. From the discovery of the first functionally relevant phosphorylation site in 1955 [4], the ability to analyze protein phosphorylation has exploded in the last five years to the point where it is now possible to quantitate changes in tens of thousands of phosphorylation sites in response to a cell receiving an external stimulus or undergoing a normal change in the physiology [5]. While phosphorylation is known to occur on histidine, aspartate, cysteine, lysine, and arginine residues, this chapter focuses on the more commonly modified and well-studied amino acids: serine, threonine, and tyrosine.

The first evidence for protein phosphorylation was uncovered in 1906 when Phoebus Levene identified phosphate in the amino acid composition of the egg yolk protein vitellin [6]. While there was evidence in the 1920s to suggest the

Analysis of Protein Post-Translational Modifications by Mass Spectrometry,
First Edition. Edited by John R. Griffiths and Richard D. Unwin.

phosphate was on the amino acid serine [7], it was not until 1932 that Levene and Fritz Lipmann isolated phosphoserine from vitellin [8]. Prior to the 1950s, research on phosphoproteins was focused mainly on abundant proteins found in egg yolk (such as vitellin) and milk (casein), and the biological function, if any, of the phosphorylation was unknown. But by the early 1950s, it was being shown that in tumor cells the phosphorus in phosphoproteins was being turned over rapidly and that tumors contained high levels of phosphoserine [9, 10], together suggesting that this modification must have some function. In 1954 Kennedy and Burnett, using labeled ATP, demonstrated that an enzyme from rat liver mitochondria was responsible for catalyzing the phosphorylation of serine on both alpha and beta casein [11]. A year later Fischer and Krebs provided the first evidence that protein phosphorylation had a biological function. They demonstrated that inactive phosphorylase b could be converted to active phosphorylase a in the presence of ATP and Mg [4], and in the next few years they identified phosphorylase kinase as the enzyme responsible for the activation and showed that it phosphorylated a specific serine residue on phosphorylase b [12].

It is now widely recognized that cascades of protein phosphorylation transmit signals from the extracellular environment to trigger a biological response within the cell. The first evidence that kinases worked in series came in 1968 with the discovery of cAMP-dependent protein kinase A (PKA) and the fact that it phosphorylated and activated phosphorylase kinase [13]. It quickly became clear that PKA had many substrates in multiple tissues [14], and the idea that protein phosphorylation was a widespread phenomenon began to take hold. Throughout the 1970s and 1980s many additional serine/threonine (S/T) protein kinases were discovered, and in 1983 Tony Hunter showed that the v-Src protein was a tyrosine kinase (TK) [15]. The difficulty in detecting phosphotyrosine in these early years arises from the fact that we now know it constitutes only a few percent of the total phosphoamino acid pool [5, 16] and that it comigrated with the much more abundant phosphothreonine in the standard electrophoretic systems used in the late 1970s to detect ^{32}P-labeled phosphoamino acids [17].

With the development by Hunter and Sefton of a two-dimensional (2D) separation method for phosphoamino acids [15], it quickly became clear that phosphorylation on tyrosine was also widespread. In 1981 the EGF receptor (EGFR) was shown to have TK activity and that stimulation of cells with EFG led to rapid tyrosine phosphorylation on multiple proteins [18, 19]. By the end of the 1980s more than 10 receptor tyrosine kinases (RTKs) had been identified. The realization that growth factor receptors had intrinsic TK activity connected intracellular signaling through (largely) serine/threonine (S/T) kinases with external signals communicated via ligand binding to transmembrane receptors. In many cases, nonreceptor tyrosine kinases (NRTK) constitute the next step in the signaling cascade, transmitting signals from the intracellular domains of the RTK to downstream S/T protein kinases [20, 21]. Vast amounts of research in the 1980s and 1990s encompassing all areas of cellular biology would discover many more kinases and their substrates and add much fine detail to the mechanism of phosphorylation-dependent signaling.

The identification of all human kinase genes was made possible with the complete sequencing of the human genome [22]. Bioinformatic analysis has identified 478 protein kinases (see Figure 2.1, right), belonging to a large super-family that shares a eukaryotic protein kinase (ePK) domain. There are an additional 40 atypical protein kinases (aPK), which have been demonstrated to have protein kinase activity, but do not share the ePK domain. Altogether the 518 protein kinases make up one of the largest families of eukaryotic genes (see Figure 2.1). All major kinase groups and most kinase families are shared across metazoans, and many are shared in yeast [23]. Protein tyrosine kinases (PTK) of which 90 have been identified are found only in metazoans [24]. More than half of these (58) are RTKs, involved in regulating the multicellular aspects of an organism via cell-to-cell communication. It is surprising how little is actually known about most of these 518 protein kinases (termed the "kinome").

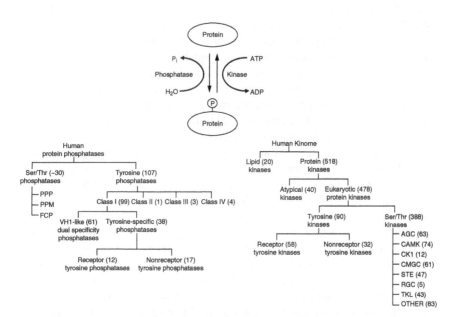

Figure 2.1 Protein phosphorylation is governed by two large superfamilies of enzymes. Protein kinases (right) add phosphate to (primarily) serine, threonine, and tyrosine residues. Protein phosphatases (left) remove the phosphate group. There are similar numbers of tyrosine kinases and phosphatases. The very small number of serine/threonine phosphatases achieves selectivity by forming combinatorial enzyme complexes with a large number of regulatory subunits.

More than 100 of the kinases have absolutely no known function, and 50% are largely uncharacterized [25]. A very small percentage of the kinome accounts for most of the published literature. This lack of knowledge about most of the human kinome is reflected in the fact that, of the twenty approved kinase therapeutics, they address only nine different kinases as their primary targets [26]. This is in spite of the fact that kinases are characterized as excellent drug targets in cancer and many other diseases. Kinase gene profiling shows distinct expression pattern differences between healthy and disease tissues for large clusters of the kinome [27].

Given the wide range of processes that are under the control of reversible protein phosphorylation and the large number of protein kinases in the metazoan genomes, it is not surprising that the extent of phosphorylation in higher-order organisms is massive. Current phosphosite databases [28, 29] list more than 150,000 sites on over 18,000 human proteins, many more than were previously predicted. The large majority of these sites have been identified in high-throughput phosphoproteomics studies utilizing MS. Large-scale phosphoproteome studies suggest that the overall phosphoamino acid composition of any cell is approximately 75–85% phosphoserine, 10–20% phosphothreonine, and 1–6% phosphotyrosine [5, 30–33]. This composition likely reflects the biology of the cell and not some bias of the mass spectrometer, as it has been shown using a large-scale synthetic phosphopeptide library that peptides containing all three types of phosphoamino acids are detected equally [32].

In 15–25% of phosphoproteins only a single site has been identified. The functional significance of these single sites is to act, in many cases, as a simple switch. Glycogen phosphorylase, for instance, contains only a single phosphoserine that drives it from the inactive to the active state [34]. The majority of proteins, however, are phosphorylated on more than one site and by more than one kinase. The spliceosome protein Srrm2 was found to contain anywhere between 177 and 300 sites [30, 33]. As might be expected, a weak but significant correlation exists between a protein's abundance and the number of sites identified in an analysis [5]. However, it is clear that multisite phosphorylation is the rule rather than the exception. It has been suggested that the multiplicity of phosphorylation on proteins might just be background noise. However, it is equally likely that given the wide variety of biological functions under the control of protein phosphorylation and the wide variety of mechanisms by which it occurs, the functional significance of most of the complex hyperphosphorylation that occurs on proteins is not yet understood. What is emerging, however, is just how intricately this multisite phosphorylation is coordinated. While some phosphorylation clusters share a common biological function, in many cases each site or a combination of sites has distinct and separable roles in that function.

The budding yeast transcription factor Pho4 controls the expression of genes needed by the organism to survive under conditions of phosphate starvation.

In a normal phosphate-rich environment, PHO4 is phosphorylated on 5 cyclin/Cdk sites and exported out of the nucleus. When yeasts are deprived of phosphate, these sites are unoccupied, and Pho4 accumulates in the nucleus and activates expression of phosphate-responsive genes. Four of the five cyclin/Cdk sites have distinct roles to play in the regulation of this function, with two being required for nuclear export, one for blocking nuclear import, and one for blocking promoter binding [35, 36]. To add complexity to this mechanism, under intermediate conditions of phosphate availability, PHO4 is phosphorylated on only one of the sites, allowing it to bind differentially to its target promoters and trigger expression of only a subset of the phosphate-responsive genes [37].

In contrast to PHO4, whose function is regulated by multisite phosphorylation via a single kinase, Sic1 is regulated by a multisite phosphorylation cascade that involves a complex dance of two different kinases. Sic1 controls the G1/S phase transition in budding yeast by inhibiting the S-phase Clb5–Cdk1 kinase. Ubiquitin-mediated destruction of Sic1 releases Clb5–Cdk1 and allows the cell to proceed to S phase (Figure 2.2a). In one of the first examples of how phosphorylation regulates ubiquitin-mediated proteolysis, Sic1 was shown to be phosphorylated on at least nine different sites and required a combination of at least three of six to trigger degradation [38]. In fact it was later shown that some phosphorylation on at least six of the nine Cdk sites is required for destruction [39]. Five of the nine Cdk-dependent sites form three pairs of high-affinity recognition elements termed phosphodegrons (see Figure 2.2b), which are recognized by ubiquitin ligases [40]. These nine sites are phosphorylated by two different cyclin/Cdks, with each showing preference for different sites. At the transition to S phase, Cln2–Cdk1 phosphorylates Sic1 on a subset of the nine sites, but with no fully formed degrons (Figure 2.2b, top). This cluster of phosphorylation sites, however, is an excellent docking platform for the slowly released Clb5–Cdk1 (Figure 2.2b, bottom), which goes on to complete phosphorylation of the residues critical for the formation of the degrons [41]. The ordered phosphorylation by two different kinases imposes a tight regulation on the G1/S transition in which Cln2–Cdk1 is not allowed to trigger the change until sufficient levels of Clb5–Cdk1 accumulate.

For both Pho4 and Sic1, phosphorylation drives the protein's biological function by regulating protein–protein interactions. In the case of Pho4, it blocks the interaction of Pho4 with nuclear import and export transport proteins and the transcriptional coactivator protein that allows promoter binding. In the case of Sic1, phosphorylation of the priming sites facilitates binding of cyclin/Cdk complexes through their regulatory subunit Cks1. Phosphorylated sites within the three degrons of Sic1 then serve as docking sites for the SCF ubiquitin ligase. Indeed while the earliest examples of the biological significance of protein phosphorylation were in the conformation-induced stimulation of enzymatic activity, it has since become clear that much of protein

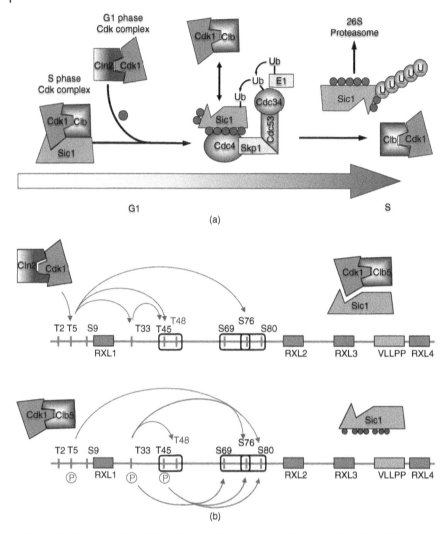

Figure 2.2 Cascades of multisite phosphorylation regulate biological function. (a) Sic1 controls the G1/S phase transition in budding yeast by inhibiting the S-phase Clb5–Cdk1 kinase. Phosphorylation-dependent ubiquitin-mediated destruction of Sic1 releases Clb5–Cdk1 and allows the cell to proceed to S phase. (b) In the first wave of phosphorylation (top), a subset of required sites are sequentially modified, but no fully formed binding sites (□) for the ubiquitination machinery are formed. These initial sites act as priming sites for the second wave of phosphorylation (bottom), which is being carried out by the slowly released Clb5–Cdk1. The now fully formed phosphodegrons bind the ubiquitination machinery, initiating destruction of Sic1. Without further sequestration of Clb5–Cdk, the cells can transition into S phase.

phosphorylation serves to either recruit or block the recruitment of other proteins. The first example of this came with the discovery of SH2 domains. The search for TK substrates in the early 1980s revealed that growth factor receptor TKs preferred themselves as substrates. This raised the question "How do RTK transmit signals to drive cellular behavior?" In 1986 Tony Pawson identified a region in the oncogenic NRTK v-Fes that was conserved in all cytoplasmic tyrosine kinases and influenced their kinase activity [42]. Termed Src homology domain 2 (SH2), it was later shown that SH2 domain-containing proteins bind other proteins, including growth factor receptors, that are phosphorylated on tyrosine [43, 44]. The recruitment of SH2 domain-containing proteins to phosphotyrosine-containing residues on growth factor receptors thus provides a mechanism by which RTKs can cascade signals into the cytoplasm. There are 120 SH2 domains on 115 proteins in the human genome. They occur on proteins that link tyrosine phosphorylation to intracellular signaling, including all NRTKs, some tyrosine phosphatases, some lipid kinases, and many adaptor proteins [45]. While the SH2 domain remains the prototype for phosphorylation-mediated protein–protein interactions, other phosphosite-dependent binding domains have since been discovered, including the PTB domain that also binds phosphotyrosine [46]. More than ten phosphoserine and phosphothreonine binding domains have also been discovered [47] including WD40 domains, which are part of the F-box proteins that act as the substrate recognition element of SCF E3 ubiquitin ligases including the one that mediated the destruction of Sic1 as described earlier.

Along with the reality that multisite phosphorylation is the norm for eukaryotic proteins, it has also now become clear that most of this phosphorylation occurs in intrinsically disordered regions of proteins [48]. Nearly all eukaryotic proteins contain disordered regions, and some proteins are predicted to be entirely disordered [49]. Intrinsically disordered proteins (IDP) play a central role in mediating protein–protein interactions and the assembly of complex protein interaction networks [50]. The disordered regions contain multiple conserved sequence motifs that serve as docking sites for other proteins, including protein kinases. The flexibility of the disordered regions makes them accessible to PTM, including but not limited to phosphorylation. With the addition of these PTMs, it is estimated that perhaps a million sequence-specific interaction motifs exist with the disordered regions of the proteome [51]. In addition to Sic1, two other well-studied examples of phosphorylation (and other PTM) clusters in disordered regions that control function are p53 [52] and RNA polymerase II [53]. The latter protein contains 52 YSPTSPS repeats in the disordered C-terminal tail that are phosphorylated on the second and fifth serines in the motif, recruiting splicing factors, chromatin modifiers, termination machinery, and other protein modules to the elongation machinery. Interestingly, the phosphorylation of intrinsically disordered regions often brings about a disordered to ordered transition in the protein structure that can either facilitate or inhibit protein–protein interactions [54, 55].

The massive amount of phosphorylation on proteins in the cell depends not only on the activity of protein kinases but also on the opposing activity of protein phosphatases. Just as there are specific kinases, which phosphorylate proteins on specific sites, there are specific phosphatases that remove phosphate from those sites. Phosphatases can be broadly divided according to their substrate specificity to include protein serine/threonine phosphatase (PSP), protein tyrosine phosphatase (PTP), and dual-specificity protein phosphatase families (Figure 2.1, left). Based on structure, rather than function, phosphatases group into several completely separate families than share no structural similarities [56]. The human genome codes for about 40 classical PTPs [57, 58], approximately half the number of TKs. Surprisingly there are only approximately 30 different PSPs, based on distinct catalytic subunits, to balance the 388 S/T kinases. The specificity of PSPs comes from their interaction with a large number of regulatory subunits [59], which together account for many more S/T phosphatases than there are S/T kinases [56].

For a protein to be regulated by phosphorylation, the activities of specific kinases and phosphatases must be in balance. A tip in the balance triggers the regulation. The intrinsic activity of S/T kinases and phosphatases are approximately equivalent, as are their intracellular concentrations [60]. In contrast to PSPs, there is compelling evidence that the activity of PTPs is several orders of magnitude higher than that of PTKs [61]. This much higher activity accounts for the overall low stoichiometry and more transient nature of tyrosine phosphorylation. The half-life of tyrosine phosphorylation on EGFR was found to be on the order of 15 s [62]. Clearly the maintenance of the balance between kinase and phosphatase goes beyond simple intrinsic enzyme activity and concentration. An equally complex system of checks and balances must regulate the regulators.

The complexity of protein phosphorylation in the cell is enormous, and the range of biological functions under the control of protein phosphorylation covers every aspect of cellular life. How all of this is regulated is also of enormous complexity. More than a 1000 protein kinase and phosphatase complexes attach and remove phosphate from the side chains of serine, threonine, and tyrosine. In all cases where the mechanisms are well studied, the addition and removal of phosphorylation are ordered and controlled. Yet most of the phosphorylation identified to date has no biological function assigned to it or how it is regulated. This understanding will require the analysis of changes in global phosphorylation patterns as well as an in-depth analysis of individual proteins. And this analysis will need to be quantitative. It is no longer sufficient to just catalog phosphorylation sites. MS has played a central role in contributing to what we have learned about protein phosphorylation, particularly in the last 10 years. It will be an indispensible tool for us going forward as we seek to understand the dynamic nature of protein phosphorylation and how it controls a cell's function. This chapter covers both the basic elements of analyzing phosphorylation on isolated proteins and the use of phosphoproteomics strategies to understand the global phosphorylation-dependent response of a cell to changes in its physiology or environment. Both of these two approaches are covered more or less from the perspective of a discovery mode. The individual or multiplexed analysis of targeted phosphorylation events is covered elsewhere [63] (also see Chapter 1).

2.2 Analysis of Protein Phosphorylation by Mass Spectrometry

Prior to the advent of "soft" ionization techniques for MS, the analysis of protein phosphorylation was commonly carried out by performing extensive, iterative enzymatic digestion of ^{32}P-labeled proteins, followed by a variety of electrophoretic and chromatographic separations [64]. Phosphoamino acids were identified from partial acid hydrolysates by comigration in a 2D separation of ^{32}P-labeled amino acids with authentic phosphoamino acid standards whose location was marked by ninhydrin staining [15]. The cellulose thin-layer plates used in these separations were often exposed for days to detect *in vivo* phosphorylated amino acids. Determining the sequence-specific location of the phosphoamino acid was accomplished by measuring radioactivity released from ^{32}P-containing peptides after iterative cycles of manual Edman degradation [65]. With the introduction of fast atom bombardment (FAB) in 1984 [66], MS would eventually become the default method for the analysis of protein phosphorylation, eliminating many of the sequential enzymatic digestion and separation steps. However, the use of ^{32}P in combination with MS would continue into the early 2000s.

FAB brought to MS the ability to ionize peptides without fragmenting them [66]. The adduction of a proton to the intact peptide produced an $[M+H]^+$ ion, which made it a trivial task to determine the molecular weight of the peptide. Because FAB worked well with simple mixtures, it was now also possible to analyze enzymatic digests of proteins and map the derived peptide masses onto the established or predicted amino acid sequence of the protein [67]. For larger proteins, the enzyme digests could be fractionated into simpler mixtures using reversed-phase (RP) HPLC. In this way the translated amino acid sequence of even very large genes could be confirmed [68, 69].

PTM of a protein results in a change in mass of the modified amino acid. This mass shift is readily observed on peptides containing the modified amino acid. Thus it was quickly realized that phosphorylated peptides might be readily observed in the FAB spectrum of a protein digest by looking for peptide masses 80 Da (HPO_3) higher than predicted. The first reported analysis of a phosphoprotein by FAB was the identification of a 23-amino acid peptide from chicken egg yolk riboflavin-binding protein that contained eight phosphorylation sites [70]. In this case the authors used 25 μg of peptide to record the spectrum and were fortunate in that the protein was homogeneously and stoichiometrically phosphorylated. The phosphorylated peptide was identified by comparison with peptides from a dephosphorylated protein. In 1986 it was demonstrated that peptides could be sequenced by tandem mass spectrometry (MS/MS) [71], and a few years later the same group sequenced three phosphopeptides from spinach chloroplast Photosystem II proteins [72]. While MS/MS was clearly an emerging technique, throughout most of the 1980s the location of specific sites of phosphorylation was still most conveniently done by measuring the release of ^{32}P during Edman sequencing [64, 73].

In 1988 two new soft ionization techniques were introduced, which would revolutionize the analysis of biomolecules. Franz Hillenkamp and Michael Karas showed that matrix-assisted laser desorption/ionization (MALDI) could ionize large proteins and produce singly charged intact protein ions that could be mass measured in a time-of-flight (TOF) mass spectrometer [74]. As early as 1988 proteins as large as an IgG (M_r = 149,190) and the tetrameric glucose isomerase (M_r = 172,460) were analyzed as intact biopolymers [75]. While MALDI shared many analytical attributes with FAB, it was more sensitive for peptides by as much as 3 orders of magnitude, was less susceptible to contaminants and incipients in the sample, and, as mentioned, had a very large mass range. It was soon shown that peptides [76] and phosphopeptides [77] could be sequenced at the sub picomole level using MALDI on a reflectron TOF mass spectrometer by recording the metastable fragment ions produced in the field-free region of the drift tube. Metastable ions produced by the loss of various phosphate-containing groups from the side chain of phosphoamino acids made it possible to distinguish tyrosine phosphorylation from serine and threonine phosphorylation on peptides [77]. An abundant $[MH-H_3PO_4]^+$ ion, accompanied by a smaller $[MH-HPO_3]^+$ ion, indicates that a peptide is most likely phosphorylated on serine or threonine. In contrast, phosphotyrosine-containing peptides generally exhibit $[MH-HPO_3]^+$ fragment ions and little, if any, $[MH-H_3PO_4]^+$.

While MALDI continues to be used as an important tool in the analysis of proteins and peptides, electrospray (ES) ionization, introduced by John Fenn in 1989 [78], has emerged as the dominant MS method for the analysis of most types of biomolecules. From an analytical perspective, ES ionization differs from MALDI largely in the fact that it is a solution-based ionization method and that it produces primarily multiply charged ions from proteins and peptides. While a description of the mechanistic features of ES ionization is outside of the scope of this chapter, it can be generalized that with the mass spectrometer operating in the positive ion mode, electrosprayed proteins and peptides will produce intact molecular ions containing approximately as many charges as there are basic groups on the molecule or for large proteins one charge for every 1000 Da of molecular mass. For proteins this usually results in a series of molecular ions, each containing a different number of charges. Most peptides are typically represented by only a few charge forms.

Perhaps the biggest advantage that ES ionization has over MALDI is that being a solution-based ionization method, it is easily interfaced to liquid chromatography, in particular RP-HPLC. Furthermore, because of voltage considerations in the ion source, ES ionization was not easily implemented on magnetic sector mass spectrometers and thus was interfaced primarily on low-voltage, less expensive quadrupole mass spectrometers, which also conveniently handled the flow rate of conventional HPLCs. Because of the multiple charging phenomenon, typical of ES, the charge envelope of a protein or peptide always falls more or less in the same mass range (m/z 500–3000), regardless of the size of the protein or peptide, and this mass range is also conveniently within the operating mass range of most quadrupole mass spectrometers. Thus for a variety of practical and theoretical reasons, ES was primarily implemented on quadrupole mass spectrometers, which made the technique much more accessible to a large research community and galvanized investigations into how to use this important new tool for biological research. The development of the triple quadrupole-tandem mass spectrometer [79] interfaced via an ES ion source made possible the direct sequencing of peptides as they flowed into the mass spectrometer [80].

Because the signal intensity produced in an ES mass spectrometer depends on concentration rather than absolute amount of the analyte, it is desirable to use the lowest flow rate possible to achieve the best sensitivity. Already by 1992 microcapillary HPLC flowing at 1–2 μL/min had been interfaced to an ES ion source [81]. In 1994 Wilm and Mann introduced the concept of nanoelectrospray [82] where they flowed peptide and protein samples at a rate of 20–30 nL/min from pulled glass capillaries [83]. This method was widely adopted and became known as static nanospray since no HPLC pumps or separations were employed. In the same year Emmett and Caprioli introduced a microelectrospray source that accommodated 300–800 nL/min, spraying directly from a capillary needle packed with C18 RP material [84]. The realization that ES was stable at these low flow rates quickly led to the development of nanoliter flow rate LC (NanoLC) coupled with ES ionization MS [85, 86]. NanoLC-MS technology, having matured over the last twenty years, when combined with modern biochemical and cell biology techniques, provides a tool that has the sensitivity and selectivity necessary for the analysis of *in vivo* biology regulated by protein phosphorylation.

To understand how phosphorylation regulates biology, the following questions need to be addressed: "Is a protein phosphorylated?" "Where is it phosphorylated?" "Is the phosphorylation regulated?" "What is the stoichiometry?"

Each of these questions poses a unique analytical challenge, and a variety of strategies utilizing MS can be applied to meet the challenges.

Unlike the riboflavin protein mentioned earlier [70], most proteins are phosphorylated at substoichiometric levels, often as low as a few percent. Therefore, in practice identifying phosphopeptides in anything other than the simplest mixtures can be quite challenging. In addition to the problem of low stoichiometry, the ion yield for peptides in general can be very different, with the result being that some peptides are difficult to detect regardless of the complexity of the sample. The ion yield for peptides is the product of both the ionization efficiency of the molecule (which in turn is related to proton affinity and gas-phase basicity) and the rate of vaporization of the molecule in the ES process. These factors are strongly dependent on the physicochemical properties of the side chains of the individual amino acids that make up the peptide. In general peptides with hydrophobic amino acids desorb more efficiently from the charged droplets created by the ES process, and basic residues enhance the ionization efficiency of the molecule. However, since both Arg and Lys are rather hydrophilic, more than a single Arg or Lys will decrease the desorption efficiency of the peptide [87, 88]. The influence of the addition of one or more phosphate groups to a peptide, and how this affects the ionization efficiency, is a matter of some debate. A commonly held belief is that phosphorylated peptides have lower ionization efficiencies than their nonphosphorylated counterparts. Conceptually this would seem to make sense. The phosphate group is negatively charged and therefore should affect the ionization efficiency under the acidic conditions employed in the positive mode. In addition, phosphoamino acids are very hydrophilic, thereby affecting the vaporization of the peptide from the charged droplets. In practice, however, it is evidently more complicated. In one of the largest studies done on this subject, a diverse set of twenty synthetic phosphopeptides (P) were mixed in two different defined ratios with their nonphosphorylated (NP) counterparts and measured by positive ion ES using low-pH RP-HPLC to introduce the samples. Surprisingly more phosphopeptides demonstrated better ionization efficiency than their nonphosphorylated counterparts, and in almost 70% of the cases the ratio of P/NP was within a factor of two [89]. Most of these peptides were nontryptic and several contained multiple basic residues, particularly arginine. In the case of very basic peptides, the ionization efficiency was strikingly charge state dependent. In contrast, a separate study examining 66 model peptides and their singly phosphorylated counterparts found that in nearly all cases the phosphopeptides had lower ionization efficiency than their nonphosphorylated counterparts [90]. But here again, the average difference in relative ionization efficiency was only a factor of two. In this study three sets of peptides were created based on a single sequence and all contained a C-terminal Lys residue. The charge and hydrophobicity in each set were varied by making fixed substitutions at specific sites within the sequence. However, there was no apparent correlation between the relative response of the peptides and their molecular weight, charge, or hydrophobicity. Interestingly, those peptides containing an Arg residue showed relative ionization efficiencies close to 1, consistent with the previous study and work in our own lab [91]. Since the most common approach to protein analysis by MS involves digestion of the protein with trypsin, it is fair to ask whether tryptic phosphopeptides have similar relative response ratios as those described previously. Nearly all of the peptides in the two studies described earlier were decidedly nontryptic. Unfortunately, no large studies using matched synthetic tryptic peptides have been conducted to assess the effect of phosphorylation on ionization efficiency. From the literature we were able to extract measured relative intensity ratios for 29 fully tryptic phosphopeptides and their counterparts [89, 92–96], and in all but one case the ratio was very nearly 1.

A less well-controlled but still informative alternative to using synthetic peptides to measure relative responses is to treat half of a phosphopeptide sample with phosphatase and then measure the response of the phosphorylated and dephosphorylated sample. Using a TiO_2-enriched phosphopeptide sample, Marcantonio [97] matched 452 tryptic phosphopeptides to their dephosphorylated counterparts and determined that on average the phosphorylated peptides had a 1.5-fold lower response. Whereas only 33% of singly phosphorylated peptides had a lower response than their dephosphorylated counterpart, multiply phosphorylated peptides were twice as likely to show a greater than twofold lower response.

From the aforementioned cases we can conclude that the ionization efficiency of phosphopeptides relative to their nonphosphorylated counterparts is difficult to determine. For singly phosphorylated peptides, they would appear, in general, to have a response similar to that of the nonphosphorylated counterpart, in most cases within a factor of two. However, there are certainly cases where this is not true and where the difference is dramatic. Unfortunately, these differences are not predictable. Even the position of the phosphorylated residue in an otherwise unaltered sequence can have a dramatic effect on the response [98]. The uncertainty in the relative response of the phosphorylated and nonphosphorylated form of a given peptide makes determining phosphorylation stoichiometry problematic. The extent of phosphorylation is determined by dividing the abundance of a phosphopeptide by the sum of the abundances of the phosphorylated and nonphosphorylated forms. For measurements made by MS, the ion intensities for all charge states of each form are used. However, this determination only holds true if the responses of the two forms of the peptide are equal. While the data presented earlier would suggest that at least for singly phosphorylated peptides any error resulting from making this assumption is likely to be small, we prefer, in the absence of any empirical evidence, to refer to determinations made in this way as "apparent" stoichiometry. The uncertainty in these apparent stoichiometries is nevertheless an uncomfortable situation, and so alternative ways to determine phosphorylation occupancy are important to consider.

The most straightforward, simplest, and by far most widely applied approach to deriving absolute stoichiometry is to synthesize the phosphorylated and nonphosphorylated forms of the peptide and measure their relative responses in a series of mixtures. If the measured responses are linear over a range of mixtures, then the response ratio (if not equal to 1) can be used to correct the apparent stoichiometry [94, 99]. If the peptides are synthesized using stable isotope-encoded amino acids, then the peptides can be added directly to the sample, providing both a measure of absolute abundance and a way to calculate stoichiometry [100]. This latter approach has the advantage that if the target peptide is produced with ragged ends due to adjacent tryptic cleavage sites, peptides, which extend across the cleavage sites, can be synthesized and added to the sample prior to digestion. In this way internal standards are produced regardless of the preference of the enzyme for the cleavage sites [100]. Peptide synthesis has become relatively inexpensive, and it is now possible to buy peptides in smaller quantities. It is still a requirement that the peptides are pure or the final determination of the absolute stoichiometry will be no more certain than the apparent stoichiometry.

Clearly this approach does not scale well as the number of sites to be determined increases. Nor is it always possible to synthesize peptides for certain long or multiply phosphorylated species that are unstable over time. As an alternative, our laboratory described an approach that uses phosphatase digestion and differential chemical labeling to determine absolute stoichiometry [101]. As shown in Figure 2.3a, an enzymatic digest is split into two equal parts, and one part is treated with phosphatase and the other undergoes a mock treatment. Both samples are then derivatized with a chemical label, one light and the other containing heavy stable isotopes. After mixing, the sample is analyzed by MS. All nonphosphorylated peptides will appear as doublets with equal intensity ratios. The presence of a doublet with a different intensity ratio followed by a singlet peak 80 Da higher in mass is an indication of a partially phosphorylated peptide. In the example shown in Figure 2.3, a sample containing the three peptides shown (Figure 2.3b) is taken through the protocol and reanalyzed (Figure 2.3c). The nonphosphorylated peptide (■) generates two peaks of equal intensity at m/z 723.81 and 726.31. The phosphorylated (○) and nonphosphorylated (●) peptide pair give rise to three peaks. The peak at m/z 706.31 arises from the half of the sample that was treated with alkaline phosphatase. It represents the total amount of the sequence present in that half of the sample (sum of phosphopeptide and nonphosphopeptide). The peak at m/z 703.81 comes from the half of the sample that was not treated with phosphatase, and it thus represents the original amount of nonphosphorylated peptide. The peak at m/z 743.81 also comes from the half of the sample that was not treated with phosphatase, and it thus represents the original amount

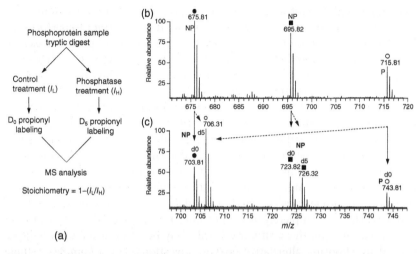

(a)

Figure 2.3 Determination of absolute stoichiometry. (a) Schematic showing the generalized protocol. Any suitable label that provides an adequate mass shift may be used. (b) Partial ES mass spectrum of a sample containing a phosphorylated (O) peptide and its nonphosphorylated counterpart (●). The apparent stoichiometry determined from this spectrum is 31%. (c) After being taken through the protocol, each peptide adds one propionyl group, either d0 or d5. The absolute stoichiometry determined from the nonphosphorylated d0/d5 cluster is 45%. See text for details.

of the phosphorylated peptide. The stoichiometry (S) is determined from the intensities of the nonphosphorylated pair using the equation $S = 1 - (I_L/I_H)$, where I_L is the intensity of the untreated sample tagged with the light label (L) and I_H is the intensity of phosphatase-treated sample tagged with the heavy label (H). In this case there is no question of different ionization efficiencies, and the increase in intensity of the nonphosphorylated peptide originating from the phosphatase-treated sample represents that portion of the sequence that was originally phosphorylated [101]. In this example the absolute stoichiometry is found to be 45%. Peptides that are stoichiometrically phosphorylated undergoing this protocol would be represented by a nonphosphorylated heavy-labeled singlet and a corresponding light-labeled phosphorylated singlet.

This basic strategy has been adapted to a variety of chemical labels [93, 102, 103] and was recently shown to be applicable on a large scale. Gygi and coworkers determined the stoichiometries for more than 5000 phosphorylation sites in asynchronous cultures of *Saccharomyces cerevisiae*. This study showed that only about 10% of the sites identified showed full or nearly full occupancy, confirming the overall low stoichiometry and suggesting that phosphorylation regulates function by influencing only a small fraction of the available protein molecules [102].

As simple and elegant as the previous approach is, there are a number of important caveats that need to be kept in mind. It is important to perform the phosphatase reaction on the peptide level, as the activity of phosphatases on proteins may be unpredictable. Differentially phosphorylated peptides will collapse down to a single nonphosphorylated species. Thus the determined stoichiometry will be for the total amount of phosphate on that peptide. It will be impossible to discriminate the stoichiometry for multiple sites of monophosphorylation on a single peptide or the contribution of mono- and diphosphorylated peptides sharing the same sequence. Finally there are two circumstances that are problematic for any method used to determine site-specific stoichiometry. The first is the ability to distinguish the relative contribution of two different sites on the same peptide, and the second is when a site occurs adjacent to an enzyme cleavage site. In the case of the latter, the nonphosphorylated peptide will have a completely different (shorter) sequence. For a single protein or a simple mixture, these situations will likely be recognized and might be resolved by choosing alternative enzyme digestions. For a complex sample such as a cellular proteome, these circumstances will be more difficult to resolve and will undoubtedly complicate the interpretation of the data.

Once a peptide has been identified as being phosphorylated, the next major analytical challenge is to localize the site of phosphorylation. Peptides are most commonly sequenced by MS/MS using collision-induced dissociation (CID). Collisions with an inert gas cause the protonated peptides to fragment at the amide bonds along the backbone. Because the proton, which directs the fragmentation, can reside at multiple locations along the backbone, a distribution of fragment ions will result. Charged fragments containing the N-terminus form a b_n ion series, while charged fragments containing the C-terminus constitute a y_n ion series (a subscripted number denotes the nth amino acid from the terminus where the fragmentation occurred). Fragmentation along the backbone is not completely random, since certain amino acids influence the localization of the proton. This influence is manifest in the different intensities of the various b and y ion fragments. In addition to the amide bond cleavages, both peptide molecular ions and fragment ions can undergo neutral loss from the amino acid side chains. With a few exceptions, phosphopeptides fragment much like nonphosphorylated peptides. A detailed description of peptide fragmentation via CID can be found elsewhere [104].

Fragmentation readily occurs at those locations where the least amount of energy is required to break a chemical bond. Fragmentation of a phosphopeptide on the side chain of a phosphorylated amino acid is an energetically favorable process that competes very effectively with cleavage at the amide bonds. The facile cleavage of H_3PO_4 (98 Da) and HPO_3 (80 Da) from the phosphopeptide precursor ion often gives rise to very abundant neutral loss ions ($M-H_3PO_4$ and $M-HPO_3$, respectively). In the case of phosphoserine- or phosphothreonine-containing peptides, the neutral loss of H_3PO_4 can dominate the spectrum, suppressing the formation of sequence-specific b and y ions. Phosphotyrosine-containing peptides on the other hand are not able to lose H_3PO_4. The aromatic ring on tyrosine stabilizes the carbon–oxygen bond. Since the oxygen–phosphorus bond is much weaker, cleavage here results in the loss of HPO_3. Because the loss of the phosphate group from tyrosine is much less facile than from either serine or threonine, the resulting neutral loss ion is much less abundant in the spectrum of phosphotyrosine-containing peptides. The charge state of the precursor and the number of basic residues (related characteristics) also strongly influence the neutral loss of phosphate. In general, higher charge state precursors tend to show a lower degree of neutral loss. A detailed examination of the neutral loss pathway [105] and the sequence-related factors that influence it [106] is described elsewhere.

The prevalence of fragment ions resulting from the neutral loss of phosphate is also dependent on two nonsequence-related factors. The collision energy imparted during the CID process and the time frame in which it takes place can have a profound effect on the production of neutral loss ions. Ion trapping type instruments use resonant excitation of cool ions to induce hundreds of collisions over a relatively long period of time. Only a few vibrational quanta of energy are transferred with each collision [107]. This slow heating process favors low-energy fragmentation pathways such as the neutral loss of phosphate. Thus ion trap MS/MS spectra of serine- and threonine-phosphorylated peptides tend to be dominated by very abundant ions resulting from the neutral loss of phosphate from the precursor (taking into account the other factors described earlier) [108]. On the other hand, instruments that are tandem in space accelerate ions into a gas-filled collision cell, where only a handful of collisions occur and much greater energy is transferred at each collision (typically 15–40 eV). CID under these conditions tends to favor backbone cleavages. Technically referred to as low-energy CID (relative to magnetic sector instruments where collision energies are in the 5–10 KeV range), this process still imparts much more energy than the very-low-energy resonance excitation process. Furthermore, the fragment ion spectrum can be readily tuned by adjusting the collision energy through increasing or decreasing the accelerating potential on the precursor ion. Recently, separate multipole collision cells have been introduced into orbitrap and quadrupole orbitrap instruments, combining the best features of true collision cell CID and the sensitivity and high performance of the orbitrap. Unfortunately termed higher-energy collision-induced dissociation (HCD), the collision energies in this regime are nevertheless similar to those of conventional low-energy CID. As expected, the phosphate neutral loss is less prevalent on these instruments, and the fragmentation pathways can be controlled to some extent by adjusting the collision energies.

The human proteome contains nearly 15% serine, threonine, and tyrosine by amino acid composition. Thus most phosphopeptides will have more than one possible location for the phosphorylation to reside. To be certain of the phosphorylation site location, an ion series of primary backbone fragments should run across the phosphorylated residue. In practice this is not so easily achieved. Phosphoserine- and phosphothreonine-containing b_n and y_n ions readily lose phosphate to produce b_n-H_3PO_4 and y_n-H_3PO_4 ions (designated b_n▲ and y_n▲, respectively). Interestingly, b_n ions are much more likely to exhibit this loss [106]. It is not clear whether this is because the neutral loss from a b_n ion forms a more stable product or that b_n ions are more likely to undergo a neutral loss. Regardless, this suggests that since y_n ions are more likely to contain an intact phosphoamino acid, this series may be more valuable in localizing the site of phosphorylation. In addition to the loss of phosphate, peptides can also readily lose H_2O (-18) and NH_3 (-17) as neutrals from both the precursor and backbone fragment ions. For peptides phosphorylated on serine or threonine, the subsequent loss of phosphate and water from b_n and y_n ions dramatically complicates the assignment of the phosphorylation site.

The spectrum shown in Figure 2.4a has been assigned to the sequence SASQSSLDKLDQELK plus two moles of phosphate. An unmodified b_2 ion (SA-) indicates no phosphorylation on S1. The next two b ions resulting from cleavage after S3 (SAS-) and Q4 (SASQ-) are represented as b_3▲ and b_4▲, suggesting S3 is phosphorylated. The b_5▲ ion representing the fragment SASQS- could result from the loss of phosphate from either S3 or S5; however, since there is no b_5▲▲, the latter is more likely. Cleavage after the next amino acid (SASQSS-) results in a b ion that losses two phosphate groups to produce b_6▲▲. This suggests that the second phosphate group is on S6 not S5. However the lack of an intact ion b_5 or b_6 ion leaves some doubt. There is an intact y ion series from y_1 to y_9 confirming the C-terminal sequence. Unfortunately the next y ion (-SLDKLDQELK) could represent either y_{10}▲ or y_{10}•, and the possibility of phosphorylation on either S5 or S6 converges after y_{10}. The appearance of y_{13}▲▲ (two losses of phosphate) reinforces the assignment of phosphate on S3. So from this spectrum the phosphorylation is assigned to S3 and S6, but the evidence is not unambiguous.

The second spectrum shown in Figure 2.4b is from a singly phosphorylated peptide with the sequence TSSIADEGTYTLDSILR. An extensive primary y ion series from y_1 to y_{14} indicates that there is no phosphorylation on the four possibilities in the C-terminus. A single N-terminal fragment, represented by an unmodified b_2 ion, suggests that the phosphate is on S3. With no other supporting evidence, point mutants were made for each the first three amino acids, and phosphorylation exclusively on S3 was confirmed using an *in vivo* activity assay [109].

The ability of MS/MS to localize sites of phosphorylation may be further compromised by the recent evidence that in the gas phase, phosphate groups can migrate to other hydroxyl-containing amino acids [110, 111]. In a study of

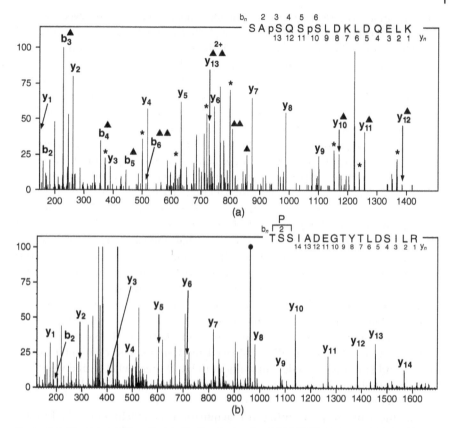

Figure 2.4 Phosphorylation site localization by tandem MS. (a) CID spectrum of a doubly phosphorylated peptide assigned by MASCOT to the sequence SASQSSLDKLDQELK. The doubly charged precursor was fragmented using HCD on a quadrupole orbitrap hybrid. Although the spectrum is not dominated by the neutral loss of phosphate from the precursor, the abundant loss of H_3PO_4 from fragment ions makes localization of the sites difficult; however there is evidence for the localization as shown. See the text for details. (b) CID spectrum of a singly phosphorylated peptide assigned to the sequence TSSIADEGTYTLDSILR. The doubly charged precursor was fragmented using low-energy CID on a quadrupole time-of-flight hybrid. Notice the residual unfragmented precursor (marked •). Neutral loss of phosphate from the precursor is a minor ion (not marked). With only a single N-terminal fragment, it was not possible to unambiguously assign the phosphorylation on any of the first three residues. See the text for details. Peptide fragment ion nomenclature is that of Biemann [104], except that b_n or y_n ions marked with either ▲ or ▲▲ refer to b_n or y_n-H_3PO_4 or b_n or y_n-$2(H_3PO_4)$, respectively. Peaks labeled with only ▲ or ▲▲ refer to $[M+2H]^{2+}$ minus one and two H_3PO_4 groups. Loss of water from the preceding b_n or y_n ion is marked *.

thirty-three fully tryptic synthetic phosphopeptides, 45% showed evidence for rearrangement of the phosphate group. These rearrangements were most prevalent in peptides that showed a significant neutral loss of phosphate from the precursor. Interestingly, these rearrangements were not observed in mass spectrometers that are tandem in space [110], the much shorter time frame of the CID process being inconsistent with the rearrangement reaction [112].

The inability to localize the site of phosphorylation due to a lack of backbone fragmentation has led to the development of methods, primarily in trapping instruments, to activate and sequence the prominent $M-H_3PO_4$ ions [113]. After a typical MS/MS (MS^2) data acquisition sequence is performed and the neutral loss ion identified in a data-dependent manner, the trap is refilled and the sequence repeated. This time however, following the initial MS^2 fragmentation, the neutral loss ion is isolated, activated, and fragmented. Since the precursor for this round of fragmentation does not contain a phosphate group, these MS^3 spectra contain primarily backbone cleavages. Phosphoserine and phosphothreonine residues will have the in-chain masses of dehydroalanine (69 Da) and dehydrobutyric acid (83 Da), respectively. A complicating issue in this approach is the recent evidence that what appear to be $M-H_3PO_4$ ions might actually derive from the loss of HPO_3 (-80) followed by the simultaneous loss of H_2O (-18) from a nearby serine or threonine [110]. In this case, the MS^3 spectrum of an M-98 ion would be incorrectly interpreted. The actual phosphoserine residue would have the in-chain mass of an unmodified serine, and the actual unmodified serine would have the in-chain mass of dehydroalanine. Even more complicating would be the situation where the M-98 ion was a mixture of $M-H_3PO_4$ and $M-HPO_3$ followed by $M-H_2O$. The combined loss of HPO_3 and H_2O was much less prominent in HCD spectra [114], and we would expect this also to be true for conventional low-energy CID tandem-in-space instruments.

An alternative to performing MS^3 acquisitions is a technique called multistage activation (MSA) [115]. In this approach the phosphopeptide precursor ion is activated and fragmented as usual, but the fragment ions are left in the trap, and the m/z at which the neutral loss ion would appear is further activated and undergoes additional fragmentation. All the ions are collected in one spectrum, which is a composite of the first and second activation processes. The main advantages of the MSA acquisition over an MS^3 acquisition are that they take much less time, that they result in only a single spectrum, and that the signal-to-noise ratio of MSA spectra is much better.

The value of performing either MS^3 or MSA is debatable. For proteome-scale studies it is not clear that the impact on the duty cycle of the experiment is offset by an increase in phosphopeptide identifications. The evidence in the literature is quite conflicting [113, 116–119]. The assumption is that while fewer spectra are acquired, the quality of these spectra will result in more identifications. MSA would seem to have an advantage here in that it does not impose as large a hit on the sequencing duty cycle. In practice MS^2 CID spectra, while not very "good looking," frequently contain sufficient backbone fragmentation to identify the sequence and in many cases localize the site of phosphorylation. The evidence that the additional stages of activation add anything substantial to the sequence data already present in an MS^2 spectrum is not compelling. In our own laboratory we find that more class 1 phosphopeptide identifications are made with MS^2 than with MSA, even if the MASCOT scores of the latter are better. In practice these approaches are not widely used in phosphoproteomics studies, but may prove more useful in the analysis of isolated phosphoproteins or very simple mixtures where the duty cycle imposed by the experiment is not an issue.

In addition to the sequence and structural features unique to phosphopeptides, certain sequence features common to all peptides complicate the localization of phosphorylation sites. Very long peptides often suffer from the fact that ions series tend to die out as the collision energy is dissipated over the molecule, leaving large gaps in the sequence information. Tryptic peptides with missed cleavages have internal basic amino acids, which hold a charge firmly and inhibit cleavages based on the mobile proton model of fragmentation. Peptides can undergo internal cleavages that lead an out-of-context sequence series. A problem unique to ion trap instruments is the lack of information at the low mass end of the spectrum due to the 1/3 cutoff rule. An unwelcome feature of resonance excitation is the loss of all ions below 1/3 the mass of the precursor. This problem is not an issue in hybrid trapping instruments that employ an HCD collision cell or in other mass spectrometers that are tandem in space.

To mitigate the labile nature of many PTMs, including phosphorylation and the impact this has on localizing the site of modification as well as circumventing some of the other less desirable sequence-specific features of CID, low-energy electron-based dissociation techniques have been developed. In electron capture dissociation (ECD) [120, 121], protonated precursor ions capture low-energy electrons (0.2 eV), leading to rapid charge neutralization and fragmentation that is very specific to the amino acid N-Cα bond. ECD spectra are dominated by backbone fragments of the c and z• (radical) type and show very little neutral loss or side chain fragmentation. ECD spectra of phosphopeptides are characterized by long runs of contiguous sequence ions, where the phosphate group stays intact. Unfortunately ECD suffers from low fragment ion yields, poor sensitivity, and the need to perform the experiment in an ion cyclotron resonance mass spectrometer (FTICR).

Electron transfer dissociation (ETD) was developed by Syka and Coon [122] a few years later as an alternative electron-based dissociation technique that could operate in an ion trap. During ETD, radical anions are formed in a separate chemical ionization source and then mixed with protonated peptides in the ion trap. Electrons are then transferred to the peptides by ion–ion reactions. Although the ionization process is different, the fragmentation mechanism in ETD is thought to be mechanistically similar to ECD, producing the same type of fragment ions and little or no neutral loss fragmentation.

Common to both types of electron-based dissociation methods is the fact that they work more efficiently on smaller peptides with a charge of 3^+ or higher, that is, peptides with a greater charge density [123]. Trypsin, the most commonly used enzyme for digestion in proteome studies, produces mostly peptides with a 2^+ charge. In a comparison of CID and ETD for tryptic peptides [124], less than 1% of the identified ETD peptides were 2^+. Interestingly, Lys-C peptides, fifty percent of which would be expected to contain an internal arginine and therefore be 3^+ or higher, performed no better than tryptic peptides [125]. This is likely due to the fact that Lys-C peptides containing an internal arginine would be longer and therefore not have an increase in charge density. To account for the fact that CID outperforms ETD for 2^+ peptides, a decision tree strategy has been developed, which allows the mass spectrometer to opt for CID or ETD depending on the charge state and m/z of the peptide [124]. Recently, it has been shown that HCD outperforms ETD in a phosphoproteome experiment for the identification of phosphopeptides and the localization of the site [32, 126], although ETD does perform better on multiply phosphorylated peptides. HCD is now in widespread use on commercial orbitrap-based mass spectrometers, providing true collision cell fragmentation, with no low mass cutoff and high-resolution mass measurement of fragment ions. The most current generation of quadrupole TOF mass spectrometers offers similar performance characteristics. While ETD remains an attractive alternative ionization technique, its use is still not widespread, and currently it does not show signs of replacing low-energy CID for the routine sequencing of phosphopeptides.

Isomeric phosphopeptides, which are identical in sequence but carry the phosphate group on different residues, can also cause difficulties in analysis. Isomeric phosphopeptides that cannot be separated by standard LC procedures will not be resolved in the MS1 precursor scan due to their identical masses. This limitation will result in a mixed fragment ion population in MS2 and subsequently confound phosphosite localization efforts [127]. Gas-phase ion mobility separation is a way to resolve this circumstance. Usually these devices operate between the LC system and the first mass analyzer and separate different ionic species based on their differential mobility through an electric field that oscillates in strength. Field asymmetric waveform ion mobility spectrometry (FAIMS) has been particularly useful in separating isomeric phosphopeptides [128], as has been described from a study of a *Drosophila* cell phosphoproteome [129].

Continuous significant improvements in MS performance over the last ten years have brought about increasingly faster data acquisition rates and dramatically improved sensitivity that has been coupled with higher mass accuracy and resolution. Despite the overall low stoichiometry, the overwhelming number of nonphosphorylated peptides, and the potential for lower ionization efficiency, high-performance tandem mass spectrometers in use today are very effective at identifying phosphorylation sites in relatively simple mixtures such as the tryptic digest of an immunopurified protein. Rather than attempting to identify phosphorylated peptides based on differences in an MS1 spectrum and targeting those peptides for sequencing, MS/MS is used to shotgun sequence as many peptides as possible and phosphorylated peptides are distinguished from nonphosphorylated peptides by sophisticated database search algorithms that use accurate mass and fragment ion spectra. In this approach phosphopeptides are identified and phosphorylation sites localized in the same event (taking into account all the caveats described earlier). The sequencing speed and high sensitivity of state-of-the-art mass spectrometers ensure very high sequence coverage (80–95%) for even very faintly stained Coomassie Blue bands on SDS-PAGE gels. While high sequence coverage is not a guarantee that all phosphorylation sites will be identified (it only guarantees you have found all the nonphosphorylated peptides), it does increase your chances. If the overall sequence coverage for the protein is low, then digestion with an alternative enzyme is a useful strategy. Alternate enzymes (e.g., lysyl endopeptidase, endoproteinase Asp-N, endoproteinase Glu-C (V8), chymotrypsin, elastase), used alone or in combination with trypsin, can improve sequence coverage and phosphorylation site identification [130–132]. Increasing the amount of starting material can make a difference, as can switching from an in-gel digest to a solution digest. If the analysis of a phosphoprotein is to be done by MALDI, then some form of enrichment (described later) will be useful.

For the identification of phosphopeptides from more complex mixtures such as multiprotein complexes or intact proteomes, the same basic shotgun sequencing approach is coupled with a phosphopeptide enrichment strategy. In most cases the global analysis of protein phosphorylation will also require a fractionation step at the peptide or protein level prior to enrichment.

2.3 Global Analysis of Protein Phosphorylation by Mass Spectrometry

With the advancement of MS instrumentation that allows for the rapid and sensitive sequencing of peptides and the concurrent development of phosphopeptide enrichment strategies, we now have the ability to analyze changes in protein phosphorylation at an unprecedented scale. These studies have led to a rethinking of the prevalence of phosphorylation throughout the proteome and have revealed the true complexity of signaling pathways in living cells.

Global phosphoproteome studies initially focused on qualitative cataloging of phosphosites in a number of model organisms. One of the earliest global studies analyzed the phosphoproteome in the budding yeast *S. cerevisiae* and identified 383 phosphosites, 365 of which had never been identified before [133]. Since that time, the yield of global phosphoproteomics studies has expanded to the point where it is not uncommon to identify >20,000 phosphosites from a single biologic sample [5, 33, 132, 134, 135]. From hundreds of studies, more than 300,000 phosphorylation sites have been identified across a multitude of both prokaryotic and eukaryotic species, with more than 200,000 of these coming from mammals alone. Greater than 95% of these sites have been identified using MS-based phosphoproteomics [29]. Prior to the onset of global phosphoproteomics, it was often estimated that about 30% of proteins could be phosphorylated [136], but these studies now suggest that up to 75% of the eukaryotic proteome is phosphorylated in at least some cell or tissues [5], and the true percentage may be even higher. Based on the natural abundance of serine, threonine, and tyrosine, there are approximately 700,000 potential sites of phosphorylation in the human genome alone [137].

For a true comparative analysis of global phosphorylation, a quantitative proteomics strategy is necessary. Quantitative phosphoproteomics allows researchers to investigate signaling pathways in different model systems to identify phosphorylation events that vary in terms of abundance and duration as a result of a given stimulation [138]. A particularly challenging aspect of signal transduction research is the identification of the protein kinase and phosphatases that are activated by a given stimulus. Global quantitative phosphoproteomics, in conjunction with a suite of bioinformatic tools, has been particularly useful in addressing this challenge [139]. The integration of kinase consensus motifs that are enriched among the phosphopeptides that are identified to be modulated by the stimuli of interest with known protein–protein interactions can lead to the initial mapping of the pathways affected (see following section on bioinformatics).

The success of these large-scale studies often relies on choosing the right biological system and the correct controls. Several different approaches have been used to map signaling pathways [138]. A common technique is to expose a cell to an extracellular agonist of a receptor of interest and measure changes to the cellular phosphoproteome (Figure 2.5a), sometimes in combination with small-molecule inhibitors to determine the roles of individual pathway components. One example of this approach is to compare starved cells treated with serum and starved cells treated with serum and a MAPK inhibitor to map the MAPK-dependent phosphoproteome [140]. A similar approach is to compare cells with aberrant signaling (i.e., as a result of expression of a constitutively active kinase) to the same cells treated with a chemical inhibitor of that pathway. This method has been used to study signaling downstream of mTOR and BRAF [141–143]. When specific inhibitors are not available, RNAi-mediated knockdown or CRISPR-based genome editing can be used to inactivate components of signaling pathways to see how the phosphoproteome is affected [144]. It is important to remember that with chemical inhibition, direct, short-term effects on signaling can be studied, whereas with genetic knockdown, the effects will be more long term and may be more likely to be indirect.

Spatial contexts can also be assessed by using subcellular fractionation to study how discrete components of the cell respond to a stimulus [127], and moreover this can help to enrich for relevant phosphorylation sites. Time-course studies can add another layer of resolution to signaling pathway maps (Figure 2.5b). By collecting and analyzing samples at several time points after a stimulus, it can be possible to map waves of phosphorylation events and more

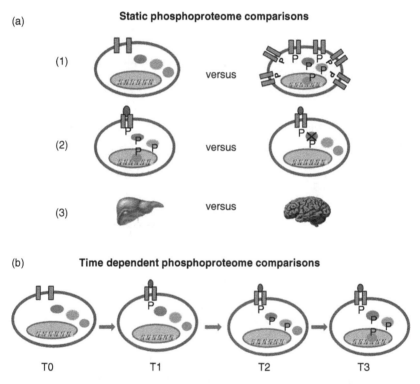

Figure 2.5 Quantitative global phosphoproteomics studies. (a) Static comparisons of phosphoproteomes can come in a variety of forms. This could include (1) a comparison of a normal cell to a cancerous one or to the normal cell where a constitutively activated kinase is overexpressed, (2) a comparison of a cell line with an activated signaling pathway to one where a component of that pathway has been either genetically or pharmacologically ablated, and (3) a comparison of 2 different cell types or tissues. (b) Time-course studies allows phosphorylation events downstream of a given stimulus to be ordered. By mapping waves of phosphorylation as a function of time, signaling pathways can be organized from stimulus to phenotype, and hypotheses about epistatic relationships between pathway members can be generated.

easily trace how early phosphorylation events lead to later ones. The choice of time points for a quantitative phosphoproteomics experiment should be carefully considered. For example, tyrosine phosphorylation of transmembrane receptors immediately downstream of external stimuli often peaks at 5–10 min after stimulus [145], but phosphorylation of transcription factors as a result of a stimulus can often last for many hours [146].

Although global phosphoproteomics studies can generate testable hypotheses and allow inferences to be made about the pathways that are modulated in a particular cell state, understanding the functional significance of any given phosphorylation site requires further validation [138], which usually involves mutation of the phosphosite. Deciding how to measure the effect of this mutation may depend on prior knowledge of the function of the protein. Disruption of the phosphosite may affect the protein structure or localization or affect interactions with other proteins. Phosphosite mutations can also have effects on other PTMs, as it has been shown in multiple studies that phosphorylation can both positively and negatively regulate other modifications such as ubiquitylation, acetylation, and glycosylation [127, 138, 147]. Finally, when a

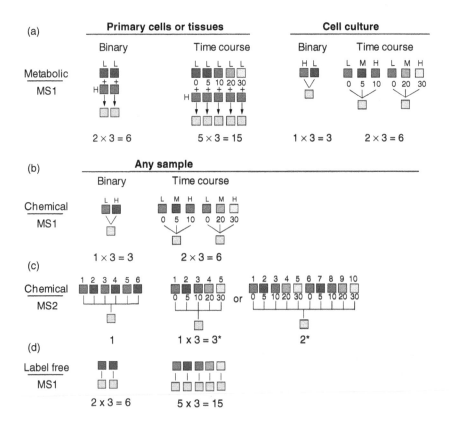

phosphosite mutation causes a loss of function, it is important to rule out that the mutation did not result in nonspecific unfolding of the protein.

Quantitative phosphoproteomics studies usually follow similar strategies to conventional global protein-centric quantitative proteomics experiments. Both label-free and stable isotope labeling strategies have been successfully applied to phosphoproteomics studies. Metabolic labeling as typified by stable isotope labeling by amino acids in cell culture (SILAC) is an efficient strategy for both expression proteomics and phosphoproteomics studies [148], particularly for cell culture experiments (Figure 2.6a, right). It has been demonstrated that whole organisms, up to the rat, can be metabolically labeled [149, 150], though this approach is time consuming and expensive. In recent years isotope-labeled chemical tags have become more common, as these approaches are more easily adapted to primary cells and tissues (Figure 2.6b and c). Chemical tags include nonisobaric tags, such as dimethyl labeling and mass tags for relative and absolute quantification (mTRAQ), where quantitation is done using the MS1 scan [151, 152], and isobaric tags, such as isobaric tags for relative and absolute quantitation (iTRAQ) and tandem mass tags (TMT), where quantitation is

---

Figure 2.6 Quantitative strategies for phosphoproteomics studies. (a) Metabolic labeling approaches as typified by SILAC are most efficiently utilized with samples from cell culture (right side). In simple binary comparisons the control (■) cells are grown in heavy label (H) and the experimental cells (■) in light (L) label. After harvesting, the populations are mixed to produce one sample (▣). Three biological replicates of the experiment would require three final samples (1×3=3). A third experimental condition can be evaluated using an intermediate or medium (M) labeled population. By using a common condition in a triple-label experiment, a 5-point time course can be evaluated with two final samples. Final number of samples for three biological replicates would be six. In cases where it is difficult or impossible to metabolically label a sample such as primary cells or tissues (left side), a labeled sample (▣) can be produced from cell culture or whole organism labeling and spiked into the experimental samples including the controls. In this case, peptides from the labeled sample act as internal references permitting comparisons across all the samples. While enabling accurate relative quantification across multiple samples, this approach has no option for multiplexing, and the number of final samples is equal to the number of experimental conditions. (B) Chemical labeling strategies where the quantitation is done using precursor ion intensities from the MS1 scans are set up similar to the way metabolically labeled cell culture experiments are performed. This is largely due to the limited number of labeling options for MS1 readouts. The major advantage of this approach is that it can be used with any type of sample. (c) Chemical labeling where the quantitation is derived from reporter ions in the MS2 spectrum have a big advantage in terms of multiplexing. Each sample, up to ten, is separately labeled with a different reporter ion (1, 2, 3, …) and the samples combined. In the simple binary comparison, all three biological replicates can be evaluated in one final 6plex sample. The five-point time course can be done as three 5plex samples, as shown (*). Additional time points (up to ten) might be evaluated in the approach, with no additional cost in instrument time. Alternatively, an additional replicate might be added and all four samples analyzed in two 10plex experiments. (d) Label-free quantitation has the advantage that the sample is analyzed with any prior manipulations. However, there are no options for multiplexing. Each sample is analyzed independently.

done on the MS2 scan [153, 154]. Although there are few studies that have directly compared these methods specifically for phosphoproteomics studies, at least one indicates that iTRAQ is a better choice than mTRAQ for these types of experiments [155]. Using a HeLa cell-derived sample, iTRAQ-based quantitation resulted in threefold more phosphopeptides being identified when compared to mTRAQ. This is likely due to less complex MS1 scans in iTRAQ experiments, as all labeled peptides are isobaric. An advantage that MS2-based isobaric labeling approaches have over the various MS1-based approaches is in multiplexing (Figure 2.6c). MS1-based approaches are typically 2plex experiments, though it is possible to perform a 3plex experiment with triple-label SILAC (Figure 2.6a, right) or differentially labeled chemical tags (Figure 2.6b). The increase from 2- to 3plex comes with a loss in sensitivity, however, as the sample becomes proportionally more complex. Isobaric tagging methods, on the other hand, are commonly available in 6, 8, and 10plex formats, with no significant loss in sensitivity since the complexity in the MS1 scan is unchanged.

On the other hand, reporter ion compression is a problem that is particular to isobaric labeling approaches. If another precursor (peptide or otherwise) is coisolated with the targeted peptide precursor ion in the MS isolation window, the resulting MS/MS spectrum will have fragment ions and reporter ions from both peptides. The targeted peptide will likely still be identifiable using this spectrum, but the reporter ion ratios may have been diluted or distorted by the reporter ions arising from the other peptide [156]. Since most peptides are present at a 1:1 ratio in the given samples being studied, ratios tend to be compressed by this interference toward 1, resulting in an underestimation of actual protein/peptide abundance differences. It has even been suggested that fragmentation of this background noise can lead to this compression problem [134]. This problem can be magnified for quantification of post-translationally modified peptides, such as phosphopeptides. For protein-level quantitation, ratio compression effects on individual peptides can be diluted out by measurements of other peptides where there is no compression. Phosphopeptide quantification, however, often relies on just one or two spectra.

Several methods have been adopted to minimize this challenging issue. Narrowing the isolation window and targeting a peptide for MS/MS when it reaches its apex of elution have both been applied to improve isobaric quantitation [157]. Reducing sample complexity using high-resolution sample fractionation has been shown to somewhat alleviate this problem, although this can significantly increase analysis time [158]. Removal of coisolated peptide ions of different charge states than the precursor improves the quantification accuracy as well; however this requires specialized instrumentation [159]. MS3 methods (described earlier) has been shown to drastically decrease the reporter ion contamination but significantly affects the sensitivity of the analysis [160]. More recently, coisolation and cofragmentation of multiple fragments have been shown to increase the reporter ion signals and increase sensitivity and accuracy of quantitation compared to standard MS3 methods [134, 161]. As of now, this feature is only available on the most state-of-the-art instrumentation, although in theory it could be applied to most instruments that are amenable to isobaric quantitation. Finally, a computational approach has been developed that estimates the degree of ratio compression for each tandem mass spectrum based on potential contaminating peaks observed in the preceding and subsequent MS1 spectra and corrects for this interference accordingly [162]. This computational approach is more amenable to protein-based quantitative studies rather than phosphoproteomics studies, however, given the variability seen with this correction on a peptide-to-peptide basis.

Label-free quantitation is an alternative to isotope labeling strategies (Figure 2.6d). For this approach, the ion intensity of a peptide is measured over its chromatographic elution profile [163]. The integrated intensities across the peak for a given peptide are compared between LC-MS runs of different samples to measure a relative abundance of the peptide between samples [164]. Sample from any biological source can be analyzed using this approach, without the significant extra cost of stable isotope labeling. While this seems like a relatively straightforward approach, it is essential that individual chromatograms can be unambiguously assigned to a given peptide to ensure accurate quantitation. In addition, ionization efficiency of a peptide is affected by the presence of coeluted peptides, and therefore changes in retention time can have dramatic effects on ionization and measured intensity [165]. Therefore, it is imperative to have a highly reproducible LC-MS system for these experiments, with narrow peak widths and robust retention time stability. High mass accuracy and resolution of the instrument can significantly help with unambiguous peptide assignment. In addition, chromatographic peak alignment software is necessary to define peptide elution profiles across multiple data files [166]. These types of experiments often require multiple technical replicates to make sure that the LC-MS and data analysis tools are robust. Unlike metabolic labeling strategies, the number of samples that can be compared is not limited. However, each sample that is added to the experiment dramatically increases instrument time, as each must be analyzed individually (Figure 2.6d). Nevertheless, this approach has been successfully utilized for a number of global quantitative phosphoproteomics studies [167, 168].

Quantitative proteomics expression studies almost always rely on quantitative data from multiple peptides that are used together to infer the relative protein abundance. This is an important distinction from quantitative phosphoproteomics studies, where usually quantitation of a phosphosite is based on a single, and at most a few, phosphopeptides. Therefore, there is usually less confidence built into a quantitative phosphosite experiment than there would be for a protein abundance experiment, since there are significantly less intensity measurements used for calculating these numbers. One can help increase the confidence in quantitative phosphoproteomics experiments by including more technical and biological replicates in the study, so that more measurements of relative abundance can be made.

A very important consideration for quantitative global phosphoproteomics studies is that any measured changes in phosphopeptide abundance not only could be the result of a change to that PTM itself but also could be due to a more general change in the protein level [127, 138]. Therefore, it is almost always required to include a quantitative proteomics experiment to measure changes in protein abundance from the same samples used in the quantitative phosphoproteomics experiment. This is particularly important for longer-term studies, such as siRNA treatments, or comparisons of different tissues. At these time scales, effects on gene and protein expression can confound measurements of phosphosite abundance. However, even for short-term manipulations such as growth factor stimulations, effects on protein stability can lead to changes on the protein level that cannot be measured in a phosphoproteomics experiment. Normalization of phosphosite abundance to protein abundance needs to be considered in all experiments of this type [169].

2.4 Sample Preparation and Enrichment Strategies for Phosphoprotein Analysis by Mass Spectrometry

A generalized phosphoproteomics strategy as used in the author's lab that can be adapted to any particular experiment is shown in Figure 2.7. Variations of this general strategy are in use in many laboratories worldwide. Employing peptide-based fractionation, phosphopeptide enrichment, and high-performance MS, numerous labs are now able to routinely detect and quantify 10,000–20,000 unique phosphorylation sites in single large-scale phosphoproteomics experiment [33, 135, 167]. Despite this effort, these numbers may still be considerably lower than the actual phosphoproteome complement of a mammalian cell [29]. Full and accurate coverage of the phosphoproteome remains a daunting and unrealized analytical challenge, with significant areas for improvement.

Protein phosphorylation is a reversible, highly dynamic PTM regulated by the interplay of opposing enzymatic activities. As such, particular care must be taken to preserve the integrity of the cellular phosphoproteome throughout the process of sample treatment and harvest, preparation of protein extracts, proteolytic digestion, and phosphopeptide capture and analysis. For example, precautions must be taken during cell harvesting to avoid induction of kinase pathways associated with changes in osmolarity, temperature, or nutrient availability. Cultured cell pellets or tissue samples should be snap-frozen in liquid nitrogen, stored frozen at −80 °C, and processed on ice. Cocktails of

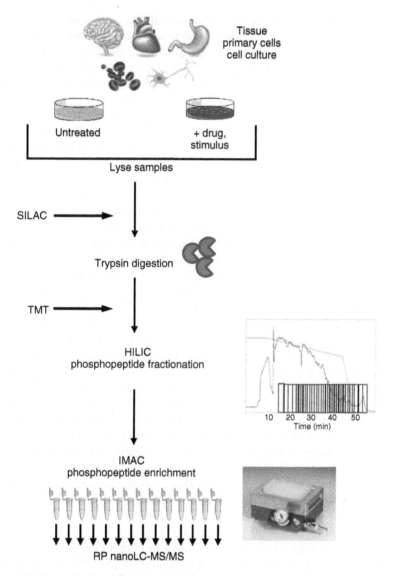

Figure 2.7 Generalized workflow for quantitative phosphoproteomics. The choice of quantitative strategy will depend upon the type of samples being analyzed. See the text for details. After lysis, (a) SILAC-labeled samples are mixed 1:1 and digested with trypsin. (b) Samples to be labeled with isobaric TMT reagents are digested first, then labeled, and mixed 1:1. HILIC chromatography of the protein digests separates phosphopeptides from the bulk of the nonphosphorylated peptides. The less complex phosphopeptide-containing fractions are enriched in 96-well plates using Fe^{3+}-IMAC. The resulting purified phosphopeptide mixtures are separated by nanoLC and analyzed directly by ES interfaced to a high-resolution, high-mass-accuracy mass spectrometer.

broad-spectrum protease and phosphatase inhibitors are routinely included during cell lysis to prevent protein degradation and dephosphorylation.

Protein lysates are routinely prepared under denaturing conditions using detergents or chaotropes to irreversibly halt protein enzymatic activity and to solubilize the largest portion of the cellular proteome possible. Because in general detergents have a deleterious effect on MS performance, proteins in detergent lysates are often separated on SDS-PAGE followed by proteolytic digestion of the gel-embedded proteins. "In-solution" digestion under denaturing conditions may be performed when proteins are extracted with strong chaotropic reagents such as urea and thiourea. A method describing filter-aided sample preparation (FASP) that facilitates buffer exchange to combine the advantages of detergent extraction with in-solution digestion has been frequently employed in MS-based proteomics [170]. Care must be taken when using this approach with the large-scale lysate requirements typical of proteomics experiments that seek to analyze PTMs [33]. Several MS-compatible detergents have been introduced for enhanced protein solubilization and digestion [171].

Although trypsin is most commonly used in the MS analysis of proteins, tryptic phosphopeptides may not always possess sequence properties suitable for phosphosite localization by MS. This is especially the case for phosphorylated residues, which lie within basic amino acid-rich protein domains. Several studies have shown that using multiple proteases on the same sample can increase proteome [172, 173] and phosphoproteome coverage significantly [130, 132, 174]. For large-scale phosphoproteomics analysis, thousands of phosphopeptide sequences were identified that would have been undetectable when using only trypsin, and fewer than a third of the phosphosites identified were detected in more than one protease data set [132]. However, using multiple protease digests drastically increases instrument time and requires a larger amount of sample. In this case, using five proteases required five times the amount of sample as would be required for a study with just trypsin and increases analysis time fivefold as well. These caveats may make this approach untenable for some studies. As with any proteomics tool, one has to weigh the purpose of the study, the benefits made by the approach, and instrument time restrictions to fully assess the best strategy.

Numerous strategies have been developed to enrich phosphopeptide from the overwhelming number of nonphosphorylated peptides in a complex sample. The most successful of these incorporate chemoaffinity and immunoaffinity strategies, either on intact phosphoproteins or phosphopeptides derived from proteolytic digests of proteins. Chemical modification and derivatization approaches have also been tested to enhance the selectivity of enrichment strategies.

Immunoaffinity approaches using antibodies have been most successfully applied to analyzing tyrosine phosphorylation. Tyrosine phosphorylation constitutes only <2% of the total cellular protein phosphorylation [175];

thus, specialized enrichment procedures are necessary to isolate phosphotyrosine-containing peptides from both unmodified peptides and the large pool of serine- and threonine-phosphorylated peptides. Antibody-based affinity purification procedures for phosphoproteomics studies involve selective enrichment of proteins or peptides that contain epitopes comprising single phosphoamino acid residues, or phosphorylated residues positioned within short linear sequence-dependent recognition motifs. Antibodies raised to specifically recognize phosphotyrosine residues in macromolecules have been in use for over 30 years [176]. Several highly specific, fully characterized, and validated antibodies are commercially available, which are well suited for large-scale study of phosphotyrosine signaling pathways in cell cultures and tissue extracts [177]. Due to the low abundance of pTyr, immunoprecipitation with anti-pTyr antibodies typically requires much larger amounts (>10 mg) of protein extract for large-scale, in-depth studies [178].

Enrichment with anti-pTyr antibodies for analysis by MS has been applied at both the protein and peptide level. Several studies have focused on quantitative elucidation of the temporal profile of phosphotyrosine-mediated signal transduction following various growth factor stimulations. Although immunoprecipitation of tyrosine-phosphorylated peptides was reported as early as 1995 [179], most of the early work applying immunopurification to tyrosine-phosphorylated proteins was done at the protein level [180–185]. This approach requires lysis under nondenaturing conditions, which facilitates copurification of protein complexes comprising phosphotyrosine-tethered scaffolds and signaling adaptors. If done quantitatively, this approach enables the elucidation of temporal recruitment and assembly dynamics of complete phosphotyrosine signaling modules. A disadvantage of protein-level enrichment is that relatively few of the resulting proteolytic peptides will contain the phosphotyrosine modification. To identify particular sites of phosphorylation, the antiphosphotyrosine immune complex can be digested with trypsin and enriched for phosphopeptides using metal affinity resins (see following text). In contrast, phosphotyrosine enrichment at the peptide level has the advantage that lysis and protein extraction can be performed under harsher conditions, allowing better solubilization of integral membrane signaling proteins and deeper coverage. Additionally, the majority of eluted peptides will be expected to contain the phosphotyrosine modification [186]. Large-scale studies employing peptide-level immunoaffinity enrichment have allowed the characterization of several hundred to thousands of phosphotyrosine sites [178, 187–189].

Many commercial antibodies are available that claim to show specificity for phosphorylated serine and phosphorylated threonine residues. The generally poor selectivity of these reagents as immunopurification tools has limited their use, though motif-specific antibodies have been effectively employed

[190–192]. In general, the global study of serine and threonine phosphorylation is better accomplished using chemoaffinity enrichment techniques. The two most frequently used strategies for global phosphopeptide enrichment are both chemoaffinity based and include various forms of immobilized metal ion affinity chromatography (IMAC) and metal oxide affinity chromatography (MOAC) where the most common reagent is TiO_2. Both have been applied with a great deal of success in a variety of permutations, alone or together using proteolytic peptides from purified proteins or crude cellular extracts, and frequently coupled with up-front chromatographic prefractionation. The binding, wash, and elution steps are most commonly performed offline from the mass spectrometer, in tubes, plates, pipette tips, or microcolumn format.

Immobilization of multivalent cations on an affinity resin support has been widely utilized in protein chemistry, exemplified by the use of Ni^{2+} for the purification of recombinant proteins engineered to contain a polyhistidine affinity tag. IMAC was introduced for the separation of phosphoproteins in 1986 [193] and later adapted and refined for phosphopeptide enrichment [194, 195]. Enrichment of phosphopeptides by IMAC exploits the affinity of phosphate groups for multivalent transition metal ions, such as Fe^{3+}, Ti^{4+}, Ga^{3+}, and Zr^{4+}. IMAC uses metal chelators such as iminodiacetic acid (IDA) and nitrilotriacetic acid (NTA) linked to solid-phase chromatographic supports to coordinate the metal ion. Fe^{3+} is the basis for the most common form of IMAC. Available coordination sites of the positively charged metal ions are presented for interaction with the negatively charged phosphate groups on peptides.

One of the major shortcomings of IMAC is its affinity for highly acidic peptides rich in glutamic and aspartic acid residues, which also coordinate well with metal complexes. An early solution to nonspecific binding was chemical derivatization to convert carboxylic acids to their corresponding O-methyl esters prior to IMAC [133]. This approach suffered from incomplete esterification and deamidation of asparagine and glutamine-containing peptides, which increased the sample complexity and complicated the sequence database search space. Rather, systematic investigations into modulation of pH, acid content, and organic modifiers led to the findings that a low-pH (2.0–2.5) loading solution improved selectivity by protonating the acidic amino acids while maintaining phosphate groups unprotonated and available for binding. Additionally, binding in the presence of trifluoroacetic acid (TFA) and acetonitrile (ACN) minimized hydrophobic interactions between peptides and the IMAC resin [196–198]. The enriched peptides are most commonly eluted with basic pH buffers such as ammonium hydroxide, but can also be eluted with phosphoric acid or EDTA. An additional shortcoming of IMAC enrichment is that it is generally intolerant of salts and detergents, which can interfere with phosphopeptide binding [199].

IMAC has been reported to bind multiply phosphorylated peptides with higher affinity than singly phosphorylated peptides, producing a bias in phosphopeptide enrichment [199]. This appears to be the case for complex, unfractionated peptide digests from whole lysates. This bias all but disappears when the sample complexity is reduced, either through prefractionation and segregation of singly and multiply phosphorylated peptide pools prior to enrichment [200] or through sequential rounds of IMAC [198, 201]. Recently, Ti^{4+}-IMAC has been reported to be superior in terms of specificity and efficiency to both IMAC using other metals or TiO_2 chromatography, particularly when applied to crude enrichment from unfractionated whole cell lysates. [202–204]. This has proven highly advantageous when characterizing cellular phosphoproteomes via performing single enrichments and subsequent MS analyses on numerous samples, as would be the case for label-free quantitative studies [167, 205].

MOAC takes advantage of the ability of some metal oxides, such as TiO_2 and ZrO_2, to form complexes with phosphate groups. It was reported by multiple groups for the selective enrichment of phosphopeptides in 2004 [206–208]. As with IMAC, nonspecific binding of nonphosphorylated peptides rich in acidic residues is a persistent challenge. To circumvent this, loading under highly acidic conditions and the inclusion of various substituted organic acids have been found to alleviate nonspecific binding. Organic acid additives such as 2,5-DHB [209] bind metal oxides more weakly than the phosphate groups, but more strongly than carboxyl groups, through a chelating bidentate mode rather than the bridging bidentate mode exhibited for phosphate group binding. Poor solubility of 2,5-DHB has prompted its substitution with more soluble and hydrophilic acids such as glycolic, phthalic, and lactic acids [199, 210, 211]. A disadvantage is that a desalting step is usually required to remove these additives prior to MS. As is the case with IMAC, the nonspecific binding of highly acidic peptides goes away when complex samples are fractionated prior to enrichment. Advantages of MOAC include the superior chemical stability of TiO_2 spherical particles. Unlike IMAC, there is no need to charge the resin with metal ions, and the support exhibits exceptional tolerance to buffer excipients such as salts, detergents, and chelating agents [199].

Is one of the affinity capture technique inherently better than the others for phosphopeptide enrichment? Numerous, often conflicting, reports have claimed the superiority of either IMAC or TiO_2 with respect to phosphopeptide capture performance, as judged by efficiency, selectivity, and recovery. The differences observed between the two resins are often so pronounced as to call into question the experimental design of the studies. Most often a lack of experience in one method or the other combined with undersampling of the enriched pools easily accounts for the very large differences observed. In fact, numerous large-scale phosphoproteomics studies using IMAC [33, 167] and MOAC [135, 212] enrichment demonstrate that, judged by overall performance, both strategies are highly effective and are capable of similar numbers of phosphosite identifications, particularly when coupled with a multidimensional separation strategy. Unlike what has been reported in earlier studies on limited sample sets, both methods capture similar numbers of singly and multiply phosphorylated peptides.

A related consideration under debate is the perceived orthogonality of the two techniques. The prevailing consensus in the field is that IMAC and TiO_2 methods are capable of enriching complementary parts of the phosphoproteome, with each technique showing particular strengths in its ability to isolate unique classes of phosphopeptides possessing distinct physicochemical properties [201, 213–216]. This would suggest that neither method alone is sufficient for a comprehensive enrichment of the phosphoproteome. Several recent studies and work in our own lab have called this assumption into question. Comparison of TiO_2 and Ti^{4+}-IMAC using a very large synthetic phosphopeptide library of known composition, and tryptic lysates of HeLa cells, indicated minimal differences between enrichment of phosphopeptides based on sequence composition, peptide length, or various physicochemical properties [205]. In a direct comparison of rat liver data sets (Figure 2.8a) obtained from different laboratories employing different phosphopeptide isolation strategies centered on either IMAC [33] or TiO_2 [31], fully 84% of the TiO_2 sites were contained within the IMAC data set [33], and there was no difference in the preference for multiply phosphorylated peptides between the two methods. Likewise, Ruprecht et al. observed that optimized Fe^{3+}-IMAC, Ti^{4+}-IMAC, and TiO_2 platforms bound the same phosphopeptide species and concluded that insufficient resin capacity, inefficient elution conditions, and the stochastic nature of data-dependent acquisition MS are the causes of the experimentally observed complementarity between platforms [217].

Unpublished data from our laboratory are in full agreement with these findings. Using isobaric labeling to quantify differences between IMAC and TiO_2 enrichment of hydrophilic interaction liquid chromatography (HILIC) fractions, we find that both resins isolate the same pool of phosphopeptides, irrespective of the number of phosphorylation sites and other physicochemical properties (Figure 2.8b). In fact, only approximately 10% of phosphopeptides were found to be enriched greater than fourfold by one technique or the other, suggesting that in an optimized platform, there are only marginal gains to be made by a combination or sequential IMAC and TiO_2 capture. Thus it is clear that both strategies can be highly effective, especially when coupled with a multidimensional separation strategy; however, it is equally clear that both IMAC and MOAC need to be optimized individually and for specific workflows.

All of the methods described earlier utilize the intact phosphate group to facilitate enrichment of phosphopeptides. Historically, several innovative methods have been introduced that use chemical modification as a way to improve the enrichment of phosphopeptides. A widely used derivatization

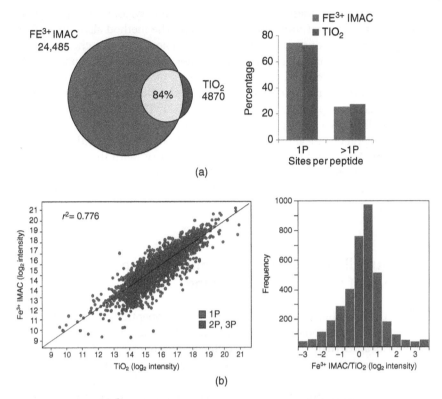

Figure 2.8 TiO$_2$ and Fe^{3+}-IMAC isolate the same phosphopeptides. (a) Rat liver phosphosites enriched by either HILIC-IMAC [33] or TiO$_2$ [31] show extensive overlap. Despite different phosphopeptide enrichment strategies conducted in two independent laboratories, 84% of phosphopeptides from the TiO$_2$ experiment are shared with the IMAC data set. Contrary to conventional thinking, no difference in the proportion of singly and multiply phosphorylated peptides recovered was observed between Fe^{3+}-IMAC and TiO$_2$. (b) Tryptic peptides from two adjacent HILIC fractions were divided into six equal parts and enriched three each on either Fe^{3+} IMAC or TiO$_2$. The six enriched pools were labeled with TMT to produce a 6plex quantitative experiment. Individual reporter ion intensities indicate the relative enrichment of a given peptide from each of the six pools. The scatterplot (left) shows very good correlation between the log$_2$ mean intensities of 2765 unique phosphopeptides enriched by either method. The histogram (right) displaying binned log$_2$ ratios for Fe^{3+} IMAC versus TiO$_2$ indicates that only approximately 10% of phosphopeptides are enriched >fourfold by one method over the other.

protocol is based on combining β-elimination of the phosphate group under strongly basic conditions, followed by a Michael addition on the resulting dehydroalanine or dehydroamino-2-butyric acid products [218–220]. The resulting side chain permits the attachment of a variety of affinity tags with the added option of being able to incorporate stable isotopes into the tag for quantification. The method is not suitable for tyrosine-phosphorylated peptides, which do not undergo β-elimination. The method also suffers from chemical side reactions and incomplete conversion. Another chemical derivatization strategy uses phosphoramidate chemistry where phosphopeptides can be linked to iodoacetyl groups immobilized on a synthetic polymer solid support or glass beads. Acid hydrolysis of the phosphoramidate bonds allows the phosphate groups to be recovered intact [221, 222]. Though amenable to pTyr as well as pSer- and pThr-containing peptides, the extensive workflow involves several chemical reactions and methyl esterification, with the associated sample losses and increased sample complexity. With the maturation of efficient chemoaffinity enrichment techniques described earlier, chemical derivatization approaches have largely fallen out of favor for phosphoproteome analysis.

2.5 Multidimensional Separations for Deep Coverage of the Phosphoproteome

Despite the availability of highly selective phosphopeptide enrichment methods and mass spectrometers with ever-increasing data acquisition rates, the complexity and dynamic range of the cellular phosphoproteome are still far too great to permit its full characterization by direct mass spectrometric analysis. Fractionation of the phosphoproteome, typically at the peptide level, using multidimensional chromatographic separations, is a necessary prerequisite for obtaining in-depth, comprehensive coverage (see Figure 2.7). Fractionation in itself is generally not sufficient to permit efficient phosphopeptide identifications from a complex mixture; rather it should be paired with a selective enrichment step. While this fractionation may occur prior to, or following the affinity enrichment, we [200] and others [223] have shown that fractionation prior to enrichment results in better selectivity in the enrichment and overall more in-depth coverage of the phosphoproteome. Phosphoproteomics prefractionation is typically performed offline to accommodate the processing of up to several milligrams of protein lysate required for large-scale, in-depth studies. Less frequently, a multidimensional separation or phosphopeptide enrichment step is fully integrated online with the mass spectrometer [224–226]. Ideally, any multidimensional separation strategy combines a chromatographic mode of fractionation highly orthogonal to the final RP separation interfaced to the MS.

Multidimensional chromatographic separations for phosphoproteomics are invariably based on exploiting the unique physicochemical properties with which the phosphorylated amino acid residue bears on modified peptides: (i) the ability of the phosphate group as a weak acid to carry up to two negative charges at physiological pH (pK_{a1} 2.1, pK_{a2} 7.2) [227] and (ii) the polarity of the phosphoamino acid ester, which imparts a strongly hydrophilic character on phosphopeptides relative to their nonphosphorylated counterparts [228].

Chromatographic enrichment methods based on ion-exchange mechanisms are widely utilized due to the difference in solution charge states between phosphorylated and nonphosphorylated peptides. In strong cation-exchange (SCX) chromatography, peptides are bound to a column containing a hydrophilic, anionic stationary phase through their positively charged functional groups. Phosphopeptide enrichment using SCX is typically conducted at low pH (2.7). Under these conditions, the N-terminal amino group as well as the side chains of arginine and lysine will be protonated. Thus, a proteotypic tryptic peptide has a net charge of +2. Addition of a negatively charged phosphate group to the peptide will reduce the charge by one, providing a mechanism for separation and enrichment of phosphopeptides from the bulk of nonphosphorylated peptides [113]. Though widely applied, it is now well known that many phosphopeptides do not elute in the 1+ fractions. Peptides containing greater than one phosphate can carry a net zero or negative charge and will elute with highly acidic nonphosphorylated peptides in the unbound fraction; this negatively impacts subsequent IMAC and MOAC binding selectivity. Other peptides that are charge reduced, including peptides derived from protein N-termini (N-acetylated) and C-termini (lacking Arg or Lys), will coelute with the +1 phosphopeptide pool. Additionally, phosphopeptides containing histidine residues, or missed tryptic cleavages introducing internal Lys or Arg residues, will also not be found in the early phosphopeptide-rich fractions. Finally, SCX is a fairly low-resolution technique and is only moderately orthogonal to reverse phase as a component of a 2D strategy [229]. Despite these considerations, SCX has been widely used as a first-dimension separation coupled with either IMAC [119, 188] or MOAC [5, 230] enrichment for large-scale phosphoproteomics studies, though it is now recognized that fractions should be taken across the entire SCX gradient to provide the greatest depth of coverage.

Strong anion-exchange (SAX) chromatography separation relies on the binding of negatively charged peptides to a positively charged stationary phase. In contrast to SCX, under neutral to mildly acidic binding conditions, the strongly acidic phosphopeptides show greater retention on SAX than nonphosphorylated counterparts, which predominate in the early eluting fractions. Using SAX, partial fractionation of the phosphoproteome based on number of phosphates is observed [231]. SAX chromatography has been increasingly applied to phosphopeptide fractionation [232–234].

HILIC is a high-resolution separation technique where the primary interaction between a peptide and the neutral, hydrophilic stationary phase is hydrogen bonding. In HILIC, retention increases with increasing polarity (hydrophilicity) of the peptide, opposite to the trends observed in RP [228]. HILIC has the highest degree of orthogonality to RP of all commonly used peptide separation modes [229], making it the ideal first-dimension separation for peptide-based proteomics. Samples are loaded at high organic solvent concentration and eluted by increasing the polarity of the mobile phase (e.g., an inverse gradient of ACN in water). Under these conditions, phosphopeptides are strongly retained because phosphoamino acids are far more hydrophilic than the twenty standard amino acids. Using an optimized gradient, phosphopeptides can be partitioned away from the bulk of nonphosphorylated peptides and further fractionated based on their hydrophobicity and the number of phosphorylated residues [200]. This greatly improves the selectivity of subsequent chemoaffinity enrichment techniques. An advantage of HILIC is that the volatile, salt-free TFA/ACN buffer system employed makes the fractions compatible with direct IMAC or MOAC capture. HILIC has been frequently applied in multidimensional phosphopeptide isolation workflows [33, 235–237].

Electrostatic repulsion hydrophilic interaction chromatography (ERLIC) is a technique closely related to HILIC. Also known as ion-pair normal phase, ERLIC employs HILIC on a weak anion-exchange (WAX) column [238]. Performing separations at low pH, acidic side chains and C-terminal carboxy groups are protonated, thus reducing their effect on retention. Basic side chains and N-terminal amines are positively charged and electrostatically repulsed by the column. The negatively charged phosphate group interacts favorably with the WAX column, leading to increased retention time. Separation is performed in high organic solvent, which superimposes a hydrophilic interaction mode on the electrostatic effects and further promotes interaction of the phosphate group with the stationary phase, such that phosphopeptides are separated from the bulk of the nonphosphorylated peptides [239]. However, since nearly half of the phosphopeptide-containing fractions also contain 50% or more nonphosphorylated peptides [239], it would seem obvious that to make the most efficient use of the MS and provide deeper coverage of the phosphoproteome, these fractions would benefit from further enrichment by IMAC or MOAC. Unfortunately, the separation of phosphopeptides by ERLIC requires 20 mM sodium methylphosphonate buffer, which would need to be removed prior to further enrichment or direct MS analysis. While not in widespread use, ERLIC has been successfully employed in a number of phosphoproteomics studies [240–242].

High-pH (HpH) reversed phase is a separation mode that is gaining ground as the first-dimension fractionation for in-depth phosphoproteome studies. Intuitively, it would seem that a multidimensional separation strategy employing two successive rounds of RP chromatography would provide only marginal improvements in practical 2D peak capacity over a single dimension of RP separation and would be inferior to all 2D separations based on different physicochemical properties (i.e., charge or hydrophilicity). However, systematic evaluation of RP–RP 2D HPLC conditions revealed that while altering ion-pairing agents and stationary phases yielded only subtle changes in orthogonality, a change in the pH of mobile phase had a more pronounced effect on peptide retention behavior [243]. This is attributed to differences in individual peptide isoelectric (pI) points, which range widely from 3 to 12 and are dependent on the number and composition of ionizable termini and side chain functionalities. Acidic peptides (pI < 5.5) are more strongly retained at pH 2.6, when the carboxylic moieties are not ionized, in contrast to basic peptides (pI > 7.5), which are more strongly retained under pH 10 conditions. Separations are routinely performed at ~pH 10 in ammonium formate- or ammonium hydroxide-containing buffers, though it has been observed that formate buffers can degrade column performance [135]. Despite RP possessing the highest resolution of routine chromatographic separation modes, even under optimal conditions, differential pH-based RP–RP separations are only semiorthogonal. To further improve orthogonality, a concatenation strategy was introduced in which systematic pooling of the early, middle, and late fractions is performed prior to analysis by low-pH RPLC-MS [244]. Several recent examples have shown this to be a highly effective first-dimension separation, performed either prior to [135] or following phosphopeptide affinity enrichment [134, 234]. Direct comparison would suggest that for phosphoproteomics analysis, HpH is superior to SCX, particularly with respect to the fractionation of singly phosphorylated peptides [135]. It is thus likely that HpH will eventually supplant SCX as a primary first-dimension separation mode.

2.6 Computational and Bioinformatics Tools for Phosphoproteomics

The maturing of proteomics technologies has led to an avalanche of data that contains potential insights into the workings of living organisms. In common with genomics and the emerging field of metabolomics, it permits researchers to study how components of a cell work in concert with one another. When merged with data from the other omics technologies, an integrated picture of how biology is regulated will develop. Processing the gigabytes of raw data from protein identification through quantitation to mapping the data onto known biological processes is now impossible to do manually. Sophisticated algorithmic and software tools are needed to process the raw data and to derive biological insight from it. Computational proteomics is an emerging field. While proteomics has much in common with genomics, particularly in the bioinformatic arenas of functional analysis and data mining, it is sufficiently different that unique computational tools are necessary to produce reliable results for downstream bioinformatics analysis. Phosphoproteomics data in particular is unique in that individual sites may be regulated, quite independent of any change in protein expression. Therefore, the site itself becomes the focus of the biology, not just the protein itself. As described in the introduction, on many proteins, multiple sites are phosphorylated and regulated independent of one another. Unfortunately most bioinformatics tools have been developed from a genomic perspective and fail to capture the intricacies of phosphorylation (or other PTM)-dependent regulations. Computational proteomics and bioinformatics tools to interrogate and integrate the phosphoproteome are still maturing but are attracting an increasing amount of interest in the computational biology community.

The identification of phosphopeptides in a phosphoproteomics experiment begins like any other proteomics experiments, with peptide sequencing via data-dependent MS/MS. This is followed, after all the data has been collected, by applying database searching tools to the MS/MS spectra. Search engine algorithms, such as Mascot [245] and Sequest [246], use an amino acid sequence to calculate a theoretical spectrum for each possible peptide sequence and output the one that most closely matches the mass spectrum and tandem mass spectrum data. For most search engines, two measures of statistical significance are usually reported for peptide identification. The first focuses on individual peptides and represents the likelihood of a match between theoretical and experimental spectra occurring by chance. The second measure is the false discovery rate (FDR) for the entire data set, which estimates the total number of incorrect matches expected from all identified peptides [247].

Analysis of phosphopeptides offers several additional layers of complexity over the identification of unmodified peptides. In the first place, for a search of phosphopeptide data, an additional variable modification must be added to the database, which exponentially increases the search space and decreases the likelihood of matching a spectrum to a sequence at a given FDR. Secondly, as described previously, localizing the site of phosphorylation can be severely difficult [248]. Localization requires identification of a fragment ion that contains the phosphate group and is unique for the phosphorylated amino acid. In practice, however, such fragment ions may not be present or may be indistinguishable from noise. In general, search engines have not been optimized to localize modification sites and only recently have a few integrated scoring measures of localization confidence. Multiply phosphorylated peptides can exacerbate this problem even further, as it can exponentially increase the number of possible site combinations within a given sequence, which in turn makes confident localization more difficult.

Several software packages are now available (Table 2.1) that calculate localization probabilities, making it possible to derive a measure of confidence in phosphosite localization for a give spectrum [248]. Such programs must consider numerous variables that can affect phosphosite identification. First, they must distinguish true peaks from chemical noise. Next, they must decide which peaks are suggestive of a modification site and how much weight each

Table 2.1 Site localization tools.

Name	Localization strategy	ion tool access	Report alternate sites?	Reference
A-Score	Probability based	http://ascore.med. harvard.edu/	No	[249]
D-Score	Score difference	No tool available	Yes (all)	[250]
LuciPhor/ LuciPhor2	Probability based	http://luciphor. sourceforge.net/	Yes (Top 2)	[251, 252]
Mascot Delta	Score difference	www.matrixscience.com/ server.html	Yes (All)	[253]
MaxQuant PTM Score	Probability based	www.maxquant.org/ downloads.htm	Yes (All)	[254]
PhosphoRS	Probability based	Available as part of Proteome Discoverer	Yes (All)	[175]
PhosphoScan	Probability based	Request from developers	Yes (All)	[255]
PhosphoScore	Score difference	https://github.com/ evansenter/ucsb/tree/ master/school/CS167/ main_project/code/ PhosphoScore	No	[256]
PhosSA	Score difference	http://helixweb.nih.gov/ ESBL/PhosSA/	Yes (All)	[257]
Protein Prospector (SLIP)	Score difference	http://prospector.ucsf. edu/prospector/ mshome.htm	Yes (All)	[258]
SLoMo/ Turbo SLoMo	Probability based	http://massspec.bham.ac. uk/slomo/	Yes (Top 2)	[259]

Adapted from "Perkins, D.N., Pappin D.J., Creasy D.M., Cottrell J.S. (1999), NCBI."

peak should have in determining the site localization. For example, a given peak may be a result of two or more possible fragments, and the software must take this into account. Finally, phosphosite identification tools must calculate a confidence metric for the particular site that is called by the analysis.

There are two main strategies that are built into phosphosite localization scoring mechanisms [248, 249]. The first involves measuring the probability of each possible phosphorylation site using ions from the fragmentation spectra. First, ions that are "site determining" are identified from the theoretical fragmentation spectrum of the identified peptide sequence. These are fragment ions that can be used to uniquely assign a particular phosphorylation site. A probability of a phosphosite match for each potential modification site is determined by comparing the number of identified theoretical site-determining ions with the number of matched ions from the theoretical spectra. This probability is calculated for each possible site, and the resulting probabilities are compared to one another to make a call on the most likely phosphosite. Several popular localization algorithms utilize this strategy, including A-Score [254], PTM score [175], PhosphoRS [251], and LuciPhor [253]. The second strategy for measuring confidence in site localization utilizes results from the search engine report. The search engine provides an identification score for each possible phosphopeptide isoform of a given sequence. The difference in the scores can be used to estimate the reliability of a given localization. Tools that rely on this strategy include Mascot Delta Score [258], SLIP [250], and D-Score [260].

It is clear that there is still quite a bit of variation in the output of these different data analysis tools. In one case, the ABRF provided 35 participants with the same phosphoproteomics data set, and they were asked to identify the phosphosites present in the sample with a 1% FDR. Phosphosites were unanimously agreed upon for 79% of the peptide spectral matches; however, there was a disagreement in site localization for the remaining 21% of data set (reviewed in [248]). In a more recent study, Kuster and colleagues created a library of >57,000 synthetic phosphopeptides with known sequences and phosphosites that was used to compare the accuracy and sensitivity of three different site localization tools: PhosphoRS, PTM Score, and Mascot Delta Score [32]. Given that sequences and sites were known, the authors were able to calculate a true false localization rate (FLR) for the data. PhosphoRS and PTM Score provided a larger number of correctly localized phosphorylated peptides than Mascot Delta Score at a true FDR of 1% and FLR of 1%. However, each of the three tools underestimated the FLR at a given probability. Importantly, the results of these different algorithms were highly complementary, in that each provided a different subset of correctly localized peptides. This observation suggests that using different localization tools on the same data set may increase the number of correctly identified phosphosites and that further optimization of these algorithms is needed.

A further complication of assigning phosphosites occurs when working with quantifying phosphopeptides from biological replicates. There are often cases where a phosphorylated peptide sequence is confidently identified in more than one replicate. However, in one replicate, the site is assigned with high confidence, whereas in the other replicates it cannot be assigned. Even within the same replicate, one may have several peptide spectral matches for the same phosphopeptide sequence, where some of the matches have a confident localization and others do not. Whether the analyst can assume these peptides are indeed the same, and thus be considered for quantification across and within the replicates, is a subject of debate. Other pieces of evidence, such as retention time, can sometimes be of use, but often isomers of the same peptide will have very similar retention times and are thus indistinguishable. Some tools, such as MaxQuant [261], assume that if the confidence in localization is at least 75% in one replicate and at least 50% in other replicates, then the peptides are likely the same isomer and should be considered as such for downstream analysis [230]. However, the phosphoproteomics community has yet to reach a consensus on how to address this problem. The reliability of site localizations is a significant issue when one considers the proliferation of proteomics databases.

The target decoy database strategy for estimating global peptide identification FDR has proven useful for comparing different identification tools in an unbiased manner. With this approach, mass spectra are searched against both a true database and a database in which the same protein sequences have been either randomized or reversed. All matches to the decoy database are counted to estimate the number of false identifications within your data set at a given score cutoff [262]. However, this approach cannot be directly extended to calculate an equivalent FLR for modification site localization. This is because an incorrect site localization is not a random match. Rather, the incorrect and correct localizations are extremely similar. Therefore, using a decoy sequence is not an accurate estimation of the true error inherent in a data set [248]. Tools have been developed that modify this decoy strategy to more accurately measure a global FLR. A majority of the most current tools to measure global FLR for phosphoproteomics involve computationally allowing the phosphate group to modify amino acids that biologically cannot bear the modifications [250, 253]. A match to one of these residues is analogous to a match to a decoy database sequence, and the number of decoy matches in a global data set allows an estimation of FLR. However, picking which residues to use as a decoy can be tricky. For an accurate estimate, the decoy residues should be present with a similar frequency to the true sites throughout the tryptic peptide proteome. In addition, the decoy residues should have a similar proximity to the true modified site as other potential true sites do, as it is more likely that a mislocalization will occur on residues that are closer to the actual modified site [248]. Incorporating these parameters can be a tricky computational task, and optimizing tools for FLR estimation is an active area of investigation.

Over the last decade, many public databases have been created to store phosphosite information [139]. Phospho.ELM [263] and PhosphoSitePlus [29] are two databases that originally focused solely on phosphopeptides but have expanded to include other PTMs. Phospho.ELM contains more than 42,000 phosphosites, 90% of which were identified from high-throughput proteomics experiments. The vast majority of these sites were obtained from human and mouse samples [263]. Each phosphosite is reported along with a conservation score, as well as an accessibility score based either on the crystal structure of the protein or as predicted by the RealSpine accessibility algorithm.

PhosphoSitePlus, a large database of PTMs initiated in 2003, has since grown to encompass data from thousands of data sets and includes unpublished results generated from in-house data. As of the end of 2014, the site includes over 330,000 nonredundant PTMs from a large cohort of model eukaryotic organisms [29], including over 240,000 nonredundant phosphosites.

Several other more general databases are also useful resources of phosphosites. The Human Protein Reference Database (HPRD), a curated source of proteomics data from human samples, includes phosphosite information obtained from human phosphoproteomics experiments [264, 265]. Other organism-specific phosphosite databases include PhosPhat for *Arabidopsis thaliana* [266] and PhosphoPep for *Drosophila melanogaster, Caenorhabditis elegans*, and *S. cerevisiae* [267]. The Universal Protein Resource (UniProt) is a more species comprehensive source of protein sequence and annotation data [268]. UniProt compiles data from a large variety of sources and curates this into a single knowledgebase. This includes known PTMs such as phosphorylation. Other PTM databases often encompass phosphorylation as well. This includes PHOSIDA, which links extensive peptide information to the modification sites [269]; SysPTM, which compiles data on over 50 different PTMs [270]; and dbPTM, which integrates both large public databases with modification site data mined from the literature [271].

Although these databases can be rich resources for hypothesis generation, the quality of the data being analyzed must be carefully considered. The vast majority of the phosphosites in these databases come from published reports of large-scale phosphoproteomics experiments, with the assumption that authors have correctly identified the sites of modification. However, much of the data were generated prior to any guidelines from journals to assess site localization, and many of these studies did not address localization at all [248]. More recently collected data can also vary in quality depending on the tool used to measure localization reliability. Researchers who use the information in phosphopeptide databases should remain aware of the degree of uncertainty inherent to these resources. It has been suggested that researchers who produce and analyze the raw data should lean toward a conservative estimate of identification/localization reliability [248].

Cataloging phosphosites may provide some insight into how a protein or set of proteins might be regulated by PTM, but quantitative phosphoproteomics experiments are often utilized to gain a deeper mechanistic understanding of the biologic pathways that differentiate two or more cell states. However, analysis of the information produced from these experiments without incorporation of other tools is rarely enlightening. Because a single phosphoproteomics experiment can only quantify a small subset of all phosphosites, a full understanding of the biologic pathways involved requires the integration of several types of data to build phosphorylation networks. There are two general nonmutually exclusive approaches to developing phosphorylation-mediated signaling networks [272]. The first focuses solely on the protein level (Figure 2.9a). All phosphoproteins that are found to contain a regulated phosphosite are mapped to known pathways or protein–protein interaction

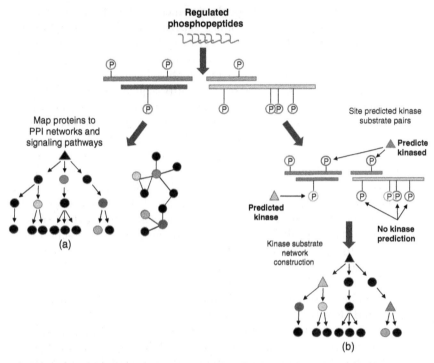

Figure 2.9 Constructing networks from global phosphoproteomics data. Significantly modulated phosphopeptides as measured from a phosphoproteomics data set are first mapped to protein sequences. (a) These protein sequences can then be directly mapped onto known signaling pathways or protein–protein interaction maps. Protein complexes or pathways that are enriched in these maps potentially participate in the response to the stimulus under study. (b) The amino acid sequence around the phosphorylation sites are used to predict active kinases that preferentially phosphorylate this sequence motif. This allows for the construction of kinase substrate phosphorylation networks. These maps enable more connections to be made between regulated phosphosites by linking them with predicted kinase activities that are modulated. Note that not all phosphorylation sites can be matched with predicted regulatory kinases, but these sites can still be mapped to proteins and included in the final network maps. These provide the basis for hypothesis generation and further biological study.

networks to identify modules that are involved in a biological process of interest. Software programs like Ingenuity Pathway Analysis (www.ingenuity. com), PANTHER [273], and KEGG [274] provide databases and algorithms to measure enrichment of given pathways in data sets. STRING can be used to identify both known and predicted interaction partners of the phosphoproteins of interest [275]. Integration of these tools can help provide a more complete understanding of the signaling pathways implicated by quantitative phosphoproteomics experiments.

These bioinformatics tools, although extremely useful, do not utilize key data obtained from phosphoproteomics experiments: the phosphorylation sites themselves (Figure 2.9b). There is a growing list of tools being developed that create kinase substrate networks by incorporating information on kinases and phosphatases that are known to act upon phosphosites identified in phosphoproteomics studies. The first step is to identify sequence motifs enriched in the phosphopeptide data that are known kinase substrates, which can provide information on modulation of kinase activity. Software programs like Scansite [276], GPS [277], and NETPHOSK [278] are used to predict kinase-specific phosphorylation sites from hundreds of kinases and identify sequence motifs that are preferred substrates. Using these tools, one can infer whether various kinases are active or repressed based on sequence motifs that are enriched in a data set of modulated phosphosites. Moreover, tools like NetworKIN [279], KinomeXplorer [280], and iGPS [281] take other contextual factors such as colocalization data and direct or indirect protein–protein interaction data into account to make connections between various active kinase modules and create large networks of potential signaling pathways, allowing the generation of testable hypotheses. In addition, phosphopeptide databases like PhosphoSitePlus, Phospho.ELM, and HPRD all incorporate tools that search for motifs and identify kinase–substrate relationships.

An important point to remember when considering these tools is that they rely heavily on prior knowledge, and so their ability to reveal less studied pathways or novel kinase–substrate interactions is limited. Because a majority of the literature focuses on a small subset of kinases and pathways, there may be overrepresentation of certain pathway components and masking of unexpected relationships that may be important for a given cell state. Tools such as SELPHI are attempting to ameliorate this problem by identifying relationships between kinases and substrates in unbiased ways [282]. Nevertheless, a researcher using these tools should still be aware of the biases that may result from using software that relies on previous knowledge to generate its reports.

There is also a growing interest in integrating phosphoproteomics data with genomic and transcriptomic data to determine how mutations in or aberrant activity of certain genes and proteins can lead to changes in phosphorylation cascades, which in turn affect gene expression. This integrative approach can sometimes be difficult because the technologies are so different, there is often inconsistency between experiments, and samples are rarely matched to the same individual [139]. Recent studies have been undertaken to determine how disease-causing genomic mutations lead to dysregulation of pathways and large-scale changes in gene expression. One recent study integrated phosphoproteomics data from various sources with cancer genomic data from The Cancer Genome Atlas (TCGA) to identify phosphosites that are frequently mutated in cancer. Of over 87,000 phosphosites, 150 were identified as frequently mutated in known or candidate cancer-driving genes [283]. Another study integrated transcriptomic, proteomics, and phosphoproteomics data from 13 non-small cell lung cancer cell lines to identify driver pathways specifically active in KRAS-dependent lines [284]. These examples illustrate how multiple methodologies and tools can be integrated to build comprehensive models of molecular signaling networks.

Despite many innovations in instrumentation and data analysis tools, phosphoproteomics experiments still suffer from incomplete coverage and a certain amount of irreproducibility across platforms [272]. In addition, single phosphopeptide measurements are insufficient for a full understanding of biological systems. Taken together, phosphorylation network signatures will serve as more informative and reproducible measures of biological system perturbations when integrated with other data types.

2.7 Concluding Remarks

Nearly 85 years after the isolation of phosphoserine, many analytical challenges remain in our pursuit of understanding how phosphorylation drives complex biological processes. As with all areas of science and natural history, the more we know, the more questions that will be raised. Phosphorylation is more widespread than could ever have been imagined. Is most of this activity noise, or is it, like intronic DNA, not yet fully understood? Uncovering the functional significance of these waves of phosphorylation is the major task at hand for at least the next decade. Confidently localizing specific sites of phosphorylation can still be problematic. The newly realized possibility of phosphate group migration during the CID process on certain types of mass spectrometers is troublesome and calls into question the validity of the database results posted from these instruments. Fortunately newer generations of mass spectrometers are less prone to these effects. Ultimately the assignment of specific phosphorylation sites must be confirmed by site-directed mutagenesis and biological testing. Strategies to enrich samples for their phosphopeptide content are now well established. However, in the absence of radioactive phosphate labeling, it is impossible to know whether all sites in a particular sample have been recovered. Localizing phosphorylation on multiply phosphorylated peptides is particularly difficult. Two phosphorylation events, differentially localized over three or more residues, with stoichiometry varying between the combinations is not an unusual circumstance.

Correlating a change in the degree of phosphosite occupancy with a biological response is the key to assigning functional significance to a given site. In practice this is less straightforward than simply measuring a change in the intensity of the phosphopeptide signal. For short duration signaling events, *de novo* protein synthesis can be ignored, but clearly both protein synthesis and protein degradation can influence the abundance of a phosphopeptide, regardless of any real change at the site. For quantitative phosphoproteomics this means having to also analyze global protein expression in a separate set of experiments.

The occurrence of phosphorylation in close proximity with other PTMs is now well documented. Measuring changes in phosphosite occupancy will be confounded by any changes occurring on the other sites of modification at the same time. In fact, understanding how various PTMs work in opposition to each other or together to influence function is a major challenge for the study of cell signaling. While the global analysis of various PTMs has much to offer in terms of understanding how these modifications affect overall cell biology, a high-resolution picture of the intricacies of the process will likely need to result from a focus on simpler systems. As a universal quantitative tool, able in principle, to identify any type of PTM, MS has and will continue to have an important role to play in this investigation.

References

1 Cohen P. The role of protein phosphorylation in human health and disease. The Sir Hans Krebs Medal Lecture. *Eur J Biochem* 2001;**268**:5001–5010.
2 Hunter T. The role of tyrosine phosphorylation in cell growth and disease. *Harvey Lect* 1998;**94**:81–119.
3 Zhu X, Lee HG, Raina AK, Perry G, Smith MA. The role of mitogen-activated protein kinase pathways in Alzheimer's disease. *Neurosignals* 2002;**11**:270–281.
4 Fischer EH, Krebs EG. Conversion of phosphorylase b to phosphorylase a in muscle extracts. *J Biol Chem* 1955;**216**:121–132.
5 Sharma K, D'Souza RC, Tyanova S, Schaab C, Wisniewski JR, Cox J, Mann M. Ultradeep human phosphoproteome reveals a distinct regulatory nature of Tyr and Ser/Thr-based signaling. *Cell Rep* 2014;**8**:1583–1594.
6 Levene PA, Alsberg CL. The cleavage products of vitellin. *J Biol Chem* 1906;**2**:127.
7 Posternak S, Posternak T. *C R Acad Sci* 1928;**187**:313.
8 Levene PA, Lipmann FA. Serinephosphoric acid obtained on hydrolysis of vitellinic acid. *J Biol Chem* 1932;**98**:109–114.
9 Araki M, Yonezawa T, Chin S, Kuga M, Shimada N, Imagi S, Ochiai Y. Investigation of phosphorus metabolism of Yoshida sarcoma with the aid of P32. I. *Gan* 1952;**43**:69–72.
10 Williams-Ashman HG, Kennedy EP. Oxidative phosphorylation catalyzed by cytoplasmic particles isolated from malignant tissues. *Cancer Res* 1952;**12**:415–421.

11 Burnett G, Kennedy EP. The enzymatic phosphorylation of proteins. *J Biol Chem* 1954;**211**:969–980.

12 Krebs EG, Fischer EH. The phosphorylase b to a converting enzyme of rabbit skeletal muscle. *Biochim Biophys Acta* 1956;**20**:150–157.

13 Walsh DA, Perkins JP, Krebs EG. An adenosine 3′,5′-monophosphate-dependant protein kinase from rabbit skeletal muscle. *J Biol Chem* 1968;**243**:3763–3765.

14 Glass DB, Krebs EG. Protein phosphorylation catalyzed by cyclic AMP-dependent and cyclic GMP-dependent protein kinases. *Annu Rev Pharmacol Toxicol* 1980;**20**:363–388.

15 Hunter T, Sefton BM. Transforming gene product of Rous sarcoma virus phosphorylates tyrosine. *Proc Natl Acad Sci U S A* 1980;**77**:1311–1315.

16 Hunter T. The genesis of tyrosine phosphorylation. *Cold Spring Harb Perspect Biol* 2014;**6**:a020644.

17 Collett MS, Erikson E, Erikson RL. Structural analysis of the avian sarcoma virus transforming protein: Sites of phosphorylation. *J Virol* 1979;**29**:770–781.

18 Chinkers M, Cohen S. Purified EGF receptor-kinase interacts specifically with antibodies to Rous sarcoma virus transforming protein. *Nature* 1981;**290**:516–519.

19 Hunter T, Cooper JA. Epidermal growth factor induces rapid tyrosine phosphorylation of proteins in A431 human tumor cells. *Cell* 1981;**24**:741–752.

20 Penhallow RC, Class K, Sonoda H, Bolen JB, Rowley RB. Temporal activation of nontransmembrane protein-tyrosine kinases following mast cell Fc epsilon RI engagement. *J Biol Chem* 1995;**270**:23362–23365.

21 Oda K, Matsuoka Y, Funahashi A, Kitano H. A comprehensive pathway map of epidermal growth factor receptor signaling. *Mol Syst Biol* 2005;**1**:2005.

22 Manning G, Whyte DB, Martinez R, Hunter T, Sudarsanam S. The protein kinase complement of the human genome. *Science* 2002;**298**:1912–1934.

23 Manning G, Plowman GD, Hunter T, Sudarsanam S. Evolution of protein kinase signaling from yeast to man. *Trends Biochem Sci* 2002;**27**:514–520.

24 Robinson DR, Wu YM, Lin SF. The protein tyrosine kinase family of the human genome. *Oncogene* 2000;**19**:5548–5557.

25 Fedorov O, Muller S, Knapp S. The (un)targeted cancer kinome. *Nat Chem Biol* 2010;**6**:166–169.

26 Knapp S, Arruda P, Blagg J, Burley S, Drewry DH, Edwards A, Fabbro D, Gillespie P, Gray NS, Kuster B, Lackey KE, Mazzafera P, Tomkinson NC, Willson TM, Workman P, Zuercher WJ. A public-private partnership to unlock the untargeted kinome. *Nat Chem Biol* 2013;**9**:3–6.

27 Kilpinen S, Ojala K, Kallioniemi O. Analysis of kinase gene expression patterns across 5681 human tissue samples reveals functional genomic taxonomy of the kinome. *PLoS One* 2010;**5**:e15068.

28 Hornbeck PV, Kornhauser JM, Tkachev S, Zhang B, Skrzypek E, Murray B, Latham V, Sullivan M. PhosphoSitePlus: A comprehensive resource for investigating the structure and function of experimentally determined post-translational modifications in man and mouse. *Nucleic Acids Res* 2012;**40**:D261–D270.

29 Hornbeck PV, Zhang B, Murray B, Kornhauser JM, Latham V, Skrzypek E. PhosphoSitePlus, 2014: Mutations, PTMs and recalibrations. *Nucleic Acids Res* 2015;**43**:D512–D520.

30 Huttlin EL, Jedrychowski MP, Elias JE, Goswami T, Rad R, Beausoleil SA, Villen J, Haas W, Sowa ME, Gygi SP. A tissue-specific atlas of mouse protein phosphorylation and expression. *Cell* 2010;**143**:1174–1189.

31 Lundby A, Secher A, Lage K, Nordsborg NB, Dmytriyev A, Lundby C, Olsen JV. Quantitative maps of protein phosphorylation sites across 14 different rat organs and tissues. *Nat Commun* 2012;**3**:876.

32 Marx H, Lemeer S, Schliep JE, Matheron L, Mohammed S, Cox J, Mann M, Heck AJ, Kuster B. A large synthetic peptide and phosphopeptide reference library for mass spectrometry-based proteomics. *Nat Biotechnol* 2013;**31**:557–564.

33 Zappacosta F, Scott GF, Huddleston MJ, Annan RS. An optimized platform for hydrophilic interaction chromatography-immobilized metal affinity chromatography enables deep coverage of the rat liver phosphoproteome. *J Proteome Res* 2015;**14**:997–1009.

34 Cohen P. The origins of protein phosphorylation. *Nat Cell Biol* 2002;**4**:E127–E130.

35 Zappacosta F, Collingwood TS, Huddleston MJ, Annan RS. A quantitative results-driven approach to analyzing multisite protein phosphorylation: The phosphate-dependent phosphorylation profile of the transcription factor Pho4. *Mol Cell Proteomics* 2006;**5**:2019–2030.

36 Komeili A, O'Shea EK. Roles of phosphorylation sites in regulating activity of the transcription factor Pho4. *Science* 1999;**284**:977–980.

37 Springer M, Wykoff DD, Miller N, O'Shea EK. Partially phosphorylated Pho4 activates transcription of a subset of phosphate-responsive genes. *PLoS Biol* 2003;**1**:E28.

38 Verma R, Annan RS, Huddleston MJ, Carr SA, Reynard G, Deshaies RJ. Phosphorylation of Sic1p by G1 Cdk required for its degradation and entry into S phase. *Science* 1997;**278**:455–460.

39 Nash P, Tang X, Orlicky S, Chen Q, Gertler FB, Mendenhall MD, Sicheri F, Pawson T, Tyers M. Multisite phosphorylation of a CDK inhibitor sets a threshold for the onset of DNA replication. *Nature* 2001;**414**:514–521.

40 Hao B, Oehlmann S, Sowa ME, Harper JW, Pavletich NP. Structure of a Fbw7-Skp1-cyclin E complex: Multisite-phosphorylated substrate recognition by SCF ubiquitin ligases. *Mol Cell* 2007;**26**:131–143.

41 Koivomagi M, Valk E, Venta R, Iofik A, Lepiku M, Balog ER, Rubin SM, Morgan DO, Loog M. Cascades of multisite phosphorylation control Sic1 destruction at the onset of S phase. *Nature* 2011;**480**:128–131.

42 Sadowski I, Stone JC, Pawson T. A noncatalytic domain conserved among cytoplasmic protein-tyrosine kinases modifies the kinase function and transforming activity of Fujinami sarcoma virus P130gag-fps. *Mol Cell Biol* 1986;**6**:4396–4408.

43 Anderson D, Koch CA, Grey L, Ellis C, Moran MF, Pawson T. Binding of SH2 domains of phospholipase C gamma 1, GAP, and Src to activated growth factor receptors. *Science* 1990;**250**:979–982.

44 Moran MF, Koch CA, Anderson D, Ellis C, England L, Martin GS, Pawson T. Src homology region 2 domains direct protein-protein interactions in signal transduction. *Proc Natl Acad Sci U S A* 1990;**87**:8622–8626.

45 Liu BA, Jablonowski K, Raina M, Arce M, Pawson T, Nash PD. The human and mouse complement of SH2 domain proteins-establishing the boundaries of phosphotyrosine signaling. *Mol Cell* 2006;**22**:851–868.

46 Kavanaugh WM, Williams LT. An alternative to SH2 domains for binding tyrosine-phosphorylated proteins. *Science* 1994;**266**:1862–1865.

47 Yaffe MB, Elia AE. Phosphoserine/threonine-binding domains. *Curr Opin Cell Biol* 2001;**13**:131–138.

48 Iakoucheva LM, Radivojac P, Brown CJ, O'Connor TR, Sikes JG, Obradovic Z, Dunker AK. The importance of intrinsic disorder for protein phosphorylation. *Nucleic Acids Res* 2004;**32**:1037–1049.

49 Wright PE, Dyson HJ. Intrinsically unstructured proteins: Re-assessing the protein structure-function paradigm. *J Mol Biol* 1999;**293**:321–331.

50 Kim PM, Sboner A, Xia Y, Gerstein M. The role of disorder in interaction networks: A structural analysis. *Mol Syst Biol* 2008;**4**:179.

51 Tompa P, Davey NE, Gibson TJ, Babu MM. A million peptide motifs for the molecular biologist. *Mol Cell* 2014;**55**:161–169.

52 Meek DW, Anderson CW. Post-translational modification of p53: Cooperative integrators of function. *Cold Spring Harb Perspect Biol* 2009;**1**:a000950.

53 Munoz MJ, de la Mata M, Kornblihtt AR. The carboxy terminal domain of RNA polymerase II and alternative splicing. *Trends Biochem Sci* 2010;**35**:497–504.

54 Bah A, Vernon RM, Siddiqui Z, Krzeminski M, Muhandiram R, Zhao C, Sonenberg N, Kay LE, Forman-Kay JD. Folding of an intrinsically disordered protein by phosphorylation as a regulatory switch. *Nature* 2015;**519**:106–109.

55 Wright PE, Dyson HJ. Intrinsically disordered proteins in cellular signalling and regulation. *Nat Rev Mol Cell Biol* 2015;**16**:18–29.

56 Mustelin T. A brief introduction to the protein phosphatase families. *Methods Mol Biol* 2007;**365**:9–22.

57 Sacco F, Perfetto L, Castagnoli L, Cesareni G. The human phosphatase interactome: An intricate family portrait. *FEBS Lett* 2012;**586**:2732–2739.

58 Kinexus: Systems Bioinformatics Company. http://www.kinexus.ca/ scienceTechnology/protein_phosphatases/protein_phosphatases.html (2015)

59 Shi Y. Serine/threonine phosphatases: Mechanism through structure. *Cell* 2009;**139**:468–484.

60 Cohen P, Cohen PT. Protein phosphatases come of age. *J Biol Chem* 1989;**264**:21435–21438.

61 Sun H, Tonks NK. The coordinated action of protein tyrosine phosphatases and kinases in cell signaling. *Trends Biochem Sci* 1994;**19**:480–485.

62 Kleiman LB, Maiwald T, Conzelmann H, Lauffenburger DA, Sorger PK. Rapid phospho-turnover by receptor tyrosine kinases impacts downstream signaling and drug binding. *Mol Cell* 2011;**43**:723–737.

63 Whiteaker JR, Zhao L, Yan P, Ivey RG, Voytovich UJ, Moore HD, Lin C, Paulovich AG. Peptide immunoaffinity enrichment and targeted mass spectrometry enables multiplex, quantitative pharmacodynamic studies of phospho-signaling. *Mol Cell Proteomics* 2015;**14**:2261–2273.

64 Gould KL, Woodgett JR, Cooper JA, Buss JE, Shalloway D, Hunter T. Protein kinase C phosphorylates pp60src at a novel site. *Cell* 1985;**42**:849–857.

65 van der Geer P, Hunter T. Phosphopeptide mapping and phosphoamino acid analysis by electrophoresis and chromatography on thin-layer cellulose plates. *Electrophoresis* 1994;**15**:544–554.

66 Barber M, Bordoli RS, Sedgwick RD, Tyler AN. Fast atom bombardment of solids (F.A.B.): A new ion source for mass spectrometry. *J Chem Soc Chem Commun* 1981;**7**:325–327.

67 Morris HR, Panico M, Barber M, Bordoli RS, Sedgwick RD, Tyler A. Fast atom bombardment: A new mass spectrometric method for peptide sequence analysis. *Biochem Biophys Res Commun* 1981;**101**:623–631.

68 Webster TA, Gibson BW, Keng T, Biemann K, Schimmel P. Primary structures of both subunits of Escherichia coli glycyl-tRNA synthetase. *J Biol Chem* 1983;**258**:10637–10641.

69 Gibson BW, Biemann K. Strategy for the mass spectrometric verification and correction of the primary structures of proteins deduced from their DNA sequences. *Proc Natl Acad Sci U S A* 1984;**81**:1956–1960.

70 Fenselau C, Heller DN, Miller MS, White HB III. Phosphorylation sites in riboflavin-binding protein characterized by fast atom bombardment mass spectrometry. *Anal Biochem* 1985;**150**:309–314.

71 Hunt DF, Yates JR III, Shabanowitz J, Winston S, Hauer CR. Protein sequencing by tandem mass spectrometry. *Proc Natl Acad Sci U S A* 1986;**83**:6233–6237.

72 Michel H, Hunt DF, Shabanowitz J, Bennett J. Tandem mass spectrometry reveals that three photosystem II proteins of spinach chloroplasts contain N-acetyl-O-phosphothreonine at their NH2 termini. *J Biol Chem* 1988;**263**:1123–1130.

73 Holmes CF, Tonks NK, Major H, Cohen P. Analysis of the in vivo phosphorylation state of protein phosphatase inhibitor-2 from rabbit skeletal muscle by fast-atom bombardment mass spectrometry. *Biochim Biophys Acta* 1987;**929**:208–219.

74 Karas M, Hillenkamp F. Laser desorption ionization of proteins with molecular masses exceeding 10,000 daltons. *Anal Chem* 1988;**60**:2299–2301.

75 Hillenkamp F, Karas M. Mass spectrometry of peptides and proteins by matrix-assisted ultraviolet laser desorption/ionization. *Methods Enzymol* 1990;**193**:280–295.

76 Spengler B, Kirsch D, Kaufmann R, Jaeger E. Peptide sequencing by matrix-assisted laser-desorption mass spectrometry. *Rapid Commun Mass Spectrom* 1992;**6**:105–108.

77 Annan RS, Carr SA. Phosphopeptide analysis by matrix-assisted laser desorption time-of-flight mass spectrometry. *Anal Chem* 1996;**68**:3413–3421.

78 Fenn JB, Mann M, Meng CK, Wong SF, Whitehouse CM. Electrospray ionization for mass spectrometry of large biomolecules. *Science* 1989;**246**:64–71.

79 Yost RA, Enke CG. Triple quadrupole mass spectrometry for direct mixture analysis and structure elucidation. *Anal Chem* 1979;**51**:1251–1264.

80 Payne DM, Rossomando AJ, Martino P, Erickson AK, Her JH, Shabanowitz J, Hunt DF, Weber MJ, Sturgill TW. Identification of the regulatory phosphorylation sites in pp42/mitogen-activated protein kinase (MAP kinase). *EMBO J* 1991;**10**:885–892.

81 Hunt DF, Henderson RA, Shabanowitz J, Sakaguchi K, Michel H, Sevilir N, Cox AL, Appella E, Engelhard VH. Characterization of peptides bound to the class I MHC molecule HLA-A2.1 by mass spectrometry. *Science* 1992;**255**:1261–1263.

82 Wilm M, Mann M. Electrospray and Taylor-Cone theory, Dole's beam of macromolecules at last? *Int J Mass Spectrom Ion Processes* 1994;**136**:167–180.

83 Wilm M, Mann M. Analytical properties of the nanoelectrospray ion source. *Anal Chem* 1996;**68**:1–8.

84 Emmett MR, Caprioli RM. Micro-electrospray mass spectrometry: Ultra-high-sensitivity analysis of peptides and proteins. *J Am Soc Mass Spectrom* 1994;**5**:605–613.

85 Davis MT, Stahl DC, Lee TD. Low flow high-performance liquid chromatography solvent delivery system designed for tandem capillary liquid chromatography-mass spectrometry. *J Am Soc Mass Spectrom* 1995;**6**:571–577.

86 Davis MT, Stahl DC, Hefta SA, Lee TD. A microscale electrospray interface for on-line, capillary liquid chromatography/tandem mass spectrometry of complex peptide mixtures. *Anal Chem* 1995;**67**:4549–4556.

87 Osaka I, Takayama M. Influence of hydrophobicity on positive- and negative-ion yields of peptides in electrospray ionization mass spectrometry. *Rapid Commun Mass Spectrom* 2014;**28**:2222–2226.

88 Cech NB, Enke CG. Relating electrospray ionization response to nonpolar character of small peptides. *Anal Chem* 2000;**72**:2717–2723.

89 Steen H, Jebanathirajah JA, Rush J, Morrice N, Kirschner MW. Phosphorylation analysis by mass spectrometry: Myths, facts, and the consequences for qualitative and quantitative measurements. *Mol Cell Proteomics* 2006;**5**:172–181.

90 Gropengiesser J, Varadarajan BT, Stephanowitz H, Krause E. The relative influence of phosphorylation and methylation on responsiveness of peptides to MALDI and ESI mass spectrometry. *J Mass Spectrom* 2009;**44**:821–831.

91 Carr SA, Huddleston MJ, Annan RS. Selective detection and sequencing of phosphopeptides at the femtomole level by mass spectrometry. *Anal Biochem* 1996;**239**:180–192.

92 Steen H, Jebanathirajah JA, Springer M, Kirschner MW. Stable isotope-free relative and absolute quantitation of protein phosphorylation stoichiometry by MS. *Proc Natl Acad Sci U S A* 2005;**102**:3948–3953.

93 Hegeman AD, Harms AC, Sussman MR, Bunner AE, Harper JF. An isotope labeling strategy for quantifying the degree of phosphorylation at multiple sites in proteins. *J Am Soc Mass Spectrom* 2004;**15**:647–653.

94 Guo L, Kozlosky CJ, Ericsson LH, Daniel TO, Cerretti DP, Johnson RS. Studies of ligand-induced site-specific phosphorylation of epidermal growth factor receptor. *J Am Soc Mass Spectrom* 2003;**14**:1022–1031.

95 Tsay YG, Wang YH, Chiu CM, Shen BJ, Lee SC. A strategy for identification and quantitation of phosphopeptides by liquid chromatography/tandem mass spectrometry. *Anal Biochem* 2000;**287**:55–64.

96 Resing KA, Ahn NG. Protein phosphorylation analysis by electrospray ionization-mass spectrometry. *Methods Enzymol* 1997;**283**:29–44.

97 Marcantonio M, Trost M, Courcelles M, Desjardins M, Thibault P. Combined enzymatic and data mining approaches for comprehensive phosphoproteome analyses: Application to cell signaling events of interferon-gamma-stimulated macrophages. *Mol Cell Proteomics* 2008;**7**:645–660.

98 Gao Y, Wang Y. A method to determine the ionization efficiency change of peptides caused by phosphorylation. *J Am Soc Mass Spectrom* 2007;**18**:1973–1976.

99 Lee KA, Craven KB, Niemi GA, Hurley JB. Mass spectrometric analysis of the kinetics of in vivo rhodopsin phosphorylation. *Protein Sci* 2002;**11**:862–874.

100 Gerber SA, Rush J, Stemman O, Kirschner MW, Gygi SP. Absolute quantification of proteins and phosphoproteins from cell lysates by tandem MS. *Proc Natl Acad Sci U S A* 2003;**100**:6940–6945.

101 Zhang X, Jin QK, Carr SA, Annan RS. N-terminal peptide labeling strategy for incorporation of isotopic tags: A method for the determination of site-specific absolute phosphorylation stoichiometry. *Rapid Commun Mass Spectrom* 2002;**16**:2325–2332.

102 Wu R, Haas W, Dephoure N, Huttlin EL, Zhai B, Sowa ME, Gygi SP. A large-scale method to measure absolute protein phosphorylation stoichiometries. *Nat Methods* 2011;**8**:677–683.

103 Bonenfant D, Schmelzle T, Jacinto E, Crespo JL, Mini T, Hall MN, Jenoe P. Quantitation of changes in protein phosphorylation: A simple method based on stable isotope labeling and mass spectrometry. *Proc Natl Acad Sci U S A* 2003;**100**:880–885.

104 Papayannopoulos I. The interpretation of collision-induced dissociation of tandem mass spectra of peptides. *Mass Spectrom Rev* 1995;**14**:49–73.

105 Palumbo AM, Tepe JJ, Reid GE. Mechanistic insights into the multistage gas-phase fragmentation behavior of phosphoserine- and phosphothreonine-containing peptides. *J Proteome Res* 2008;**7**:771–779.

106 Brown R, Stuart SS, Houel S, Ahn NG, Old WM. Large-scale examination of factors influencing phosphopeptide neutral loss during collision induced dissociation. *J Am Soc Mass Spectrom* 2015;**26**:1128–1142.

107 March RE. An introduction to quadrupole Ion trap mass spectrometry. *J Mass Spectrom* 1997;**32**:351–369.

108 DeGnore JP, Qin J. Fragmentation of phosphopeptides in an ion trap mass spectrometer. *J Am Soc Mass Spectrom* 1998;**9**:1175–1188.

109 Karcher RL, Roland JT, Zappacosta F, Huddleston MJ, Annan RS, Carr SA, Gelfand VI. Cell cycle regulation of myosin-V by calcium/calmodulin-dependent protein kinase II. *Science* 2001;**293**:1317–1320.

110 Palumbo AM, Reid GE. Evaluation of gas-phase rearrangement and competing fragmentation reactions on protein phosphorylation site assignment using collision induced dissociation-MS/MS and MS3. *Anal Chem* 2008;**80**:9735–9747.

111 Cui L, Reid GE. Examining factors that influence erroneous phosphorylation site localization via competing fragmentation and rearrangement reactions during ion trap CID-MS/MS and -MS(3.) *Proteomics* 2013;**13**:964–973.

112 Yague J, Paradela A, Ramos M, Ogueta S, Marina A, Barahona F, Lopez de Castro JA, Vazquez J. Peptide rearrangement during quadrupole ion trap fragmentation: Added complexity to MS/MS spectra. *Anal Chem* 2003;**75**:1524–1535.

113 Beausoleil SA, Jedrychowski M, Schwartz D, Elias JE, Villen J, Li J, Cohn MA, Cantley LC, Gygi SP. Large-scale characterization of HeLa cell nuclear phosphoproteins. *Proc Natl Acad Sci U S A* 2004;**101**:12130–12135.

114 Cui L, Yapici I, Borhan B, Reid GE. Quantification of competing H3PO4 versus HPO3 + H2O neutral losses from regioselective 18O-labeled phosphopeptides. *J Am Soc Mass Spectrom* 2014;**25**:141–148.

115 Schroeder MJ, Shabanowitz J, Schwartz JC, Hunt DF, Coon JJ. A neutral loss activation method for improved phosphopeptide sequence analysis by quadrupole ion trap mass spectrometry. *Anal Chem* 2004;**76**:3590–3598.

116 Ulintz PJ, Yocum AK, Bodenmiller B, Aebersold R, Andrews PC, Nesvizhskii AI. Comparison of MS(2)-only, MSA, and MS(2)/MS(3) methodologies for phosphopeptide identification. *J Proteome Res* 2009;**8**:887–899.

117 Villen J, Beausoleil SA, Gygi SP. Evaluation of the utility of neutral-loss-dependent MS3 strategies in large-scale phosphorylation analysis. *Proteomics* 2008;**8**:4444–4452.

118 Olsen JV, Mann M. Improved peptide identification in proteomics by two consecutive stages of mass spectrometric fragmentation. *Proc Natl Acad Sci U S A* 2004;**101**:13417–13422.

119 Gruhler A, Olsen JV, Mohammed S, Mortensen P, Faergeman NJ, Mann M, Jensen ON. Quantitative phosphoproteomics applied to the yeast pheromone signaling pathway. *Mol Cell Proteomics* 2005;**4**:310–327.

120 Stensballe A, Jensen ON, Olsen JV, Haselmann KF, Zubarev RA. Electron capture dissociation of singly and multiply phosphorylated peptides. *Rapid Commun Mass Spectrom* 2000;**14**:1793–1800.

121 Zubarev RA, Horn DM, Fridriksson EK, Kelleher NL, Kruger NA, Lewis MA, Carpenter BK, McLafferty FW. Electron capture dissociation for structural characterization of multiply charged protein cations. *Anal Chem* 2000;**72**:563–573.

122 Syka JE, Coon JJ, Schroeder MJ, Shabanowitz J, Hunt DF. Peptide and protein sequence analysis by electron transfer dissociation mass spectrometry. *Proc Natl Acad Sci U S A* 2004;**101**:9528–9533.

123 Good DM, Wirtala M, McAlister GC, Coon JJ. Performance characteristics of electron transfer dissociation mass spectrometry. *Mol Cell Proteomics* 2007;**6**:1942–1951.

124 Swaney DL, McAlister GC, Coon JJ. Decision tree-driven tandem mass spectrometry for shotgun proteomics. *Nat Methods* 2008;**5**:959–964.

125 Molina H, Horn DM, Tang N, Mathivanan S, Pandey A. Global proteomic profiling of phosphopeptides using electron transfer dissociation tandem mass spectrometry. *Proc Natl Acad Sci U S A* 2007;**104**:2199–2204.

126 Frese CK, Zhou H, Taus T, Altelaar AF, Mechtler K, Heck AJ, Mohammed S. Unambiguous phosphosite localization using electron-transfer/higher-energy collision dissociation (EThcD). *J Proteome Res* 2013;**12**:1520–1525.

127 Junger MA, Aebersold R. Mass spectrometry-driven phosphoproteomics: Patterning the systems biology mosaic. *Wiley Interdiscip Rev Dev Biol* 2014;**3**:83–112.

128 Shvartsburg AA, Singer D, Smith RD, Hoffmann R. Ion mobility separation of isomeric phosphopeptides from a protein with variant modification of adjacent residues. *Anal Chem* 2011;**83**:5078–5085.

129 Bridon G, Bonneil E, Muratore-Schroeder T, Caron-Lizotte O, Thibault P. Improvement of phosphoproteome analyses using FAIMS and decision tree fragmentation. application to the insulin signaling pathway in *Drosophila melanogaster* S2 cells. *J Proteome Res* 2012;**11**:927–940.

130 Bian Y, Ye M, Song C, Cheng K, Wang C, Wei X, Zhu J, Chen R, Wang F, Zou H. Improve the coverage for the analysis of phosphoproteome of HeLa cells by a tandem digestion approach. *J Proteome Res* 2012;**11**:2828–2837.

131 Wisniewski JR, Mann M. Consecutive proteolytic digestion in an enzyme reactor increases depth of proteomic and phosphoproteomic analysis. *Anal Chem* 2012;**84**:2631–2637.

132 Giansanti P, Aye TT, Van Den Toorn H, Peng M, van Breukelen B, Heck AJ. An augmented multiple-protease-based human phosphopeptide atlas. *Cell Rep* 2015;**11**:1834–1843.

133 Ficarro SB, McCleland ML, Stukenberg PT, Burke DJ, Ross MM, Shabanowitz J, Hunt DF, White FM. Phosphoproteome analysis by mass spectrometry and its application to *Saccharomyces cerevisiae. Nat Biotechnol* 2002;**20**:301–305.

134 Erickson BK, Jedrychowski MP, McAlister GC, Everley RA, Kunz R, Gygi SP. Evaluating multiplexed quantitative phosphopeptide analysis on a hybrid quadrupole mass filter/linear ion trap/orbitrap mass spectrometer. *Anal Chem* 2015;**87**:1241–1249.

135 Batth TS, Francavilla C, Olsen JV. Off-line high-pH reversed-phase fractionation for in-depth phosphoproteomics. *J Proteome Res* 2014;**13**:6176–6186.

136 Cohen P. The regulation of protein function by multisite phosphorylation--a 25 year update. *Trends Biochem Sci* 2000;**25**:596–601.

137 Ubersax JA, Ferrell JE Jr. Mechanisms of specificity in protein phosphorylation. *Nat Rev Mol Cell Biol* 2007;**8**:530–541.

138 Roux PP, Thibault P. The coming of age of phosphoproteomics--from large data sets to inference of protein functions. *Mol Cell Proteomics* 2013;**12**:3453–3464.

139 Liu Y, Chance MR. Integrating phosphoproteomics in systems biology. *Comput Struct Biotechnol J* 2014;**10**:90–97.

140 Courcelles M, Fremin C, Voisin L, Lemieux S, Meloche S, Thibault P. Phosphoproteome dynamics reveal novel ERK1/2 MAP kinase substrates with broad spectrum of functions. *Mol Syst Biol* 2013;**9**:669.

141 Hsu PP, Kang SA, Rameseder J, Zhang Y, Ottina KA, Lim D, Peterson TR, Choi Y, Gray NS, Yaffe MB, Marto JA, Sabatini DM. The mTOR-regulated phosphoproteome reveals a mechanism of mTORC1-mediated inhibition of growth factor signaling. *Science* 2011;**332**:1317–1322.

142 Yu Y, Yoon SO, Poulogiannis G, Yang Q, Ma XM, Villen J, Kubica N, Hoffman GR, Cantley LC, Gygi SP, Blenis J. Phosphoproteomic analysis identifies Grb10 as an mTORC1 substrate that negatively regulates insulin signaling. *Science* 2011;**332**:1322–1326.

143 Old WM, Shabb JB, Houel S, Wang H, Couts KL, Yen CY, Litman ES, Croy CH, Meyer-Arendt K, Miranda JG, Brown RA, Witze ES, Schweppe RE, Resing KA, Ahn NG. Functional proteomics identifies targets of phosphorylation by B-Raf signaling in melanoma. *Mol Cell* 2009;**34**:115–131.

144 Kim JY, Welsh EA, Oguz U, Fang B, Bai Y, Kinose F, Bronk C, Remsing Rix LL, Beg AA, Rix U, Eschrich SA, Koomen JM, Haura EB. Dissection of TBK1 signaling via phosphoproteomics in lung cancer cells. *Proc Natl Acad Sci U S A* 2013;**110**:12414–12419.

145 Johnson H, Lescarbeau RS, Gutierrez JA, White FM. Phosphotyrosine profiling of NSCLC cells in response to EGF and HGF reveals network specific mediators of invasion. *J Proteome Res* 2013;**12**:1856–1867.

146 Osinalde N, Sanchez-Quiles V, Akimov V, Guerra B, Blagoev B, Kratchmarova I. Simultaneous dissection and comparison of IL-2 and IL-15 signaling pathways by global quantitative phosphoproteomics. *Proteomics* 2015;**15**:520–531.

147 Rust HL, Thompson PR. Kinase consensus sequences: A breeding ground for crosstalk. *ACS Chem Biol* 2011;**6**:881–892.

148 Ong SE, Blagoev B, Kratchmarova I, Kristensen DB, Steen H, Pandey A, Mann M. Stable isotope labeling by amino acids in cell culture, SILAC, as a simple and accurate approach to expression proteomics. *Mol Cell Proteomics* 2002;**1**:376–386.

149 Kruger M, Moser M, Ussar S, Thievessen I, Luber CA, Forner F, Schmidt S, Zanivan S, Fassler R, Mann M. SILAC mouse for quantitative proteomics uncovers kindlin-3 as an essential factor for red blood cell function. *Cell* 2008;**134**:353–364.

150 McClatchy DB, Liao L, Park SK, Venable JD, Yates JR. Quantification of the synaptosomal proteome of the rat cerebellum during post-natal development. *Genome Res* 2007;**17**:1378–1388.

151 Hsu JL, Huang SY, Chow NH, Chen SH. Stable-isotope dimethyl labeling for quantitative proteomics. *Anal Chem* 2003;**75**:6843–6852.

152 DeSouza LV, Taylor AM, Li W, Minkoff MS, Romaschin AD, Colgan TJ, Siu KW. Multiple reaction monitoring of mTRAQ-labeled peptides enables absolute quantification of endogenous levels of a potential cancer marker in cancerous and normal endometrial tissues. *J Proteome Res* 2008;7:3525–3534.

153 Ross PL, Huang YN, Marchese JN, Williamson B, Parker K, Hattan S, Khainovski N, Pillai S, Dey S, Daniels S, Purkayastha S, Juhasz P, Martin S, Bartlet-Jones M, He F, Jacobson A, Pappin DJ. Multiplexed protein quantitation in *Saccharomyces cerevisiae* using amine-reactive isobaric tagging reagents. *Mol Cell Proteomics* 2004;**3**:1154–1169.

154 Thompson A, Schafer J, Kuhn K, Kienle S, Schwarz J, Schmidt G, Neumann T, Johnstone R, Mohammed AK, Hamon C. Tandem mass tags: A novel quantification strategy for comparative analysis of complex protein mixtures by MS/MS. *Anal Chem* 2003;**75**:1895–1904.

155 Mertins P, Udeshi ND, Clauser KR, Mani DR, Patel J, Ong SE, Jaffe JD, Carr SA. iTRAQ labeling is superior to mTRAQ for quantitative global proteomics and phosphoproteomics. *Mol Cell Proteomics* 2012;**11**:M111.

156 Chahrour O, Cobice D, Malone J. Stable isotope labelling methods in mass spectrometry-based quantitative proteomics. *J Pharm Biomed Anal* 2015;**113**:2–20.

157 Savitski MM, Sweetman G, Askenazi M, Marto JA, Lang M, Zinn N, Bantscheff M. Delayed fragmentation and optimized isolation width settings for improvement of protein identification and accuracy of isobaric mass tag quantification on Orbitrap-type mass spectrometers. *Anal Chem* 2011;**83**:8959–8967.

158 Ow SY, Salim M, Noirel J, Evans C, Wright PC. Minimising iTRAQ ratio compression through understanding LC-MS elution dependence and high-resolution HILIC fractionation. *Proteomics* 2011;**11**:2341–2346.

159 Wenger CD, Lee MV, Hebert AS, McAlister GC, Phanstiel DH, Westphall MS, Coon JJ. Gas-phase purification enables accurate, multiplexed proteome quantification with isobaric tagging. *Nat Methods* 2011;**8**:933–935.

160 Ting L, Rad R, Gygi SP, Haas W. MS3 eliminates ratio distortion in isobaric multiplexed quantitative proteomics. *Nat Methods* 2011;**8**:937–940.

161 McAlister GC, Nusinow DP, Jedrychowski MP, Wuhr M, Huttlin EL, Erickson BK, Rad R, Haas W, Gygi SP. MultiNotch MS3 enables accurate, sensitive, and multiplexed detection of differential expression across cancer cell line proteomes. *Anal Chem* 2014;**86**:7150–7158.

162 Savitski MM, Mathieson T, Zinn N, Sweetman G, Doce C, Becher I, Pachl F, Kuster B, Bantscheff M. Measuring and managing ratio compression for accurate iTRAQ/TMT quantification. *J Proteome Res* 2013;**12**:3586–3598.

163 Bantscheff M, Lemeer S, Savitski MM, Kuster B. Quantitative mass spectrometry in proteomics: Critical review update from 2007 to the present. *Anal Bioanal Chem* 2012;**404**:939–965.

164 Bondarenko PV, Chelius D, Shaler TA. Identification and relative quantitation of protein mixtures by enzymatic digestion followed by capillary reversed-phase liquid chromatography-tandem mass spectrometry. *Anal Chem* 2002;**74**:4741–4749.

165 Sun W, Wu S, Wang X, Zheng D, Gao Y. An analysis of protein abundance suppression in data dependent liquid chromatography and tandem mass spectrometry with tryptic peptide mixtures of five known proteins. *Eur J Mass Spectrom (Chichester, Eng)* 2005;**11**:575–580.

166 Neilson KA, Ali NA, Muralidharan S, Mirzaei M, Mariani M, Assadourian G, Lee A, van Sluyter SC, Haynes PA. Less label, more free: Approaches in label-free quantitative mass spectrometry. *Proteomics* 2011;**11**:535–553.

167 de Graaf EL, Giansanti P, Altelaar AF, Heck AJ. Single-step enrichment by Ti^{4+}-IMAC and label-free quantitation enables in-depth monitoring of phosphorylation dynamics with high reproducibility and temporal resolution. *Mol Cell Proteomics* 2014;**13**:2426–2434.

168 Oliveira AP, Ludwig C, Zampieri M, Weisser H, Aebersold R, Sauer U. Dynamic phosphoproteomics reveals TORC1-dependent regulation of yeast nucleotide and amino acid biosynthesis. *Sci Signal* 2015;**8**:rs4.

169 Wu R, Dephoure N, Haas W, Huttlin EL, Zhai B, Sowa ME, Gygi SP. Correct interpretation of comprehensive phosphorylation dynamics requires normalization by protein expression changes. *Mol Cell Proteomics* 2011;**10**:1–12.

170 Wisniewski JR, Zougman A, Nagaraj N, Mann M. Universal sample preparation method for proteome analysis. *Nat Methods* 2009;**6**:359–362.

171 Chen EI, Cociorva D, Norris JL, Yates JR III. Optimization of mass spectrometry-compatible surfactants for shotgun proteomics. *J Proteome Res* 2007;**6**:2529–2538.

172 Guo X, Trudgian DC, Lemoff A, Yadavalli S, Mirzaei H. Confetti: A multiprotease map of the HeLa proteome for comprehensive proteomics. *Mol Cell Proteomics* 2014;**13**:1573–1584.

173 Swaney DL, Wenger CD, Coon JJ. Value of using multiple proteases for large-scale mass spectrometry-based proteomics. *J Proteome Res* 2010;**9**:1323–1329.

174 Gauci S, Helbig AO, Slijper M, Krijgsveld J, Heck AJ, Mohammed S. Lys-N and trypsin cover complementary parts of the phosphoproteome in a refined SCX-based approach. *Anal Chem* 2009;**81**:4493–4501.

175 Olsen JV, Blagoev B, Gnad F, Macek B, Kumar C, Mortensen P, Mann M. Global, in vivo, and site-specific phosphorylation dynamics in signaling networks. *Cell* 2006;**127**:635–648.

176 Frackelton J, Ross AH, Eisen HN. Characterization and use of monoclonal antibodies for isolation of phosphotyrosyl proteins from retrovirus-transformed cells and growth factor-stimulated cells. *Mol Cell Biol* 1983;**3**:1343–1352.

177 Tinti M, Nardozza AP, Ferrari E, Sacco F, Corallino S, Castagnoli L, Cesareni G. The 4G10, pY20 and p-TYR-100 antibody specificity: Profiling by peptide microarrays. *N Biotechnol* 2012;**29**:571–577.

178 Rush J, Moritz A, Lee KA, Guo A, Goss VL, Spek EJ, Zhang H, Zha XM, Polakiewicz RD, Comb MJ. Immunoaffinity profiling of tyrosine phosphorylation in cancer cells. *Nat Biotechnol* 2005;**23**:94–101.

179 Kassel DB, Consler TG, Shalaby M, Sekhri P, Gordon NNT. Direct coupling of an automated 2-dimensional microcolumn affinity chromatography-capillary HPLC system with mass spectrometry for biomolecule nalysis. *Tech Protein Chem* 1995;**VI**:39–46.

180 Gold MR, Yungwirth T, Sutherland CL, Ingham RJ, Vianzon D, Chiu R, van Oostveen I, Morrison HD, Aebersold R. Purification and identification of tyrosine-phosphorylated proteins from B lymphocytes stimulated through the antigen receptor. *Electrophoresis* 1994;**15**:441–453.

181 Pandey A, Podtelejnikov AV, Blagoev B, Bustelo XR, Mann M, Lodish HF. Analysis of receptor signaling pathways by mass spectrometry: Identification of Vav-2 as a substrate of the epidermal and platelet-derived growth factor receptors. *Proc Natl Acad Sci U S A* 2000;**97**:179–184.

182 Steen H, Kuster B, Fernandez M, Pandey A, Mann M. Tyrosine phosphorylation mapping of the epidermal growth factor receptor signaling pathway. *J Biol Chem* 2002;**277**:1031–1039.

183 Ibarrola N, Kalume DE, Gronborg M, Iwahori A, Pandey A. A proteomic approach for quantitation of phosphorylation using stable isotope labeling in cell culture. *Anal Chem* 2003;**75**:6043–6049.

184 Blagoev B, Ong SE, Kratchmarova I, Mann M. Temporal analysis of phosphotyrosine-dependent signaling networks by quantitative proteomics. *Nat Biotechnol* 2004;**22**:1139–1145.

185 Kruger M, Kratchmarova I, Blagoev B, Tseng YH, Kahn CR, Mann M. Dissection of the insulin signaling pathway via quantitative phosphoproteomics. *Proc Natl Acad Sci U S A* 2008;**105**:2451–2456.

186 Boersema PJ, Foong LY, Ding VMY, Lemeer S, Van Breukelen B, Philp R, Boekhorst J, Snel B, Hertog JD, Choo ABH, Heck AJR. In-depth qualitative and quantitative profiling of tyrosine phosphorylation using a combination of phosphopeptide immunoaffinity purification and stable isotope dimethyl labeling. *Mol Cell Proteomics* 2010;**9**:84–99.

187 Rikova K, Guo A, Zeng Q, Possemato A, Yu J, Haack H, Nardone J, Lee K, Reeves C, Li Y, Hu Y, Tan Z, Stokes M, Sullivan L, Mitchell J, Wetzel R, MacNeill J, Ren JM, Yuan J, Bakalarski CE, Villen J, Kornhauser JM, Smith B, Li D, Zhou X, Gygi SP, Gu TL, Polakiewicz RD, Rush J, Comb MJ. Global survey of phosphotyrosine signaling identifies oncogenic kinases in lung cancer. *Cell* 2007;**131**:1190–1203.

188 Villen J, Beausoleil SA, Gerber SA, Gygi SP. Large-scale phosphorylation analysis of mouse liver. *Proc Natl Acad Sci U S A* 2007;**104**:1488–1493.

189 Kumar N, Wolf-Yadlin A, White FM, Lauffenburger DA. Modeling HER2 effects on cell behavior from mass spectrometry phosphotyrosine data. *PLoS Comput Biol* 2007;**3**:e4.

190 Grunborg M, Kristiansen TZ, Stensballe A, Andersen JS, Ohara O, Mann M, Jensen ON, Pandey A. A mass spectrometry-based proteomic approach for identification of serine/threonine-phosphorylated proteins by enrichment with phospho-specific antibodies: Identification of a novel protein, Frigg, as a protein kinase A substrate. *Mol Cell Proteomics* 2002;**1**:517–527.

191 Zhang H, Zha X, Tan Y, Hornbeck PV, Mastrangelo AJ, Alessi DR, Polakiewicz RD, Comb MJ. Phosphoprotein analysis using antibodies broadly reactive against phosphorylated motifs. *J Biol Chem* 2002;**277**:39379–39387.

192 Matsuoka S, Ballif BA, Smogorzewska A, McDonald ER III, Hurov KE, Luo J, Bakalarski CE, Zhao Z, Solimini N, Lerenthal Y, Shiloh Y, Gygi SP, Elledge SJ. ATM and ATR substrate analysis reveals extensive protein networks responsive to DNA damage. *Science* 2007;**316**:1160–1166.

193 Andersson L, Porath J. Isolation of phosphoproteins by immobilized metal (Fe^{3+}) affinity chromatography. *Anal Biochem* 1986;**154**:250–254.

194 Neville DCA, Rozanas CR, Price EM, Gruis DB, Verkman AS, Townsend RR. Evidence for phosphorylation of serine 753 in CFTR using a novel metal- ion affinity resin and matrix-assisted laser desorption mass spectrometry. *Protein Sci* 1997;**6**:2436–2445.

195 Posewitz MC, Tempst P. Immobilized gallium(III) affinity chromatography of phosphopeptides. *Anal Chem* 1999;**71**:2883–2892.

196 Kokubu M, Ishihama Y, Sato T, Nagasu T, Oda Y. Specificity of immobilized metal affinity-based IMAC/C18 tip enrichment of phosphopeptides for protein phosphorylation analysis. *Anal Chem* 2005;**77**:5144–5154.

197 Tsai CF, Wang YT, Chen YR, Lai CY, Iin PY, Pan KT, Chen JY, Khoo KH, Chen YJ. Immobilized metal affinity chromatography revisited: PH/acid control toward high selectivity in phosphoproteomics. *J Proteome Res* 2008;**7**:4058–4069.

198 Ye J, Zhang X, Young C, Zhao X, Hao Q, Cheng L, Jensen ON. Optimized IMAC-IMAC protocol for phosphopeptide recovery from complex biological samples. *J Proteome Res* 2010;**9**:3561–3573.

199 Jensen SS, Larsen MR. Evaluation of the impact of some experimental procedures on different phosphopeptide enrichment techniques. *Rapid Commun Mass Spectrom* 2007;**21**:3635–3645.

200 McNulty DE, Annan RS. Hydrophilic interaction chromatography reduces the complexity of the phosphoproteome and improves global phosphopeptide isolation and detection. *Mol Cell Proteomics* 2008;**7**:971–980.

201 Thingholm TE, Jensen ON, Robinson PJ, Larsen MR. SIMAC (Sequential Elution from IMAC), a phosphoproteomics strategy for the rapid separation of monophosphorylated from multiply phosphorylated peptides. *Mol Cell Proteomics* 2008;**7**:661–671.

202 Zhou H, Ye M, Dong J, Han G, Jiang X, Wu R, Zou H. Specific phosphopeptide enrichment with immobilized titanium ion affinity chromatography adsorbent for phosphoproteome analysis. *J Proteome Res* 2008;**7**:3957–3967.

203 Yu Z, Han G, Sun S, Jiang X, Chen R, Wang F, Wu R, Ye M, Zou H. Preparation of monodisperse immobilized Ti^{4+} affinity chromatography microspheres for specific enrichment of phosphopeptides. *Anal Chim Acta* 2009;**636**:34–41.

204 Zhou H, Ye M, Dong J, Corradini E, Cristobal A, Heck AJR, Zou H, Mohammed S. Robust phosphoproteome enrichment using monodisperse microsphere-based immobilized titanium(IV) ion affinity chromatography. *Nat Protoc* 2013;**8**:461–480.

205 Matheron L, Van Den Toorn H, Heck AJR, Mohammed S. Characterization of biases in phosphopeptide enrichment by Ti^{4+}-immobilized metal affinity chromatography and TiO_2 using a massive synthetic library and human cell digests. *Anal Chem* 2014;**86**:8312–8320.

206 Pinkse MWH, Uitto PM, Hilhorst MJ, Ooms B, Heck AJR. Selective isolation at the femtomole level of phosphopeptides from proteolytic digests using 2D-NanoLC-ESI-MS/MS and titanium oxide precolumns. *Anal Chem* 2004;**76**:3935–3943.

207 Kuroda I, Shintani Y, Motokawa M, Abe S, Furuno M. Phosphopeptide-selective column-switching RP-HPLC with a titania precolumn. *Anal Sci* 2004;**20**:1313–1319.

208 Sano A, Nakamura H. Titania as a chemo-affinity support for the column-switching HPLC analysis of phosphopeptides: Application to the characterization of phosphorylation sites in proteins by combination with protease digestion and electrospray ionization mass spectrometry. *Anal Sci* 2004;**20**:861–864.

209 Larsen MR, Thingholm TE, Jensen ON, Roepstorff P, Jorgensen TJD. Highly selective enrichment of phosphorylated peptides from peptide mixtures using titanium dioxide microcolumns. *Mol Cell Proteomics* 2005;**4**:873–886.

210 Thingholm TE, Jorgensen TJD, Jensen OL, Larsen MR. Highly selective enrichment of phosphorylated peptides using titanium dioxide. *Nat Protoc* 2006;**1**:1929–1935.

211 Sugiyama N, Masuda T, Shinoda K, Nakamura A, Tomita M, Ishihama Y. Phosphopeptide enrichment by aliphatic hydroxy acid-modified metal oxide chromatography for nano-LC-MS/MS in proteomics applications. *Mol Cell Proteomics* 2007;**6**:1103–1109.

212 Wilson-Grady JT, Haas W, Gygi SP. Quantitative comparison of the fasted and re-fed mouse liver phosphoproteomes using lower pH reductive dimethylation. *Methods* 2013;**61**:277–286.

213 Bodenmiller B, Mueller LN, Mueller M, Domon B, Aebersold R. Reproducible isolation of distinct, overlapping segments of the phosphoproteome. *Nat Methods* 2007;**4**:231–237.

214 Lai ACY, Tsai CF, Hsu CC, Sun YN, Chen YJ. Complementary Fe^{3+}- and Ti^{4+}-immobilized metal ion affinity chromatography for purification of acidic and basic phosphopeptides. *Rapid Commun Mass Spectrom* 2012;**26**:2186–2194.

215 Tsai CF, Hsu CC, Hung JN, Wang YT, Choong WK, Zeng MY, Lin PY, Hong RW, Sung TY, Chen YJ. Sequential phosphoproteomic enrichment through complementary metal-directed immobilized metal ion affinity chromatography. *Anal Chem* 2014;**86**:685–693.

216 Zhou H, Low TY, Hennrich ML, van der Toorn H, Schwend T, Zou H, Mohammed S, Heck AJ. Enhancing the identification of phosphopeptides from putative basophilic kinase substrates using Ti(IV) based IMAC enrichment. *Mol Cell Proteomics* 2011;**10**:M110.

217 Ruprecht B, Koch H, Medard G, Mundt M, Kuster B, Lemeer S. Comprehensive and reproducible phosphopeptide enrichment using iron immobilized metal ion affinity chromatography (Fe-IMAC) columns. *Mol Cell Proteomics* 2015;**14**:205–215.

218 Oda Y, Nagasu T, Chait BT. Enrichment analysis of phosphorylated proteins as a tool for probing the phosphoproteome. *Nat Biotechnol* 2001;**19**:379–382.

219 Goshe MB, Conrads TP, Panisko EA, Angell NH, Veenstra TD, Smith RD. Phosphoprotein isotope-coded affinity tag approach for isolating and quantitating phosphopeptides in proteome-wide analyses. *Anal Chem* 2001;**73**:2578–2586.

220 McLachlin DT, Chait BT. Improved beta elimination-based affinity purification strategy for enrichment of phosphopeptides. *Anal Chem* 2003;**75**:6826–6836.

221 Zhou H, Watts JD, Aebersold R. A systematic approach to the analysis of protein phosphorylation. *Nat Biotechnol* 2001;**19**:375–378.

222 Tao WA, Wollscheid B, O'Brien R, Eng JK, Li XJ, Bodenmiller B, Watts JD, Hood L, Aebersold R. Quantitative phosphoproteome analysis using a dendrimer conjugation chemistry and tandem mass spectrometry. *Nat Methods* 2005;**2**:591–598.

223 Nuhse TS, Stensballe A, Jensen ON, Peck SC. Large-scale analysis of in vivo phosphorylated membrane proteins by immobilized metal ion affinity chromatography and mass spectrometry. *Mol Cell Proteomics* 2003;**2**:1234–1243.

224 Washburn MP, Wolters D, Yates JR. Large-scale analysis of the yeast proteome by multidimensional protein identification technology. *Nat Biotechnol* 2001;**19**:242–247.

225 Wolters DA, Washburn MP, Yates JR III. An automated multidimensional protein identification technology for shotgun proteomics. *Anal Chem* 2001;**73**:5683–5690.

226 Pinkse MWH, Mohammed S, Gouw JW, Van Breukelen B, Vos HR, Heck AJR. Highly robust, automated, and sensitive online TiO$_2$-based phosphoproteomics applied to study endogenous phosphorylation in *Drosophila melanogaster*. *J Proteome Res* 2008;**7**:687–697.

227 Goldberg RN, Kishore N, Lennen RM. Thermodynamic quantities for the ionization reactions of buffers. *J Phys Chem Ref Data* 2002;**31**:231–370.

228 Alpert AJ. Hydrophilic-interaction chromatography for the separation of peptides, nucleic acids and other polar compounds. *J Chromatogr A* 1990;**499**:177–196.

229 Gilar M, Olivova P, Daly AE, Gebler JC. Orthogonality of separation in two-dimensional liquid chromatography. *Anal Chem* 2005;**77**:6426–6434.

230 Olsen JV, Vermeulen M, Santamaria A, Kumar C, Miller ML, Jensen LJ, Gnad F, Cox J, Jensen TS, Nigg EA, Brunak S, Mann M. Quantitative

phosphoproteomics reveals widespread full phosphorylation site occupancy during mitosis. *Sci Signal* 2010;**3**:ra3.

231 Han G, Ye M, Zhou H, Jiang X, Feng S, Jiang X, Tian R, Wan D, Zou H, Gu J. Large-scale phosphoproteome analysis of human liver tissue by enrichment and fractionation of phosphopeptides with strong anion exchange chromatography. *Proteomics* 2008;**8**:1346–1361.

232 Dai J, Wang LS, Wu YB, Sheng QH, Wu JR, Shieh CH, Zeng R. Fully automatic separation and identification of phosphopeptides by continuous pH-gradient anion exchange online coupled with reversed-phase liquid chromatography mass spectrometry. *J Proteome Res* 2009;**8**:133–141.

233 Nie S, Dai J, Ning ZB, Cao XJ, Sheng QH, Zeng R. Comprehensive profiling of phosphopeptides based on anion exchange followed by flow-through enrichment with titanium dioxide (AFET). *J Proteome Res* 2010;**9**:4585–4594.

234 Ficarro SB, Zhang Y, Carrasco-Alfonso MJ, Garg B, Adelmant G, Webber JT, Luckey CJ, Marto JA. Online nanoflow multidimensional fractionation for high efficiency phosphopeptide analysis. *Mol Cell Proteomics* 2011;**10**:6996–7005.

235 Albuquerque CP, Smolka MB, Payne SH, Bafna V, Eng J, Zhou H. A multidimensional chromatography technology for in-depth phosphoproteome analysis. *Mol Cell Proteomics* 2008;**7**:1389–1396.

236 Engholm-Keller K, Birck P, Størling J, Pociot F, Mandrup-Poulsen T, Larsen MR. TiSH - a robust and sensitive global phosphoproteomics strategy employing a combination of TiO_2, SIMAC, and HILIC. *J Proteomics* 2012;**75**:5749–5761.

237 Zhou H, Di Palma S, Preisinger C, Peng M, Polat AN, Heck AJR, Mohammed S. Toward a comprehensive characterization of a human cancer cell phosphoproteome. *J Proteome Res* 2013;**12**:260–271.

238 Alpert AJ. Electrostatic repulsion hydrophilic interaction chromatography for isocratic separation of charged solutes and selective isolation of phosphopeptides. *Anal Chem* 2008;**80**:62–76.

239 Alpert AJ, Hudecz O, Mechtler K. Anion-exchange chromatography of phosphopeptides: Weak anion exchange versus strong anion exchange and anion-exchange chromatography versus electrostatic repulsion-hydrophilic interaction chromatography. *Anal Chem* 2015;**87**:4704–4711.

240 Bennetzen MV, Larsen DH, Bunkenborg J, Bartek J, Lukas J, Andersen JS. Site-specific phosphorylation dynamics of the nuclear proteome during the DNA damage response. *Mol Cell Proteomics* 2010;**9**:1314–1323.

241 Gan CS, Guo T, Zhang H, Lim SK, Sze SK. A comparative study of electrostatic repulsion-hydrophilic interaction chromatography (ERLIC) versus SCX-IMAC-based methods for phosphopeptide isolation/enrichment. *J Proteome Res* 2008;**7**:4869–4877.

242 Hao P, Guo T, Sze SK. Simultaneous analysis of proteome, phospho- and glycoproteome of rat kidney tissue with electrostatic repulsion hydrophilic interaction chromatography. *PLoS One* 2011;**6**:E16884.

243 Gilar M, Olivova P, Daly AE, Gebler JC. Two-dimensional separation of peptides using RP-RP-HPLC system with different pH in first and second separation dimensions. *J Sep Sci* 2005;**28**:1694–1703.

244 Song C, Ye M, Han G, Jiang X, Wang F, Yu Z, Chen R, Zou H. Reversed-phase-reversed-phase liquid chromatography approach with high orthogonality for multidimensional separation of phosphopeptides. *Anal Chem* 2010;**82**:53–56.

245 Perkins DN, Pappin DJ, Creasy DM, Cottrell JS. Probability-based protein identification by searching sequence databases using mass spectrometry data. *Electrophoresis* 1999;**20**:3551–3567.

246 Eng JK, McCormack AL, Yates JR. An approach to correlate tandem mass spectral data of peptides with amino acid sequences in a protein database. *J Am Soc Mass Spectrom* 1994;**5**:976–989.

247 Elias JE, Gygi SP. Target-decoy search strategy for increased confidence in large- scale protein identifications by mass spectrometry. *Nat Methods* 2007;**4**:207–214.

248 Chalkley RJ, Clauser KR. Modification site localization scoring: Strategies and performance. *Mol Cell Proteomics* 2012;**11**:3–14.

249 Lee DC, Jones AR, Hubbard SJ. Computational phosphoproteomics: From identification to localization. *Proteomics* 2015;**15**:950–963.

250 Baker PR, Trinidad JC, Chalkley RJ. Modification site localization scoring integrated into a search engine. *Mol Cell Proteomics* 2011;**10**: M111.

251 Taus T, Kocher T, Pichler P, Paschke C, Schmidt A, Henrich C, Mechtler K. Universal and confident phosphorylation site localization using phosphoRS. *J Proteome Res* 2011;**10**:5354–5362.

252 Fermin D, Avtonomov D, Choi H, Nesvizhskii AI. LuciPHOr2: Site localization of generic post-translational modifications from tandem mass spectrometry data. *Bioinformatics* 2015;**31**:1141–1143.

253 Fermin D, Walmsley SJ, Gingras AC, Choi H, Nesvizhskii AI. LuciPHOr: Algorithm for phosphorylation site localization with false localization rate estimation using modified target-decoy approach. *Mol Cell Proteomics* 2013;**12**:3409–3419.

254 Beausoleil SA, Villen J, Gerber SA, Rush J, Gygi SP. A probability-based approach for high-throughput protein phosphorylation analysis and site localization. *Nat Biotechnol* 2006;**24**:1285–1292.

255 Wan Y, Cripps D, Thomas S, Campbell P, Ambulos N, Chen T, Yang A. PhosphoScan: A probability-based method for phosphorylation site prediction using MS2/MS3 pair information. *J Proteome Res* 2008;**7**:2803–2811.

256 Ruttenberg BE, Pisitkun T, Knepper MA, Hoffert JD. PhosphoScore: An open-source phosphorylation site assignment tool for MSn data. *J Proteome Res* 2008;**7**:3054–3059.

257 Saeed F, Pisitkun T, Hoffert JD, Rashidian S, Wang G, Gucek M, Knepper MA. PhosSA: Fast and accurate phosphorylation site assignment algorithm for mass spectrometry data. *Proteome Sci.* 2013;**11**:S14.

258 Savitski MM, Lemeer S, Boesche M, Lang M, Mathieson T, Bantscheff M, Kuster B. Confident phosphorylation site localization using the Mascot Delta Score. *Mol Cell Proteomics* 2011;**10**:M110.

259 Bailey CM, Sweet SM, Cunningham DL, Zeller M, Heath JK, Cooper HJ. SLoMo: Automated site localization of modifications from ETD/ECD mass spectra. *J Proteome Res* 2009;**8**:1965–1971.

260 Vaudel M, Breiter D, Beck F, Rahnenfuhrer J, Martens L, Zahedi RP. D-score: A search engine independent MD-score. *Proteomics* 2013;**13**:1036–1041.

261 Cox J, Mann M. MaxQuant enables high peptide identification rates, individualized p.p.b.-range mass accuracies and proteome-wide protein quantification. *Nat Biotechnol* 2008;**26**:1367–1372.

262 Nesvizhskii AI. A survey of computational methods and error rate estimation procedures for peptide and protein identification in shotgun proteomics. *J Proteomics* 2010;**73**:2092–2123.

263 Dinkel H, Chica C, Via A, Gould CM, Jensen LJ, Gibson TJ, Diella F. Phospho.ELM: A database of phosphorylation sites--update 2011. *Nucleic Acids Res* 2011;**39**:D261–D267.

264 Goel R, Harsha HC, Pandey A, Prasad TS. Human protein reference database and human proteinpedia as resources for phosphoproteome analysis. *Mol Biosyst* 2012;**8**:453–463.

265 Keshava Prasad TS, Goel R, Kandasamy K, Keerthikumar S, Kumar S, Mathivanan S, Telikicherla D, Raju R, Shafreen B, Venugopal A, Balakrishnan L, Marimuthu A, Banerjee S, Somanathan DS, Sebastian A, Rani S, Ray S, Harrys Kishore CJ, Kanth S, Ahmed M, Kashyap MK, Mohmood R, Ramachandra YL, Krishna V, Rahiman BA, Mohan S, Ranganathan P, Ramabadran S, Chaerkady R, Pandey A. Human protein reference database--2009 update. *Nucleic Acids Res* 2009;**37**:D767–D772.

266 Durek P, Schmidt R, Heazlewood JL, Jones A, MacLean D, Nagel A, Kersten B, Schulze WX. PhosPhAt: The *Arabidopsis thaliana* phosphorylation site databaseAn update. *Nucleic Acids Res* 2010;**38**:D828–D834.

267 Bodenmiller B, Campbell D, Gerrits B, Lam H, Jovanovic M, Picotti P, Schlapbach R, Aebersold R. PhosphoPep--a database of protein phosphorylation sites in model organisms. *Nat Biotechnol* 2008;**26**:1339–1340.

268 Bateman A, Martin MJ, O'Donovan C, Magrane M, Apweiler R, Alpi E, Antunes R, Arganiska J, Bely B, Bingley M, Bonilla C, Britto R, Bursteinas B, Chavali G, Cibrian-Uhalte E, Silva AD, De Giorgi M, Dogan T, Fazzini F, Gane P, Castro LG, Garmiri P, Hatton-Ellis E, Hieta R, Huntley R, Legge D, Liu W, Luo J, MacDougall A, Mutowo P, Nightingale A, Orchard S, Pichler K, Poggioli D, Pundir S, Pureza L, Qi G, Rosanoff S, Saidi R, Sawford T,

Shypitsyna A, Turner E, Volynkin V, Wardell T, Watkins X, Zellner H, Cowley A, Figueira L, Li W, McWilliam H, Lopez R, Xenarios I, Bougueleret L, Bridge A, Poux S, Redaschi N, Aimo L, Argoud-Puy G, Auchincloss A, Axelsen K, Bansal P, Baratin D, Blatter MC, Boeckmann B, Bolleman J, Boutet E, Breuza L, Casal-Casas C, de Castro E, Coudert E, Cuche B, Doche M, Dornevil D, Duvaud S, Estreicher A, Famiglietti L, Feuermann M, Gasteiger E, Gehant S, Gerritsen V, Gos A, Gruaz-Gumowski N, Hinz U, Hulo C, Jungo F, Keller G, Lara V, Lemercier P, Lieberherr D, Lombardot T, Martin X, Masson P, Morgat A, Neto T, Nouspikel N, Paesano S, Pedruzzi I, Pilbout S, Pozzato M, Pruess M, Rivoire C, Roechert B, Schneider M, Sigrist C, Sonesson K, Staehli S, Stutz A, Sundaram S, Tognolli M, Verbregue L, Veuthey AL, Wu CH, Arighi CN, Arminski L, Chen C, Chen Y, Garavelli JS, Huang H, Laiho K, McGarvey P, Natale DA, Suzek BE, Vinayaka C, Wang Q, Wang Y, Yeh LS, Yerramalla MS, Zhang J. UniProt: A hub for protein information. *Nucleic Acids Res* 2015;**43**:D204–D212.

269 Gnad F, Gunawardena J, Mann M. PHOSIDA 2011: The post-translational modification database. *Nucleic Acids Res* 2011;**39**:D253–D260.

270 Li J, Jia J, Li H, Yu J, Sun H, He Y, Lv D, Yang X, Glocker MO, Ma L, Yang J, Li L, Li W, Zhang G, Liu Q, Li Y, Xie L. SysPTM 2.0: An updated systematic resource for post-translational modification. *Database* 2014;**2014**:bau025.

271 Lu CT, Huang KY, Su MG, Lee TY, Bretana NA, Chang WC, Chen YJ, Chen YJ, Huang HD. DbPTM 3.0: An informative resource for investigating substrate site specificity and functional association of protein post-translational modifications. *Nucleic Acids Res* 2013;**41**:D295–D305.

272 Liu Z, Wang Y, Xue Y. Phosphoproteomics-based network medicine. *FEBS J* 2013;**280**:5696–5704.

273 Mi H, Muruganujan A, Thomas PD. PANTHER in 2013: Modeling the evolution of gene function, and other gene attributes, in the context of phylogenetic trees. *Nucleic Acids Res* 2013;**41**:D377–D386.

274 Kanehisa M, Goto S, Sato Y, Furumichi M, Tanabe M. KEGG for integration and interpretation of large-scale molecular data sets. *Nucleic Acids Res* 2012;**40**:D109–D114.

275 Szklarczyk D, Franceschini A, Wyder S, Forslund K, Heller D, Huerta-Cepas J, Simonovic M, Roth A, Santos A, Tsafou KP, Kuhn M, Bork P, Jensen LJ, von Mering C. STRING v10: Protein-protein interaction networks, integrated over the tree of life. *Nucleic Acids Res* 2015;**43**:D447–D452.

276 Obenauer JC, Cantley LC, Yaffe MB. Scansite 2.0: Proteome-wide prediction of cell signaling interactions using short sequence motifs. *Nucleic Acids Res* 2003;**31**:3635–3641.

277 Xue Y, Ren J, Gao X, Jin C, Wen L, Yao X. GPS 2.0, a tool to predict kinase-specific phosphorylation sites in hierarchy. *Mol Cell Proteomics* 2008;**7**:1598–1608.

278 Blom N, Sicheritz-Ponten T, Gupta R, Gammeltoft S, Brunak S. Prediction of post-translational glycosylation and phosphorylation of proteins from the amino acid sequence. *Proteomics* 2004;**4**:1633–1649.

279 Linding R, Jensen LJ, Pasculescu A, Olhovsky M, Colwill K, Bork P, Yaffe MB, Pawson T. NetworKIN: A resource for exploring cellular phosphorylation networks. *Nucleic Acids Res* 2008;**36**:D695–D699.

280 Horn H, Schoof EM, Kim J, Robin X, Miller ML, Diella F, Palma A, Cesareni G, Jensen LJ, Linding R. KinomeXplorer: An integrated platform for kinome biology studies. *Nat Methods* 2014;**11**:603–604.

281 Song C, Ye M, Liu Z, Cheng H, Jiang X, Han G, Songyang Z, Tan Y, Wang H, Ren J, Xue Y, Zou H. Systematic analysis of protein phosphorylation networks from phosphoproteomic data. *Mol Cell Proteomics* 2012;**11**:1070–1083.

282 Petsalaki E, Helbig AO, Gopal A, Pasculescu A, Roth FP, Pawson T. SELPHI: Correlation-based identification of kinase-associated networks from global phospho-proteomics data sets. *Nucleic Acids Res* 2015;**43**(W1):W276–W282.

283 Reimand J, Wagih O, Bader GD. The mutational landscape of phosphorylation signaling in cancer. *Sci Rep* 2013;**3**:2651.

284 Balbin OA, Prensner JR, Sahu A, Yocum A, Shankar S, Malik R, Fermin D, Dhanasekaran SM, Chandler B, Thomas D, Beer DG, Cao X, Nesvizhskii AI, Chinnaiyan AM. Reconstructing targetable pathways in lung cancer by integrating diverse omics data. *Nat Commun* 2013;**4**:2617.

3

Analysis of Protein Glycosylation by Mass Spectrometry

David J. Harvey

Glycobiology Institute, Department of Biochemistry, University of Oxford, Oxford, UK

3.1 Introduction

It is estimated that about half of all proteins are glycosylated with the glycan portion responsible for many of the biophysical properties of the molecules [1, 2]. An enormous amount of work has been devoted to their analysis and much of this is covered in recent reviews [3–22]. This chapter summarizes the structure of these compounds and describes the main mass spectrometric methods used in their analysis.

3.2 General Structures of Carbohydrates

Carbohydrates are the most abundant and structurally diverse compounds found in nature. They can exist as small monosaccharides or they can link together to give polymeric compounds such as cellulose (D-Glcβ-1 → 4)- linear repeat) and chitin D-GlcNAcβ(1 → 4)- linear) with molecular weights that can exceed 1 MDa. Unlike linear biopolymers such as proteins and nucleic acids, oligomeric (2–10 monosaccharide residues) and polymeric carbohydrates typically form branched structures because linkage of the constituent monosaccharides can occur at any of the hydroxyl groups of the adjacent residue. Consequently, very large numbers of isomers are possible. For example, it has been calculated that for a simple hexasaccharide, there are more than 10^{12} possible isomeric structures [23], presenting what has been referred to as an "isomer barrier" to the analyst. It is doubtful, therefore, that any single analytical technique will ever provide all of the structural information necessary to characterize such a large number of oligosaccharides. Fortunately, however, this hypothetical situation does not occur in nature because the biosynthetic

Analysis of Protein Post-Translational Modifications by Mass Spectrometry,
First Edition. Edited by John R. Griffiths and Richard D. Unwin.
© 2017 John Wiley & Sons, Inc. Published 2017 by John Wiley & Sons, Inc.

pathways creating most complex carbohydrates are specific and limited by the available glycosyltransferases. Consequently, only a very few of the theoretically possible isomers are ever encountered, a property that can be utilized structurally if the source of the glycan is known.

The basic building blocks of oligo- and polysaccharides are monosaccharides with the general formula $C_nH_{2n}O_n$ where $n = 3$ (trioses) to 9 (nonoses) with $n = 6$ (hexoses) being the most common. The monosaccharides contain $n - 1$ hydroxy groups and one aldehyde (reducing sugars) or keto group and can exist in a number of isomeric forms as the result of equilibration between linear and cyclic forms (typically 6- (pyranose) or 5- (furanose) membered rings). Rings can also exist in different conformations, for example, chair or boat forms. Cyclization produces an additional chiral center giving rise to α- and β-anomers. Formation of a bond to the anomeric carbon prevents ring opening and fixes the ring size and conformation of that monosaccharide. Many modified monosaccharides exist such as those with missing hydroxyl groups, primary hydroxyl groups oxidized to carboxylic acids, or hydroxy groups replaced by, for example, a primary amine or *N*-acylamino group. Hydroxy groups can also be methylated or acetylated (see Kennedy [24] for more information on carbohydrate structure).

3.2.1 Protein-Linked Glycans

Three main types of attached glycans are recognized. N-linked glycans are attached to asparagine in an Asn-Xxx-Ser (or Thr) consensus sequence where Xxx is any amino acid except proline. O-linked glycans are generally smaller and more varied in structure [25] and are attached to either serine or threonine but with no directing consensus sequence. A third type comprises glycans that form part of a glycosylphosphatidylinositol lipid anchor. Recently a few glycoproteins have been reported in which cysteine replaces serine or threonine in the N-linked consensus sequence [26]. O-linked glycans from mammalian systems tend to have relatively diverse structures and are generally classified according to their core structures as outlined in Table 3.1. It is common to find several O-linked glycosylation sites in close proximity such that proteolysis rarely produces fragments with single sites as with the N-glycans. Consequently, peptides containing several glycans are often examined either as intact molecules or after stripping the glycans to their attached GalNAc residue. A database of O-glycan structures is available [28]. N-glycans, although often larger than O-linked structures, contain a trimannosyl-chitobiose pentasaccharide core that is attached to the protein by the reducing-terminal GlcNAc residue with an amide link and have well-defined overall structures.

Because of the complexity of many of these glycans, it is customary in this field to depict the structures in a picture format. Symbols such as circles or squares are chosen to represent the individual monosaccharide constituents,

Table 3.1 Structures of the cores of O-glycans 27.

Core	Structure
Core 1	Galβ1 → 3GalNAcα-Protein
Core 2	GlcNAcβ1 → 6(Galβ1 → 3)GalNAcα-Protein
Core 3	GlcNAcβ1 → 3GalNAcα-Protein
Core 4	GlcNAcβ1 → 6(GlcNAcβ1 → 3)GalNAcα-Protein
Core 5	GalNAcα1 → 3GalNAcα-Protein
Core 6	GlcNAcβ1 → 6GalNAcα-Protein
Core 7	GalNAcα1 → 6GalNAcα-Protein
Core 8	Galα1 → 3GalNAcα-Protein

Source: Adapted from "Varki, A., Cummings, R. D., Esko, J. D., Freeze, H. H., Stanley, P., Bertozzi, C. R., Hart, G. W., Etzler, M. E.: Essentials of Glycobiology, Second Edition. Cold Spring Harbor Laboratory Press, (2008).

which are then linked in the sequence of the glycan. In former times, most laboratories chose their own symbols, but a more consistent system was introduced by the Consortium for Functional Glycomics (CFG) and is outlined in Figure 3.1.

Mass-different monosaccharides were differentiated by shape, and the different types of, for example, hexoses were differentiated by color. Linkage was shown by written notes on the connecting bonds. Although widely adopted, this system suffers from major drawbacks, not least of which is the manner of indicating linkage. An alternative system, proposed in 2009, overcomes this problem by indicating the linkage positions by the angle of the lines connecting the monosaccharide symbols with full lines denoting β-bonds and broken lines showing α-bonds. The requirement for color was overcome by using shapes to show the different isobaric monosaccharides such as mannose (circle), galactose (diamond), and glucose (square) with various additions such as a solid fill to show the presence of an *N*-acetylamino group (as in GlcNAc (filled square)) or inclusion of a dot to show the absence of an OH group (e.g., a diamond with a dot to indicate fucose (deoxy-L-galactose). This system has the added advantage that it can be extended to include other monosaccharides, such as tyvelose (dideoxy-mannose (circle with two dots)), without the need to invent entirely new symbols [29]. To make this system more familiar to users of the CFG system, recently, for N-linked glycans, the CFG colors have been incorporated for depicting N-glycans. This system is used in this chapter.

The biosynthesis of N-glycans [30] is outlined in Figure 3.2. Briefly, the glycan Glc$_3$Man$_9$GlcNAc$_2$ (**1**; Figure 3.2) is attached to the protein during transcription in the endoplasmic reticulum and then degraded by enzymatic removal of the three glucose residues (**2**). The intermediates in this process are

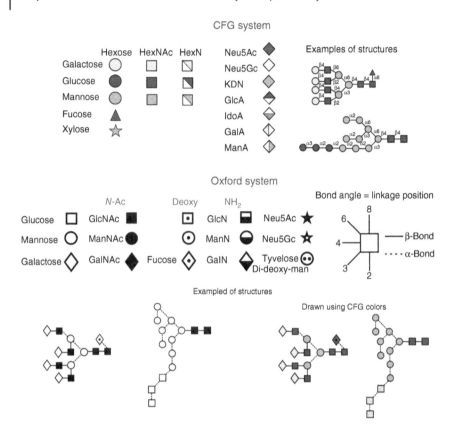

Figure 3.1 Methods for drawing N-glycans using symbols. Two systems are in common use, the Consortium for Functional Glycomics (CFG) and the Oxford systems. The CFG system uses color to differentiate hexoses, whereas the Oxford system uses shapes. Furthermore, the Oxford system identifies modifications to the basic monosaccharides by simply adding features such as a fill (*N*-acetyl) or dot (deoxy), whereas the CFG system often needs additional symbols. Linkage is indicated by text on the bonds linking the symbols in the CFG system, whereas the Oxford system uses the angle of the bonds to denote the linkage position. A new version of the CFG system will incorporate this latter feature as an option. The figure shows some of the symbols and examples of structures drawn with the different systems. Two versions of the Oxford system are shown: black and white and the same structures drawn with inclusion of the CFG colors.

used by the cell to ensure correct folding of the protein. The central (d2) mannose is then removed (**3**), and the glycoprotein is transferred to the Golgi apparatus where the two α-linked mannose residues on the 3-antenna (d1 arm) and that on the 6-branch of the 6-antenna are removed to leave Man$_5$GlcNAc$_2$ (**4**). These glycans are all referred to as "high-mannose" glycans. A GlcNAc residue is then attached at the 2-position of the 3-linked mannose (**7**), after which several pathways propagate. Galactose, followed (**8**) by sialic acid, linked

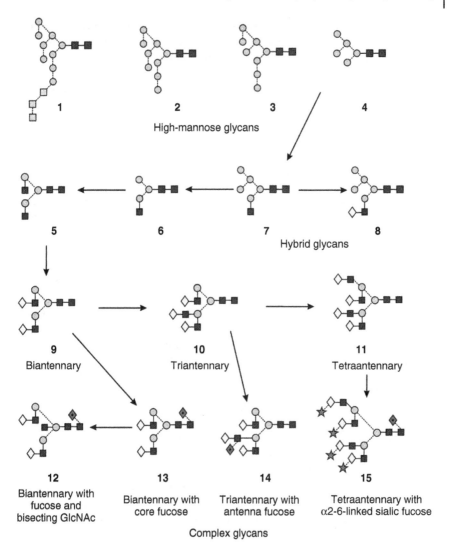

Figure 3.2 Biosynthesis of N-linked glycans. The structures (drawn in the Oxford format) that are shown are only a few of those possible. Arrows are illustrative and do not necessarily depict intact biochemical pathways. The antennae of the complex glycans usually terminate in *N*-acetylneuraminic acid in either α2-3- or α2-6-linkage.

either α2 → 3 or α2 → 6, can be attached to the 4-position of the GlcNAc to give what is known as a hybrid glycan. Alternatively, the two remaining mannose residues attached to the 6-linked core mannose residue are removed (**6**), and then, typically for mammalian systems, either of the core mannoses can have one or two Gal-GlcNAc (**9**) and finally Neu5Ac-Gal-GlcNAc antennae attached to the 2- or 4-positions of the 3-mannose or at the 2- and 6-positions of the 6-mannose (**9, 10, 11**). Glycans with two such antennae (**9**) are known as biantennary complex glycans, those with three antennae (**10**) are triantennary glycans, and those with four antennae (**11**) are tetraantennary glycans. Furthermore, these antennae can be extended with addition of further Gal-GlcNAc groups. All of these glycans are known as "complex glycans." Other common modifications to these structures are the addition of fucose to the core GlcNAc (**12, 13**) or to the galactose or GlcNAc residues (**14**) of the antennae, the addition of GlcNAc to the 4-position of the core mannose (known as a "bisecting" GlcNAc—**12**) or the addition of further sialic acid residues to the ends of the antennae (**15**). Different types of N-glycan characterize various species: thus N-glycans from birds tend to have antennae that lack galactose, those from fungi are mainly high-mannose glycans, whereas plants and insects produce truncated glycans known as "paucimannosidic" glycans, often containing xylose with fucose attached to the 3- rather than to the 6-position of the reducing-terminal GlcNAc that is common in mammalian glycans. The N-glycans in all of these species contain the same trimannosyl-chitobiose core, and only archaea appear to synthesize different types of N-glycan.

Glycoproteins can contain one or many sites for glycan attachment. Each of these sites may be fully occupied with one or many glycans (over 100 in some cases), or they may be partially occupied or vacant. Occupancy of several sites by many glycans can result in large numbers of individual glycoproteins (known as glycoforms). It has been estimated, for example, that *Desmodus rotundus* salivary plasminogen activator, with its two N- and four O-linked sites, contains in excess of 330,000 individual molecular species if the glycans are randomly distributed [31, 32].

Structural determination of these compounds, therefore, requires not only the structure of the glycans but also identification of the site(s) to which they are attached. Not all possible sites are glycosylated and some may be partially glycosylated. Thus, the degree of glycosylation (site occupancy) is also a feature to be determined. Where molecules, such as IgG, contain several glycosylated protein chains, it is also important to determine the relationship of glycosylation between the chains.

3.3 Isolation and Purification of Glycoproteins

Many glycoproteins are present in low concentrations in complex biological matrices. Concentrations in human plasma, for example, are estimated to span 10 orders of magnitude. Thus, for the less abundant glycoproteins, the first stage of analysis is to extract them in a sufficient state of purity. Immunodepletion of the major proteins such as albumin, α1-acid glycoprotein, α1-antitrypsin, apolipoprotein A, fibrinogen, haptoglobin, IgA, IgG, IgM, and transferrin is frequently used, and kits for this process are available commercially. A large number of specific methods have been developed and space only permits a brief summary of the more important ones.

3.3.1 Lectin Affinity Chromatography

Lectins are proteins that have affinity for glycans and have been used extensively for extractions and for profiling glycans. Over 160 lectins have been described with some 60 being commercially available [33, 34]. The affinity of these proteins tends to be specific for certain features of the glycans; for example, concanavalin A (ConA) mainly recognizes α-mannose and glucose, *Lens culinaris* agglutinin (LCA) recognizes hybrid or bi- and triantennary complex structures that are core fucosylated, peanut agglutinin (PNA) recognizes Galβ1-3GalNAc and β-linked Gal, whereas wheat germ agglutinin (WGA) recognizes GlcNAc and sialic acid. The lectins are usually immobilized on silica-based materials or agarose but magnetic nanoparticles have also been used [35–37].

Although these lectins individually are not suitable for total glycoprotein extractions, they are invaluable when certain sets of glycan, such as high-mannose glycans, are required. For use with a wider range of glycoproteins, multilectin [38] or sequential lectin columns [39] can be used. Madera et al. [40] have compared the two approaches for extraction of glycoproteins from human blood serum. Four silica-bound lectins (ConA, *Sambucus nigra* agglutinin (SNA-I), *Ulex europaeus* agglutinin I (UEA-I), and *Phaseolus vulgaris* lectin (PHA-L)) were used individually or mixed. 108 glycoproteins were found with the individual columns, whereas only 67 were recovered from the mixed column. This situation may not, of course, apply to all glycoprotein mixtures.

A list of suitable lectins is available in the review by Alley et al. [3]. It has been pointed out that although the affinity of any particular lectin for specific carbohydrates may be known for the carbohydrate itself, this may not apply to the glycan when it is bound to the protein.

3.3.2 Boronate-Based Compounds

Boronic acids bind to *cis*-diol groups and, thus, are appropriate for extracting a wide range of glycans and glycopeptides. The cyclic compounds that are formed are not particularly stable (binding constants in the region of 10^3 M), and the bound glycans can easily be released by treatment with mild acids. Like lectins, these boronic acids can be bound to magnetic nanoparticles [41–43] or used to construct affinity columns [44, 45]. In order to increase the binding ability of lectins, Lu et al. [46] have produced boronate-decorated lectins, termed BAD-lectins (ConA, SNA, and *Aleuria aurantia* lectin (AAL) were used) and conjugated them to magnetic nanoparticles. A 2- to 60-fold increase in detection sensitivity and enrichment efficiency was observed. 296 N-linked glycopeptides were found in whole-cell lysates with only a 6% overlap between the three lectins.

3.3.3 Hydrazide Enrichment

This method is popular for extracting glycoproteins or glycopeptides, but because the glycan is modified, it is not suitable for structural studies on this part of the molecule. The technique involves oxidation of vicinal diols with periodate to give the corresponding dialdehydes, which are then captured on various media carrying hydrazide groups [47–49]. The proteins are then released, usually with the enzyme protein *N*-glycosidase F (PNGase F) for further analysis. The oxidized glycan remains attached to the support medium.

3.3.4 Titanium Dioxide Enrichment of Sialylated Glycoproteins

Titanium dioxide is frequently used for enrichment of phosphorylated glyco-proteins, but if these are dephosphorylated by treatment with alkaline phosphatase, the titanium dioxide binds highly efficiently to any sialic acids [50, 51]. Nonspecific binding of acidic amino acids can be blocked with 5% trifluoroacetic acid (TFA) and 1 M glycolic acid. The bound glycoproteins can then be released by treatment with aqueous ammonia at pH 11.

3.4 Mass Spectrometry of Intact Glycoproteins

Most high-performance mass spectrometers are now capable of resolving gly-coforms of glycoproteins of considerable molecular weight, provided that the number of glycosylation sites is limited. Thus, for example, Thompson et al. have resolved glycoforms of half antibodies using an orbitrap-based instrument (*m/z* of the 18+ charge state in the region of 4.2 kDa) [52], and the same group have resolved 49 glycoforms of chicken ovalbumin using similar detection [53]. Differences in measured mass and that of the protein can give an indication of the glycans present. More complex glycoproteins usually give unresolved broad mass peaks, particularly by matrix-assisted laser desorption/ionization (MALDI), and further analysis requires either proteolysis or removal of the intact carbohydrates.

3.5 Site Analysis

Site analysis involves the determination of the glycan occupancy at each consensus sequence site; several methods are available. Usually, the protein is digested with a protease such as trypsin, which cleaves peptide bonds

C-terminal to lysine and arginine leaving an amino group (basic) at each terminus, thus promoting the formation of doubly charged ions by mass spectrometry (MS). The idea is to isolate each glycosylation site to a different glycopeptide with analysis usually by liquid chromatography/mass spectrometry (LC/MS). Missed cleavages are common, particularly when the cleavage site is near a glycan, and some proteins such as α1-acid glycoprotein [54] are difficult to digest. However, the use of a new surfactant "RapiGest SF^{TM}" greatly aids digestion as does the use of a microwave oven [55]. Problems arise when it is not possible to produce peptides containing only one glycosylation site and when the peptides are too large for efficient mass spectral analysis. Other proteases are available but are not used as frequently. The most common are Glu-C, which cleaves peptide bonds C-terminal to aspartic or glutamic acids; Lys-C, which cleaves C-terminal to lysine; and Arg-C, which cleaves C-terminal to arginine.

Mass spectrometric detection of glycopeptides in the presence of peptides is often difficult because of suppression effects [56]. However, glycopeptides can be identified by the presence of oxonium ions as described in Section 3.8.3.1 [57]. Site analysis has been greatly aided in recent years by the advent of electron capture dissociation (ECD) and electron transfer dissociation (ETD) fragmentation, which preferentially fragments the peptide chain rather than the glycan as is common with collision-induced dissociation (CID) [58–60].

Removal of glycans from either the intact glycoprotein or a glycopeptide with the endoglycosidase PNGase F leaves aspartic acid in place of the asparagine at the N-linked site of the protein. The concomitant increase in mass by 1 Da can be detected by MS to identify the site. The percent occupancy at the site can also be deduced. In cases where the peptide sequence is not known, the aspartic acid can be identified by partial ^{18}O incorporation if the digestion is performed in 40% ^{18}O-enriched water [61]. In other methods, glycans have been removed with the endoglycosidase endoH to leave a GlcNAc residue at the linkage site, and Lee et al. [62] have achieved the same effect by heating glycopeptides with TFA in a microwave oven to degrade the glycans (from horseradish peroxidase) to the same residue. A method for locating O-glycosylation sites described by Müller et al. [63] also involves partial deglycosylation, this time with trifluoromethanesulfonic acid to the level of core-GlcNAc residues. The glycoprotein was then cleaved with an Arg-C-specific endopeptidase, clostripain, to yield tandem repeat icosapeptides, which were analyzed by MALDI/postsource decay (PSD) from α-cyano-4-hydroxycinnamic acid (CHCA) matrix.

An alternative technique uses nonspecific protease such as the cocktail of enzymes derived from *Streptomyces griseus*, known as pronase to cleave the protein, leaving glycans containing asparagine or a very short peptide chain [64–66]. Problem with this method is that it can produce several peptide fragments from each site, thus complicating the analysis or degrading the protein to just asparagine, thus removing any data on the site of attachment. By using extended hydrolysis periods, Schiel et al. [67] have produced both N- and O-glycans containing only asparagine at the reducing terminus. Hua et al. [68] have evaluated the use of seven nonspecific proteases with ribonuclease B as a test compound as a method for site analysis. The proteinases elastase, papain, pepsin, pronase, proteinase K, subtilisin, or thermolysin gave small peptides from which those containing the glycosylation site, with its attached glycans, could be obtained. The authors claim that rather than being nonspecific, these proteinases are multispecific and able to hydrolyze proteins at a large but finite number of sites, thus having the ability to leave asparagine or small peptides (normally two to six amino acids) attached to the glycan. Dodds et al. [69] have described immobilized pronase that retains its activity after repeated use for at least 6 weeks.

Attempts to locate O-glycans on peptide chains by fragmentation sometimes fail [70] because of the preferential loss of the glycans catalyzed by proton migration from $[M+nH]^{n+}$ ions. Elimination of the proton has been proposed as a way of overcoming this reaction and has been achieved by ECD in an ion cyclotron resonance (ICR) instrument [58]. It has been argued that ionization by charge localization could achieve the same result by eliminating the ionizing protons. Consequently, Czeszak et al. [71] derivatized glycopeptides at their amino terminus with a phosphonium group and showed that the resulting ions, when studied by MALDI/PSD, undergo predictable a-type fragmentation of the peptide chain without loss of the attached glycans. In contrast, CID on the doubly charged protonated phosphonium cation gave predominant loss of the sugar moiety. Experiments were conducted with only GalNAc attached to the peptide, but the method may be applicable to peptides carrying larger glycans. Alternatively, the O-glycans could be degraded to GalNAc as in the method described by Müller et al. [63].

A method named the GlycoFilter has been described for rapid analysis of glycosylation sites and glycan structure [72]. Glycoproteins were trapped in a spin filter, reduced, and alkylated and glycans were then released with PNGase F. Recovery was by a second centrifugation and analysis was performed by MALDI after permethylation. Finally, the residual protein was hydrolyzed with trypsin in $H_2^{18}O$, and the peptides were recovered by a third centrifugation step. Each enzymatic step could be accelerated by heating in a domestic microwave oven. A total of 865 and 295 N-glycosites were identified from urine and plasma samples, respectively.

3.6 Glycan Release

It is often difficult to perform detailed structural analysis of the glycan when it is attached to the protein or peptide. And, consequently, such analyses are usually conducted on released glycans. Glycans can be released from glycoproteins either chemically using hydrazine or by β-elimination or enzymatically using a variety of endoglycosidases.

3.6.1 Use of Hydrazine

Although popular in the past, hydrazine use has declined greatly in recent years to be replaced by milder enzymatic methods. However, hydrazine does have the advantage of being nonselective with respect to the glycans that are released, a feature not shared by the majority of enzymes. Anhydrous hydrazine is a reagent that cleaves amide bonds and, thus, causes complete disruption of the protein chain. It can be used to release both N- [73–75] and O-linked [74] glycans. O-linked glycans are specifically released at 60 °C, whereas 95 °C is needed to release the N-linked sugars. A major advantage of the method is that the sugars are released with an intact reducing terminus that can subsequently be used to label the glycan with various tags such as the fluorescent tags used for analysis by high-performance liquid chromatography (HPLC). On the other hand, because hydrazine cleaves amide bonds, it removed the acetyl groups from N-acetylamino sugars requiring a re-N-acetylation step in the release procedure. A recent method employs a carbon column for simultaneous reacetylation and hydrazine removal [76]. In addition, formation of artifacts can reduce the yield of glycans. Thus, it has been calculated [77] that as much as 25% of the total glycans are modified at the reducing terminus and that these compounds can never be converted into the parent sugar. Another disadvantage of hydrazine release is that the protein is completely destroyed, resulting in loss of all information regarding glycan linkage position.

3.6.2 Use of Reductive β-Elimination

Reductive amination exploits the action of a strong base to release the glycan, but because of an extensive concomitant degradation reaction, known as peeling, the reaction is performed in the presence of an excess of the reducing agent sodium cyanoborohydride [78]. Although performed mainly in solution, the reaction has also been used to release glycans from within SDS-PAGE gels [79, 80]. β-Elimination is used extensively for the release of O-glycans but has the disadvantage of producing glycans without a reducing terminus, thus precluding derivatization at this site. In response to this, several investigators have attempted the use of milder release reagents in the hope that peeling can be prevented and reduction avoided. Ammonia [81] has been used extensively in this context. It leaves the protein intact but converts the serine and threonine residues that were linked to the glycans in their dehydro forms. These subsequently react with excess ammonia to add an NH_2 label that can be detected by MS to provide linkage information. Ammonia in the presence of ammonium carbonate has also been used [82, 83]. Glycan release appeared to be good, and cleanup of the product was minimal as all reagents were volatile. Because the hydroxide ion appears to cause unfavorable peeling reactions, Miura et al. [84] have investigated the use of the ammonium salt ammonium carbamate for glycan release and have reported efficient release with little peeling. The release was performed by addition of powdered ammonium carbamate and incubation for 20 h at 60 °C.

Although the reaction with ammonia was reported to produce quantitative release of O-glycans, a recent study [85] with human IgA1 has found incomplete liberation of O-glycans. MALDI time-of-flight (MALDI-TOF) MS analysis revealed that only one of the six glycosylated sites was susceptible to β-elimination under the conditions used. It was proposed that resistance to β-elimination was due to very close proximity of proline to the glycosylated serine or threonine residues. The author commented that the findings may have implications for similarly O-glycosylated peptides and proteins and possibly for other chemical methods that are used to carry out β-eliminations of O-glycans.

The reaction has been investigated in detail by Yu et al. [86] for O-glycan chains with β1,3-linked cores. In contrast to β1,4-linkages of the N-glycan type, which were shown to be stable under the ammonium-based alkaline conditions, the β1,3-linkage was found to be labile and to give considerable peeling. The results indicated that complete prevention of peeling under nonreducing alkali-catalyzed hydrolysis conditions remains difficult.

Zheng et al. [87] have compared ammonia, methylamine, and dimethylamine at 55 °C for 6 h for the release of GalNAc from a small glycopeptide. The O-glycosylated Thr residue was converted into a stable derivative with various amines. β-Elimination with dimethylamine and methylamine resulted in the conversion of the glycopeptide to 69.2% of the dimethylamine derivative and 61.5% of the methylamine derivative, respectively. However, the incubation of the glycopeptide with ammonia only resulted in 8% production of the product. The authors concluded that elimination with dimethylamine was the most efficient for the release of O-linked glycans. In spite of these developments, the classical β-elimination reaction with sodium hydroxide and sodium borohydride remains the most popular method for releasing O-glycans. For glycoproteins that contain both N- and O-linked glycans, the N-linked glycans are usually released first enzymatically, followed by O-linked glycan release by β-elimination.

3.6.3 Use of Enzymes

Several enzymes are available for releasing N-glycans. The most popular is peptide-*N*-glycosidase F (PNGase F) [88], an amidase that releases the intact glycan as the corresponding glycosylamine. This process converts the attached asparagine to aspartic acid, effectively labeling the site of glycan attachment with a one mass unit increase in molecular weight. The released glycosylamines are relatively unstable and readily hydrolyze to the glycan. Commonly, treatment with dilute organic acid is used to speed the reaction. Without this treatment, some glycosylamine can be retained, particularly if the glycan release has been performed in the presence of ammonium-containing buffers [89]. Various commercial preparations of PNGase F are available, most containing various additives to aid release. Some of these additives are incompatible with MS. Glycerol, for example, can be difficult to remove and can inhibit crystal formation in MALDI analysis. Dithiothreitol is sometimes present as a reagent for denaturing the protein. This compound can decompose releasing H_2S, which competes with water for reacting with the released glycosylamines. This reaction adds SH rather than OH to the reducing terminus of the glycan. The nominal mass difference between these moieties is 16, which is the same as that of oxygen, giving the impression that the glycan contains one more oxygen atom than is actually the case [90].

PNGase F releases most N-glycans except those containing fucose $\alpha1 \rightarrow 3$-linked to the reducing-terminal GlcNAc [91]. Proteins need to be denatured for efficient glycan release, but, even so, reactions can sometimes take longer than the typical overnight procedure. Various techniques such as microwave irradiation [92, 93] or high pressure [94] can be used to reduce the reaction time. Use of these methods has produced reaction times of less than 1 h. Glycans can be released in solution or by infusing the enzyme into isoelectric focusing [95] or SDS-PAGE gels [96–101], the latter technique being particularly useful when only small amounts of glycoprotein are available. After release, glycans are usually recovered from the surrounding solution although dissolution of the gel has also been used [102].

In situations where PNGase F fails to release glycans, PNGase A is usually effective. This enzyme is larger than PNGase F (75.5 as compared to 35 KDa), does not readily penetrate gels, and only releases glycans from smaller peptides [103]. Consequently, the glycoprotein should be digested with, for example, trypsin prior to incubation [104]. Like PNGase F, PNGase A releases the entire glycan, but another popular group of endoglycosidases, exemplified by endoH and endoS, cleaves the glycan between the two GlcNAc residues of the chitobiose core. Although this reaction effectively leaves the reducing terminal attached to the protein as a label for glycan attachment, the released glycan has lost the information regarding the attachment of groups such as fucose to the core GlcNAc. These endoglycosidases are usually more specific in their substrate specificity than the PNGases. EndoH, for example, only releases high-mannose and some hybrid glycans, whereas endoS from *Streptococcus pyogenes* releases some complex glycans but not high-mannose or hybrid structures [105].

3.7 Analysis of Released Glycans

3.7.1 Cleanup of Glycan Samples

Although MALDI and, to a lesser extent, electrospray ionization (ESI) are comparatively tolerant to the presence of contaminants, it is important to desalt samples before analysis in order to produce acceptable spectra. Suitable methods include the use of resins [96], aminopropyl silica in a hydrophilic interaction liquid chromatography (HILIC) microelution plate [106], or graphitized carbon [107]. Resins are usually packed into microcolumns or can be added directly to the MALDI target [108]. Alternatively, membranes such as low molecular cutoff dialysis membranes or Nafion 117 [109] can be used. Peptidic material can be removed from glycan samples with C18 such as that incorporated into ZipTips. Among other cleanup methods is the use of ZIC-HILIC and cotton wool [110] and oxylamino-containing polymers to bind the reducing terminus of the glycan followed by release with mild acid [111].

3.7.2 Derivatization

3.7.2.1 Derivatization at the Reducing Terminus

Because glycans do not fluoresce or absorb satisfactorily at UV frequencies, derivatization of the reducing terminus by addition of fluorescent tags such as 2-aminobenzamide (2-AB) [112], 2-aminobenzoic acid (2-AA) [113–115], or 2-aminopyridine (2-AP) [116, 117], usually by reductive amination (Figure 3.3), is routine for the detection of glycans. Such derivatives have also found use in mass spectral analyses for enhancement of signals or for modification of fragmentation patterns.

Because of the presence of amine groups linking the derivative to the glycan, these derivatives can increase the proton affinity of the molecules when $[M+H]^+$ ions are formed in preference to the more usual $[M+Na]^+$ ions. With ESI, for example, they have been reported to enhance signal strength by 30- to 100-fold depending on the derivative [118]. Recently procainamide (N-(2-diethylamino)ethyl-4-aminobenzamide), first introduced in 2000 [119], has received considerable interest in this context and is available in a kit from ProZyme. Derivatization with aminobenzoic acid alkyl esters has been used to increase hydrophobicity, which results in an increase in sensitivity [120, 121], particularly for fast atom bombardment (FAB) ionization. Considerable increases in sensitivity have also been achieved by using derivatives containing a constitutive cationic charge such as those prepared using trimethyl(4-aminophenyl)ammonium chloride (TMAPA) [118] or Girard's reagent T [122]. Derivatives containing a bromine atom have been used as labeling reagents because of the distinctive bromine isotope pattern that allows all fragment ions containing the derivatized end of the molecule to be identified [123, 124].

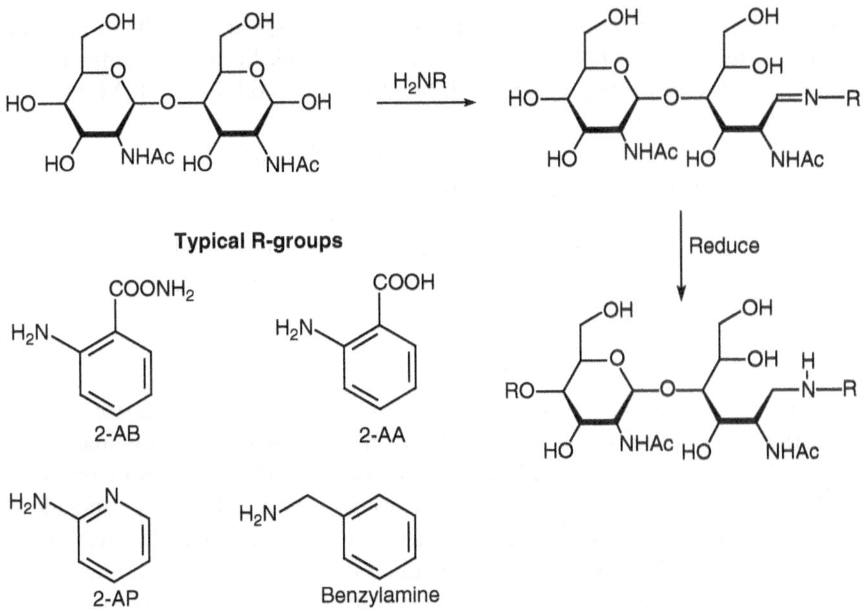

Figure 3.3 Reductive amination reaction for derivatization of the reducing terminus of carbohydrates (shown for the chitobiose core of N-glycans) and some of the typical amines used for the reaction.

The reductive amination reaction with 2-AP can be reversed to recover the carbohydrate [125]. Küster et al. [126] have studied the effect of a number of derivatives on the high-energy fragmentation of N-glycans; more cross-ring fragments were seen than with underivatized compounds. An ^{0,2}A-ion (see following text) from the reducing terminus (loss of the derivative) was often the most abundant ion in the spectrum.

An alternative derivatization strategy that avoids some of the problematic cleanup procedures necessitated by reductive amination includes the production of hydrazone and hydrazide derivatives as illustrated by reactions with phenylhydrazine [127–129]. An advantage of this reagent is that reactions are performed under slightly basic conditions causing minimal loss of sialic acid. Also, the derivatives form prominent $[M+H]^+$ ions, making them derivatives suitable for the detection of glycopeptides in the presence of peptides [130]. Thus the glycan profile and site occupancy (Asn-to-Asp conversion) could be monitored in a single spectrum. Basic and quaternary ammonium-containing derivatives linked via hydrazine chemistry have also been used for this purpose [131].

Also of interest are the derivatives containing a cationic charge [122, 131, 132]. Of these, Girard's T reagent (carboxymethyltrimethylammonium chloride hydrazide) [122, 132] has received the most interest and has been used for quantitative studies of N-glycans in positive ion mode because it eliminates the differences in ionization efficiency inherent in the production of $[M+H]^+$ and $[M+Na]^+$ ions [133, 134].

Because PNGase F releases the glycans as the glycosylamines, it is possible to prepare derivatives directly by reaction with carbonyl reagents. Chen and Novotny [135] have prepared such derivatives from 2-methyl-3-oxo-4-phenyl-2,3-dihydrofuran-2-yl acetate immediately after release.

A large number of other reducing-terminal derivatives have been investigated and are reported in recent reviews [136–138].

3.7.2.2 Derivatization of Hydroxyl Groups: Permethylation

Permethylation of carbohydrates is the most widely used technique for increasing thermal stability and reducing polarity. It is used extensively for analyses by FAB, MALDI, and combined gas chromatography/mass spectrometry (GC/MS). Reduction in polarity increases sensitivity in MALDI analyses and allows small amounts of very large glycans to be examined [139]. Furthermore it stabilizes the sialic acids as discussed later. One of their main advantages over other alkylation and acylation techniques is the relatively small (14 mass units) mass increment attending derivatization, which is important because of the large number of hydroxyl groups in carbohydrates.

The synthetic method introduced by Hakomori in 1964 [140] has been widely adopted for permethylation of carbohydrates and consists in reacting the carbohydrate, in dimethyl sulfoxide (DMSO), with methyl iodide catalyzed by the methylsulfinyl carbanion, prepared from sodium hydride. This reaction replaces active hydrogen in hydroxyl, carboxy, and amino groups by a methyl group. In order to retain information on the occurrence of natural O-Me groups, $[^2H3]$-methyl iodide can be used as the methylating reagent. The reaction requires considerable cleanup (reviewed by Levery in 1997 [141]), which can limit its usefulness for the analysis of small amounts of glycans. A simpler method that uses finely divided sodium hydroxide rather than the methylsulfinyl carbanion was introduced in 1984 by Ciucanu and Kerek [142] and has been widely adopted. A few problems have been reported but generally overcome by such modifications as adding a trace of water or using N,N-dimethylacetamide as the solvent [143]. Peeling reactions (base-catalyzed removal of monosaccharide residues from the reducing terminus) have been minimized by addition of acetic acid to the final reaction mixture and by keeping the mixture at 0 °C [144].

A recent modification that is useful for derivatization of very small amounts of carbohydrate is the use of small microspin columns or fused-silica capillaries (500 μm i.d.) packed with sodium hydroxide powder to which are added the analytes, mixed with methyl iodide in DMSO containing traces of water. Reactions were said to take less than one minute, and the procedure minimized oxidative degradation and peeling reactions and avoided the need of excessive cleanup. Picomole amounts of linear and branched, sialylated, and neutral glycan samples were rapidly and efficiently permethylated by this approach [145, 146], and a high-throughput extension of the method utilizing spin columns packed with sodium hydroxide beads has recently been described [147]. A potential problem is the appearance in mass spectra of a series of ions 30 Da larger than those from the fully methylated carbohydrate. These "overmethylation" ions appear to arise following reaction of the glycan with iodomethyl methyl ether that is formed as a side reaction [148]. Various aspects of permethylation procedures have been described in reviews by Jay [149] and Levery [141].

A major application of the permethylation reaction is in the determination of glycan structure by fragmentation. Glycosidic cleavages leave OH groups at the original sites of glycan attachment, effectively labeling them [150–152]. Cross-ring cleavage reactions produce different fragments attached to the nonreducing terminus, depending on the position of the linked residues. Thus 2-linkages produce a 74 Da increment, 4-linkages an 88 Da increment, and 6-linkages a combination of 60 and 88 Da increments, but for 3-linkages, no related ions are created.

3.7.2.2.1 Linkage Analysis

A major application of the permethylation reaction is linkage or "methylation analysis" for determination of carbohydrate linkages by GC/MS. It was developed in the late 1960s [153, 154] and is still in use today. The method has been the subject of many articles and reviews [141, 155–161], and several modifications have been developed. Basically, the procedure is one in which all hydroxyl groups in a polysaccharide are first permethylated after which the molecule is hydrolyzed to generate monosaccharides that contain free hydroxyl groups at the sites of linkage. These monosaccharides are then reduced in order to avoid the production of two peaks on chromatographic analysis due to α- and β-anomers, and, finally, the resulting alditols are derivatized with a different reagent such as acetic anhydride to produce partially methylated alditol acetates, known as PMAAs. The positions of the various substituents are subsequently located by GC/MS in order to determine which of the hydroxyl groups (those derivatized as OAc in this example) were originally involved in bonding. Ring size is also reflected in the results because for hexoses, additional free OH groups will be generated at C4 and C5 for furanoses and pyranoses, respectively. However, the additional OH groups can cause problems with certain sugars such as 4-linked hexopyranoses and 5-linked hexofuranoses, both of which will produce 1,4,5-triacetoxy products. However, carbohydrates derived from mammalian glycoproteins appear to contain only pyranose structures, and, consequently this difficulty should not present problems in this area. Although these procedures will identify which hydroxyl groups on a given sugar ring are involved in linkage, information as to which sugar is attached at that site is not available and must be obtained by other means.

Data on the retention times and mass spectra of these derivatives have been published in several reviews [141, 159, 160]. Fragmentation occurs primarily along the carbon chain following charge localization on a nitrogen (amino sugar) or oxygen atom. Secondary fragments are the result of the further elimination of neutral fragments such as acetic acid (60 u) ketene (42 u), formaldehyde (30 u), or methanol (32 u). The base peak is usually at m/z 43 (CH_3CO^+).

3.7.2.3 Derivatization of Sialic Acids

Glycans containing sialic acids are relatively unstable, particularly under MALDI conditions, and readily eliminate sialic acid. Methods for overcoming this problem are discussed in Section 3.8.1.3. One of these methods involves methyl ester or amide formation to remove the labile acidic protons that are responsible for the sialic acid losses. Permethylation not only achieves the same purpose but also blocks the hydroxyl groups preventing fragmentation in negative ion mode. More details can be found later in this chapter.

3.7.3 Exoglycosidase Digestions

An important method for N- and O-glycan analysis, and one that provides information both on the nature of the monosaccharide constituents and on their linkage, is disassembly by use of exoglycosidase digestion [162–164]. Glycan mixtures are profiled, most commonly by HPLC, although MS can be used as well and then the mixture is digested with a series of enzymes (see Dwek et al. [165] for suitable enzymes) that remove successive monosaccharide residues from the reducing terminus. At each stage, the mixture is examined to determine the number of residues that have been removed. Although a popular method for glycan analysis, the technique suffers from several drawbacks, predominant of which is reliance on a supply of suitable enzymes and the absence of enzymes that reveal specific structural features such as the presence of bisecting GlcNAc residues. Also, there is no indication as to the site of attachment of the monosaccharide residues that are removed at each stage of the disassembly.

Several investigators have developed methods for glycan release [166] and exoglycosidase sequencing directly on a MALDI target. The method developed by Küster et al. [167] for on-target sequencing involves sequential incubation of the glycans with the relevant enzyme, spectral recording, and removal of the matrix (2,5-dihydroxybenzoic acid (DHB)) prior to analysis of the next digest. Starting with 100 pmoles of glycan, it was possible to conduct three successive enzyme digestions before the amount of sample became insufficient to give a MALDI signal. Other investigators [166, 168] have developed similar methods using mixtures of exoglycosidase (exoglycosidase arrays) on different target spots in order to avoid removal of the matrix at each stage. The neutral matrix 6-aza-2-thiothymine (ATT), rather than the acidic DHB, has been used so that enzyme digests could be performed in the presence of a matrix that did not affect the enzyme activities [169, 170].

3.7.4 HPLC and ESI

The compatibility of ESI with HPLC has resulted in LC/MS being extensively used for carbohydrate analysis as described later on in the section on total methods for analysis. LC/MS [171] and microseparation methods suitable for glycan analysis, particularly when coupled with MS, have recently been reviewed [172, 173]. Normal-phase (NP) separations provide good correlation between structure, molecular weight, and retention time [174], allowing some structural information to be obtained directly from the elution profile. Reversed-phase HPLC systems have also been used as illustrated by a study of the 2-AB derivatives of small oligosaccharides and N-linked glycans [175]. Porous graphitized carbon is also popular for carbohydrate separations [176, 177], and graphitized carbon nanoflow columns (0.6 µL/min) have been reported to increase sensitivity by 10-fold over conventional columns with a detection limit in the low femtomole range for N-linked glycans [178]; cyclo-dextrin-based columns have also been used [179]. The somewhat limited resolving power of earlier HPLC columns has now largely been overcome by the use of ultra-high-pressure column chromatography and monolithic silica-based capillary HILIC columns [180, 181]. "Chip-based" methods employing arrays of small sprayers built into silicon chips enable the ESI techniques to be automated and have been exploited in the carbohydrate field [182–184].

3.8 Mass Spectrometry of Glycans

3.8.1 Aspects of Ionization for Mass Spectrometry Specific to the Analysis of Glycans

3.8.1.1 Electron Impact (EI)
Electron impact ionization is used mainly for small glycans and is used in GC/MS studies of permethylated monosaccharides as discussed in Section 3.7.2.2.1. The method relies on creating volatile derivatives of the carbohydrates and is generally restricted to carbohydrates with a maximum of about 11 residues. Larger carbohydrates need temperatures in excess of 400 °C [185–190] for analysis. Such temperatures cause extensive decomposition of the compounds, and, in addition, few GC/MS instruments are capable of operation at these temperatures. Thus, N-glycans must be ionized by other techniques.

3.8.1.2 Fast Atom Bombardment (FAB)

FAB, introduced in 1981 [191], was used extensively for ionization of N-glycans until the advent of MALDI and ESI techniques but is now mainly of historical interest. Glycans needed to be derivatized, usually by permethylation, in order to form a hydrophobic layer over the surface of the matrix that was necessary for efficient ionization. Thioglycerol was widely used as the preferred matrix for N-glycans, but the spectra suffered from a high background that limited sensitivity and the spectra contained extensive fragmentation. Nevertheless, much excellent work was performed with the technique, particularly by the Imperial College group (e.g., Refs [192–195].

3.8.1.3 Matrix-Assisted Laser Desorption/Ionization (MALDI)

The use of MALDI to ionize carbohydrates was first reported by Karas and Hillenkamp in one of their early papers [196] and was applied to N-linked glycans by Mock et al. in 1991 [197]. It has since become one of the most popular techniques for glycan analysis, mainly because it tends to produce only one ion from carbohydrates, thus providing a means to easily profile glycan mixtures (for reviews, see [11, 198–204]. Many matrices have been developed (see Table 3.2 for examples), but one of the earliest, DHB [205], is still the most popular.

Although the dried droplet method of sample preparation works reasonably well, targets prepared in this way with DHB have large crystals pointing from the periphery toward the center. Studies by Kussmann et al. [206] have suggested that there, at least for peptides, is considerable fractionation of cations between the crystals and noncrystalline areas, giving very different signals from different areas of the target. More homogeneous targets and, consequently, better performance have been achieved by techniques such as mixing the DHB with various additives including 2-hydroxy-5-methoxybenzoic acid [207] (the mixture is commonly known as super-DHB), 1-amino*iso*quinoline

Table 3.2 Matrices for MALDI mass spectrometry of carbohydrates.

Matrix and (abbreviation)	Molecular Weight	Structure
3-Amino-4-hydroxybenzoic acid	153.1	
3-Aminoquinoline (3-AQ)	144.2	

Table 3.2 (Continued)

Matrix and (abbreviation)	Molecular Weight	Structure
Arabinosazone	328.4	
6-Aza-2-thiothymine (ATT)	143.2	
5-Chloro-2-mercaptobenzothiazole (CMBT)	201.7	
α-Cyano-4-hydroxycinnamic acid (CHCA)	189.2	
2,6-Dihydroxyacetophenone (DHAP)	152.2	
2,5-Dihydroxybenzoic acid (gentisic acid, DHB)	154.1	
DHB/2-hydroxy-5-methoxybenzoic acid (super-DHB)	154.1/168.1	DHB + CH$_3$O

(*Continued*)

Table 3.2 (Continued)

Matrix and (abbreviation)	Molecular Weight	Structure
DHB/1-hydroxy-*iso*-quinoline (DHB/HIQ)	154.1/145.2	
6,7-Dihydroxycoumarin (Esculetin)	178.2	
Harmane derivatives	182.2	
2-(4'-Hydroxyphenylazo)benzoic acid (HABA)	242.2	
Sinapinic acid (3,5-dimethoxy-4-hydroxycinnamic acid)	224.2	
2,4,6-Trihydroxyacetophenone (THAP)	167.2	

(HIQ) [208], fucose [209], or spermine [210]. Recrystallization of the initially dried sample spot from ethanol [211] is also popular in producing a more homogeneous target surface and better mixing of sample and matrix. Strong signals that rapidly fade on firing the laser at a particular sample spot suggest that the carbohydrates coat the surface of the target crystals rather than being incorporated into the crystal, as is thought to be the case with peptides. Typical MALDI-TOF spectra of N-glycans are shown in Figure 3.4a.

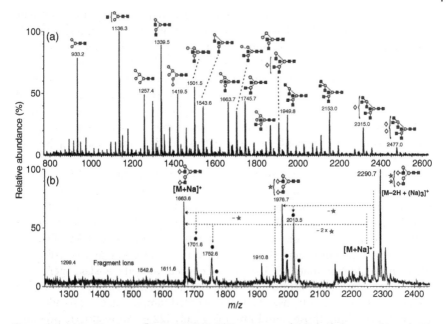

Figure 3.4 (a) Positive ion reflectron MALDI-TOF spectrum of N-linked glycans released with hydrazine from chicken ovalbumin. Major ions are [M+Na]⁺. (b) Positive ion reflectron MALDI-TOF spectrum of a disialylated biantennary N-linked glycan showing loss of sialic acid residues to give focused ions and PSD ions (marked with an asterisk).

Unlike neutral glycans, sialylated glycans are relatively unstable under MALDI conditions and readily eliminate sialic acid. Operation of MALDI-TOF instruments in linear mode restricts the observation of fragmentation occurring in the flight tube and is preferred by some investigators [212]. In reflectron mode, such decompositions within the instrument produce broad metastable ions (Figure 3.4b) whose mass can be predicted by the formula

$$M_c = M_a \left[\frac{1 + \left(\dfrac{M_b}{M_a} \right) r}{(1+r)} \right]^2$$

where M_c is the mass of the metastable ion, M_a is the mass of the parent ion, M_b is the mass of the fragment, and r is a constant that is dependent on the instrument [213]. Migration of the labile acidic proton of the carboxy group is responsible for sialic acid loss, but this can be prevented by derivatization such as methyl ester formation [214] (Figure 3.5a), permethylation [215], conversion to amides [216, 217], or ion pairing with quaternary ammonium or phosphonium salts [218]. These techniques not only stabilize the sialic acids but also prevent negative ion and salt formation, thus allowing quantitative glycan profiling to be made in the positive ion mode. Methyl ester formation of sialic acids with methanol catalyzed by 4-(4,6-dimethoxy-1,3,5-triazin-2-yl)-4-methylmorpholinium chloride (DMT-MM) is particularly useful because a2 → 6-linked sialic acids form methyl esters, whereas a2 → 3-linked sialic acids form lactones (Figure 3.5b). The 32 mass unit difference is easily detectable by MS and provides a quick method for determination of sialic acid linkages [219].

Use of different matrices can also be used to reduce sialic acid loss. Thus, 2,4,6-trihydroxyacetophenone (THAP) has been reported to cause less sialic acid loss, particularly when mixed with ammonium citrate [220], and ATT has also shown promise [220]. With a standard nitrogen laser, "softness" of the matrix is roughly in the order CHCA ≫ DHB > sinapinic acid (SA) ~ THAP > ATT > hydroxypicolinic acid (HPA) [221]. Infrared lasers with glycerol or 3-nitrobenzyl alcohol matrices can also be used to reduce sialic acid loss [222–224] but are generally not found on commercial instruments. Ice has also been used as the matrix by use of a Peltier-cooled sample stage to provide conditions approaching those found physiologically [225]. Atmospheric pressure ion sources have also found application in this context [226, 227].

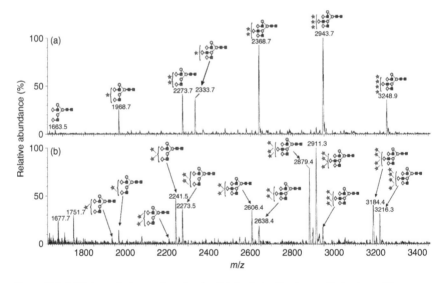

Figure 3.5 (a) Released glycans from bovine fetuin derivatized as methyl esters with methyl iodide. (b) The same glycans derivatized with methanol in the presence of DMT-MM. α2 → 6-linked sialic acids form methyl esters, whereas α2 → 3-linked acids produce lactones. The mass difference between these species allows the linkage to be determined.

Sulfated glycans are also unstable under MALDI conditions and eliminate sulfate easily [228]. Their sodium salts, however, are more stable [97, 229] and can be differentiated from phosphorylated glycans (phosphate and sulfate have the same nominal mass) by the observation that whereas phosphorylated glycans ionize as free acids, sulfated glycans are invariably seen in the positive ion spectra only as sodium salts [229]. In order to stabilize sulfates, ion pairing with the tripeptide Lys–Lys–Lys has been used [230] with the added advantage that it enables isobaric phosphates and sulfates to be differentiated by their fragmentation spectra [231]. Whereas the sulfates showed preferential cleavage of the oxygen–sulfur bond, the phosphates preferred to eliminate the ligand. Fragmentation has also helped in the location of phosphate groups on high-mannose glycans; whereas 6-linked phosphates are stable, 1-linked phosphates undergo cleavage in their PSD spectra [232]. Derivatization has also been used to examine sulfated glycans [233]. Samples were first permethylated and then subjected to methanolytic cleavage of the sulfate groups (which do not methylate) to reveal the linking hydroxyl group. The desulfated permethylated glycans were then subjected to further permethylation using deuteromethyl iodide to label the newly exposed hydroxyl groups.

In positive ion mode, MALDI produces mainly $[M+Na]^+$ ions from neutral glycans [197, 198, 234] (Figure 3.4a) although other cations can be introduced by doping the matrix with an appropriate salt. Sialylated glycans produce a mixture of $[M+Na]^+$ and $[M-nH+(n+1)Na]^+$ ions in positive mode and $[M-H]^-$ ions under negative ion conditions. Negative ions of neutral carbohydrates have only been reported occasionally from specific matrices such as β-carbolines [235, 236]. However, doping the matrix with salts, particularly chlorides, has yielded abundant $[M+adduct]^-$ ions. THAP doped with ammonium nitrate [237], harmine doped with ammonium chloride [238], and 9-aminoacridine doped with sodium iodide or ammonium chloride [239] are recent examples.

MALDI is mistakenly not generally regarded as a quantitative technique. However, if experiments are performed appropriately, good quantitative data can be produced. It is important that target inhomogeneities are overcome by acquiring spectra with many laser shots directed to several target positions [211]. One method for reducing target inhomogeneities is to use an ionic liquid matrix [240] such as that prepared from a standard MALDI matrix and an organic base such as butylamine [241]. The abundance of the $[M+Na]^+$ ions normally formed by carbohydrates appears to show little variation with N-glycan structure [211, 234, 242, 243] unlike the case with peptides where ionization by protonation reflects the proton affinity of the constituent amino acids.

3.8.1.4 Electrospray Ionization (ESI)
Unlike MALDI, ionization of carbohydrates by electrospray can lead to the production of several types of ions, depending on the ion source conditions

and additives to the solvent. Multiple charging is common, particularly in negative ion mode with polyacidic glycans. The resulting spectra, therefore, do not accurately reflect the glycan profile. Furthermore, signal suppression can be a problem, particularly when compounds are present that ionize in preference to the carbohydrates. Consequently, ESI is generally inferior to MALDI for profiling mixtures of neutral glycans but can be advantageous with sialylated and sulfated glycans because of their lower tendency to fragment. $[M+Na]^+$ ions, as in MALDI, can be obtained by use of relatively high cone voltages [244], but these ions are usually accompanied by a considerable amount of in-source fragmentation. Sensitivity generally falls as a function of increasing mass such that it is difficult to record singly charged ions from the larger glycans [119, 150, 244–248]. Part of the reason for this is the increasing tendency for the larger glycans to form doubly charged ions of the type $[M+cation_2]^{2+}$ or $[M+anion_2]^{2-}$. Both $[M+H]^+$ and $[M+2H]^{2+}$ ions can be formed under milder ion source conditions, particularly from derivatized glycans that contain an amine group. In the negative ion mode, carbohydrates form $[M-H]^-$ ions or adducts with various anions [238, 249–254]. Anion addition stabilizes the ions and increases sensitivity. Nanoelectrospray appears to be particularly appropriate for glycan analysis as it has been reported not to suffer from loss of signal at high mass to the same extent as classical electrospray [255].

3.8.2 Glycan Composition by Mass Spectrometry

Because most N- and O-linked glycans contain only a limited number of mass-different monosaccharide constituents, their measured mass can lead directly to their monosaccharide composition. Table 3.3 lists masses of the common monosaccharide constituents. Several laboratories have written software for this task, and there are applications on the web such as that in GlycoWorkbench (http://download.cnet.com/windows/glycoworkbench/3260-20_4-10238922-1. html). It is tempting to submit such compositions to a database and accept the glycan structure that is returned but, because of the possible existence of isomeric structures, this method is not rigorous enough. Such structures must be confirmed by other techniques. Even mass spectral fragmentation is insufficient because it only produces sequence and linkage information and does not yield any information on the nature of the constituent monosaccharides. Such information must be obtained by orthogonal methods such as exoglycosidase digestion or methylation analysis by GC/MS.

3.8.3 Fragmentation

Controlled fragmentation of glycans is normally produced by CID in a collision cell, but fragments can also be formed elsewhere in the instrument. Fragments can arise from decomposition in the ion source (in-source decay

Table 3.3 Residue masses of common monosaccharides.

Monosaccharide	Residue formula	Residue Mass[a]
Pentose	$C_5H_8O_4$	132.042
		132.116
Deoxyhexose	$C_6H_{10}O_4$	146.078
		146.143
Hexose	$C_6H_{10}O_5$	162.053
		162.142
Hexosamine	$C_6H_{11}NO_4$	161.069
		161.157
HexNAc	$C_8H_{13}NO_5$	203.079
		203.179
Methylhexose	$C_7H_{12}O_5$	194.079
		194.185
Hexuronic acid	$C_6H_8O_6$	176.032
		176.126
N-Acetylneuraminic acid	$C_{11}H_{17}NO_8$	291.095
		291.258
N-Glycolylneuraminic acid	$C_{11}H_{17}NO_9$	307.090
		307.257

[a] Top figure = monoisotopic mass (based on C = 12.000000, H = 1.007825, N = 14.003074, O = 15.994915). Lower figure = average mass (based on C = 12.011, H = 1.00794, N = 14.0067, O = 15.9994. The masses of the intact glycans can be obtained by the addition of the residue masses given above, plus the mass of the terminal group (H_2O for an unmodified glycan)—18.011 (monoisotopic) and 18.152 (average)—and the mass of any reducing-terminal or other derivative.

(ISD) or prompt fragmentation) and within the flight tube of time-of-flight (TOF) instruments (PSD) [256]. High-quality fragmentation spectra can now be obtained from both electrosprayed and MALDI-generated ions [246, 257–259] by use of Q-TOF-type instruments. The review by Zaia [20] provides a good overview of carbohydrate fragmentation. Fragmentation of carbohydrates occurs by two main pathways: glycosidic cleavage involving the breaking of a bond between sugar rings and cross-ring cleavage involving the bonds comprising the rings. Glycosidic cleavages from even-electron ions of the type $[M+Na]^+$ result in the loss of neutral molecules and are accompanied by hydrogen migrations. A third type of fragmentation that was proposed to involve a six-membered transition state and the transfer of a carbon-attached hydrogen atom has recently been suggested [260]; the fragment ions effectively eliminate two oxygen functions to leave the expelled neutral particle with a carbonyl group.

3.8.3.1 Nomenclature of Fragment Ions

The accepted nomenclature for naming fragment ions from carbohydrates is that introduced by Domon and Costello in 1988 [261] (Figure 3.6). Ions that retain the charge at the reducing terminus are designated X (cross ring), Y, and Z (glycosidic), whereas ions with the charge at the nonreducing terminus are A (cross ring), B, and C (glycosidic). Sugar rings are numbered from the nonreducing end for A, B, and C ions and from the reducing end for the others. Ions are designated by a subscript number that follows the letter to show the bond that is broken. However, for discussion purposes, in order to avoid the A, B, and C subscripts changing for glycans that have different length chains, we have introduced a modification in which the subscript numbers have been replaced by the subscript R (for reducing-terminal GlcNAc), R-1 (penultimate GlcNAc), and so on. [262]. Fragments from branched-chain glycans are distinguished by Greek letters following the subscript, with α representing the largest chain. The bonds that are broken in cross-ring cleavages are shown by superscripts preceding the letter.

3.8.3.2 In-Source Decay (ISD) Ions

ISD fragments are the result of very fast decomposition of the molecular ions and occur within the ion source. These ions can be the result of Y-type glycosidic cleavages that give rise to compositions indistinguishable from those of native glycans, which will distort a glycan profile. This is a common problem with FAB spectra and to the spectra of glycans that contain sialic acid.

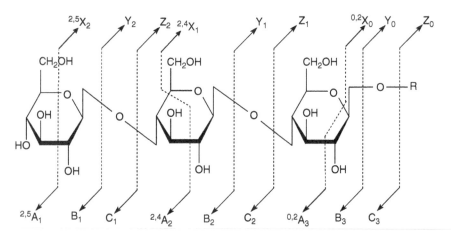

Figure 3.6 Nomenclature introduced by Domon and Costello [261] for describing the fragmentation of carbohydrates.

3.8.3.3 Postsource Decay (PSD) Ions

Ions that decompose more slowly can give rise to PSD fragments if decomposition occurs between the ion source and the detector as first noted by Huberty et al. in 1993 [263] for sialylated glycans. The observation has developed into a general technique for obtaining fragmentation spectra with reflectron TOF instruments [256]. However, in order to record a complete spectrum using instruments without a curved-field reflectron, the reflectron needs to be stepped across the mass range and each subspectrum stitched together to give the final result. The abundance relationship of ions in different sections of the spectra can, thus, be distorted. Ion kinetics dictates that the abundance of these PSD ions depends on the relative time that the parent ions spend in the ion source compared with the time involved in traversing the flight tube. Long in-source delay times, thus, can have an adverse effect on the abundance of PSD ions. Kaufmann et al. [264] have proposed that an additional loss of PSD fragments can be due to a reduced collisional activation in delayed-extraction sources. Unlike the production of CID fragments, PSD ion production is relatively difficult to control and generally needs reasonably large amounts (e.g., up to 100 pmoles) of carbohydrate. To some extent, judicious matrix selection can be used to control fragmentation. Thus, glycans derivatized by reductive amination have been shown to ionize by MALDI to produce mainly $[M+H]^+$ ions, which subsequently fragment to give mainly Y-type ions. Ionization with DHB, on the other hand, produces only $[M+Na]^+$ ions, which fragment to give a complex mixture of B and Y ions together with some cross-ring fragments [118].

3.8.3.4 Collision-Induced Dissociation (CID)

CID fragments are usually produced in collision cells filled with an inert gas, and because fragmentation is isolated from ion production, it can be controlled by application of varying voltages on the collision cell. In positive ion mode, low-energy collisions with glycans produce mainly glycosidic cleavages and low amounts of cross-ring cleavages, mainly of the A type. Higher-energy collisions of the type produced with TOF/TOF instruments or magnetic sector instruments fitted with an orthogonal TOF analyzer [245, 265] produce spectra with a larger proportion of X-type cleavage ions [266]. Glycopeptides tend to produce most fragment ions from the glycan moiety because of the weaker bonds in this part of the molecule [267–269]. Kolli and Dodds [270] have used energy-resolved spectra to show with an N-linked glycopeptide from *Erythrina cristagalli* lectin (ECL) (17 amino acids) that, at a collision cell voltage of 17.5 V, the fragmentation spectrum of the triply protonated ion contained only glycosidic cleavage ions, whereas at 37.5 V mainly b- and y-type peptide fragments were observed. By switching between the two energies, a composite spectrum was obtained. Similar results were obtained with the $Man_5GlcNAc_2$-SRNLTK glycopeptide from ribonuclease B. The presence of glycans on the peptides is revealed by the production of oxonium ions such as those from hexose (m/z 163), HexNAc (m/z 204), Neu5Ac (m/z 292), and HexHexNAc (m/z 366).

3.8.3.5 Electron Transfer Dissociation (ETD)

ETD spectra are generated by the transfer of an electron from a radical ion to the target molecule. The radical ion is generated from an aromatic hydrocarbon such as fluoranthene that has accepted the electron thermalized by a methane buffer gas. Unlike CID spectra of glycopeptides, ETD causes fragmentation of the peptide chain by cleavage of the N–Cα bonds to produce c- and z-type ions, unlike the b and y ions that dominate the CID spectra [58–60]. Little or no loss of the glycans is observed. The method requires the production of multiply charged ions and is, thus, particularly appropriate for ESI-generated glycopeptide spectra. Glycans, on the other hand, typically produce only singly charged ions such as $[M+Na]^+$ or $[M+H]^+$ but can be induced to produce doubly charged ions, which is suitable for ETD reactions by adducting them with cations such as Ca^{2+} or Mg^{2+} [271]. Fragmentation of such species generates extensive cross-ring fragmentation, mainly of the X type, with Mg^{2+} reported to be the best cation. ECD is similar to ETD but is performed in an ICR cell.

3.8.3.6 Infrared Multiphoton Dissociation (IRMPD)

Infrared multiphoton dissociation (IRMPD) of $[M+Li]^+$ and $[M+Na]^+$ ions from glycans implemented in ICR instruments appears to give very similar fragmentation to that observed by CID [272] and enables fragmentation of larger N-linked glycans to be observed when it could not be obtained by CID [273]. IRMPD of $[M+nH]^{n+}$ ions of large glycopeptides, on the other hand, show preferential fragmentation of the peptide chain [274]. Positive ion IRMPD spectra of these O-linked alditols, as with the N-glycans, are very similar to low-energy CID spectra [275].

3.8.3.7 MS^n

Sequential fragmentation or "disassembly" has also been used to obtain structural information on N-glycans. Several successive stages of fragmentation can be observed in an ion trap instrument, particularly where cleavages can occur adjacent to GlcNAc residues [144, 276]. The technique is particularly valuable when interfaced with HPLC [175]. MS^n spectra have shown that many of the fragment ions in the MS^2 spectra from complex glycans arise from several pathways, a property that can present problems for the interpretation of "unknown" spectra [262].

Ashline et al. [277] have studied the fragmentation of small permethylated oligosaccharides as alkali metal adducts and identified ions that are characteristic of, for example, the nature of the terminal monosaccharide residue. They [278] have also used the technique with fragmentation stages up to MS^7 to investigate the occurrence of isomers from N-glycans released from glycoproteins such as chicken ovalbumin and IgG. 2-AP-labeled N-linked glycans have been studied by Ojima et al. [279] using a MALDI-QIT-TOF instrument. The work suggested that some isomer information is available in the form of peak intensity in the positive ion MS^3 spectra. Information on anomeric configuration has been obtained with energy-resolved spectra at various stages of MS^n [280]. It was noted that for the $[M+Na]^+$ ions from linear oligosaccharides, α-linkages fragment at a lower energy than β-ones. However, the authors were not sure if the relationship would apply to all types of carbohydrate.

MS^n experiments have provided information on both the glycan and peptide moieties of glycopeptides. Thus, for example, Deguchi et al. [281] have chosen both peptide- and carbohydrate-derived ions from the MS^2 spectra of these compounds and fragmented each at the MS^3 stage in both positive and negative ion modes to obtain data on both halves of the molecule. In another approach for the analysis of glycoproteins, the compounds were first digested in-gel with trypsin followed by MS^n experiments in an ion trap-TOF instrument. Information on the carbohydrate was obtained in the early stages of fragmentation until finally all of the sugar had been removed, allowing the peptide to be sequenced by a series of y ions [282].

Although acquisition of MS^n spectra is usually performed with trapping instruments, it is possible to use in-source fragmentation to provide the MS^2 spectrum and then to fragment the resulting fragments in a collision cell in the conventional manner to provide the MS^3 spectrum [283]. However, the method is only satisfactory for single compounds when there is no ambiguity as to the source of the MS^2 fragments.

3.8.3.8 Fragmentation Modes of Different Ion Types

Carbohydrates produce a variety of ion types depending on the ionization method used and the structure of the carbohydrate. $[M+Na]^+$ are ubiquitous in MALDI spectra of neutral glycans, but $[M+H]^+$ ions can be formed if there is a site for protonation, as with many derivatized glycans. In negative ion mode, relatively unstable $[M-H]^-$ or $[M-nH]^{n-}$ ions are frequently encountered, particularly from glycans containing anionic groups, but more stable ions can be formed by adduction with various anions.

3.8.3.8.1 $[M+H]^+$ Ions

Protonated molecules decompose much more readily than metal-cationized species and yield mainly B- and Y-type glycosidic cleavage ions with very little or no cross-ring fragments [244] (Figure 3.7a). Thus, they are of limited use for determining the detailed structure of unknown carbohydrates. Another problem is the tendency for rearrangement reactions to occur [284–288], particularly when the carbohydrate has been derivatized at the reducing terminus [289, 290]. Derivatization by reductive amination introduces a secondary amine at the reducing terminus that readily attracts a proton to give the $[M+H]^+$ ion. Charge localization at this site produces a spectrum that contains mainly Y-type glycosidic fragments.

3.8.3.8.2 *[M+Metal]⁺ Ions*

In contrast to the fragmentation of $[M+H]^+$ ions, fragmentation of [M+alkali metal]$^+$ ions (Figure 3.7b) generally requires more energy, but the rearrangements reported from the $[M+H]^+$ ions are absent [291]. The ease with which various ions decompose follows the order $H = Li^+ > Na^+ > K^+ > Cs^+$. $[M+Cs]^+$ ions do not fragment other than to give Cs^+ [244, 292, 293]. Internal fragments (losses from two or more sites) are common and can cause difficulties with spectral interpretation. Cross-ring fragments, usually of the A type [244, 294], are often present, but when low energies are used for fragmentation, as with CID on Q-TOF-type instruments [150, 244, 246], these ions are not usually abundant. More abundant cross-ring fragment ions, particularly X type, can be produced at higher energies of the type found in TOF/TOF-type mass spectrometers [266, 295–298] or formed by high-energy photodissociation [299]. The $[M+Na]^+$ ions of permethylated carbohydrates have been examined with TOF/TOF instruments. Their spectra also contained many A- and X-type cross-ring fragments, but, in general B and Y ions formed by cleavage adjacent to GlcNAc residues dominated the spectra [151, 152, 300].

For some monosaccharides, $[M+Ag]^+$ ions have been used to produce characteristically different spectra relative to adducts of group I metals allowing them to be differentiated, whereas $[M+Cu]^+$ ions did not [301]. With N-linked glycans, silver appears to be particularly good at cleaving glycosidic bonds but does not offer any particular advantage for structural determination [302]. Among doubly charged cations, calcium appears to be particularly effective at fragmenting carbohydrates and for producing high sensitivity detection. Doubly charged ions are the major products formed with these metals although copper has a tendency to form singly charged ions [303].

3.8.3.8.3 *[M–H]⁻ and [M+Anion]⁻ Ions*

CID spectra of negative ions from neutral N-glycans tend to be simpler than the corresponding positive ion spectra, but they carry considerably more specific structural information [254, 304–306] (Figure 3.7c). Negative ion spectra are associated with $[M–H]^-$ ions, and indeed these are formed from the glycans but tend to be rather prone to fragmentation in ESI ion sources. In order to produce stable ions, the glycans can be adducted with anions such as chloride, bromide, iodide, phosphate, sulfate, or nitrate by addition of the appropriate ammonium salt to the electrospray solvent. Normally, samples derived from biological sources naturally contain phosphate and some chloride anions, but in order to avoid splitting the signal, it is advantageous to convert all glycans to their phosphate adducts. These adducts, together with the chloride and

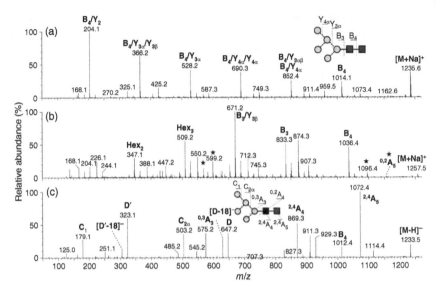

Figure 3.7 (a) CID spectrum of the $[M+H]^+$ ion from the high-mannose N-linked glycan, $Man_5GlcNAc_2$. (b) CID spectrum of the $[M+Na]^+$ ion from $Man_5GlcNAc_2$. Ions marked with a star are cross-ring products. (c) CID spectrum of the $[M-H]^-$ ion from $Man_5GlcNAc_2$.

nitrate adducts, all fragment similarly [238, 250, 254, 307–309] by first eliminating the anion with a proton to give what is essentially the $[M-H]^-$ ion. Bromide adducts fragment poorly, but sulfate and iodide adducts do not fragment at all. Doubly charged negative ions of the type $[M+(H_2PO_4)_2]^{2-}$ from neutral carbohydrates behave similarly [310].

It is the abstraction of a proton from different hydroxyl groups that is responsible for the diagnostic fragments, most of which are A-type cross-ring fragmentation products. Glycosidic fragments are also formed but often tend to be less abundant. In Q-TOF-type instruments, these tend to be mainly of the C rather than the B or Y type [311–314], whereas, more B- and Y-type cleavages appear to be produced in TOF/TOF instruments [297]. These fragment ions are very diagnostic of specific structural features such as the branching pattern of the glycan, the location of fucose residues, and the presence of a bisecting GlcNAc residue (Figure 3.8), which are difficult to determine by techniques such as exoglycosidase digestion.

One of the most prominent fragments in these spectra is a $^{2,4}A_R$ cleavage of the reducing-terminal GlcNAc residue (m/z 1154 in Figure 3.8 as an example), which can be used to indicate the presence or absence of 6-linked fucose on the core GlcNAc because fucose is eliminated in the neutral fragment (Figure 3.8c). The proposed mechanism is initiated by abstraction of the proton from the OH group in the 3-position. If a fucose residue occupies this position, the $^{2,4}A_R$

ion is missing. This ion is also absent from the spectra of glycans derivatized at the reducing terminus by reductive amination, because of the open nature of the GlcNAc ring. Abstraction of the corresponding proton from the 3-position of the penultimate GlcNAc residue leads to the production of the $^{2,4}A_{R-1}$ ion, and cleavage between the two GlcNAc residues produces a prominent B_{R-1} ion.

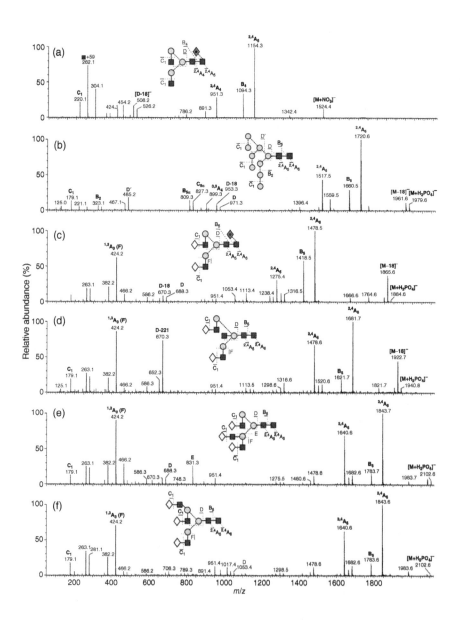

These three ions are typical of all N-linked glycans with a trimannosyl-chitobiose core and confirm the $\beta 1 \rightarrow 4$ linkage between them.

Two of the most useful ions are named D and D-18 and are formed by loss of the chitobiose core and 3-antenna, followed, in the case of the D-18 ion, by loss of water. These ions specify the composition of the 6-antenna and are usually accompanied by $^{0,3}A$ and $^{0,4}A$ cross-ring fragments of the core mannose residue (see Table 3.4 for masses).

When a bisecting GlcNAc residue is present, the D ion, which contains the bisecting GlcNAc residue, eliminates this GlcNAc as a neutral molecule (221 u) to give what is usually a very abundant ion (Figure 3.8c). Antenna composition of complex glycans is revealed by a cross-ring cleavage of the outer mannose residues to give an ion consisting of the antenna plus $-O-CH=CH-O^-$ (59 u), for example, Gal-GlcNAc-O-CH=CH-O$^-$ (m/z 424), and, consequently, the presence of substituents such as fucose or α-galactose can easily be spotted. Common fragments of this type are listed in Table 3.4. The monosaccharide residue terminating the chain is revealed by the mass of the C_1 fragment as listed in Table 3.4. Another useful feature of these negative ion spectra is their ability to distinguish the branching pattern of triantennary glycans [315]. In glycans containing branching of the 3-antenna (with Gal-GlcNAc groups), a $^{0,4}A$ cleavage of the mannose residue gives a prominent fragment containing both the 2- and 4-linked chains (m/z 831). The triantennary glycan branched on the 6-antenna contains its branches at the 2- and 6-positions; consequently, this ion is missing. The structures of these glycans can be confirmed by prominent D, D-18, and D-36 ions. An additional advantage of negative ion fragmentation is that isomeric compounds yield ions with differing m/z values that are much more useful than the differences in abundance of the same mass ion that frequently characterize isomers in positive ion spectra [256].

Figure 3.8 Negative ion MS/MS spectra of selected N-glycans. (a) The position of the core fucose is defined by the masses of the $^{2,4}A_4$ and $^{2,4}A_5$ ions. The C_1 ion at m/z 220 indicates GlcNAc at the ends of the antennae, and the D and [D-18]$^-$ ions at m/z 526 and 508, respectively, show that the 6-antenna contains only mannose and GlcNAc. (b) The masses of the $^{2,4}A_6$ and $^{2,4}A_5$ ions show the absence of fucose at the core. The C_1 ion at m/z 179 shows hexose (mannose) at the nonreducing termini, and the D, D-18, $^{0,3}A_3$ ions show the presence of the four mannose residues on the 6-antenna. Furthermore, the D' ion at m/z 585 shows two of these mannose residues on the 6-branch. (c) The C_1 ion at m/z 179 in this and spectra d–f indicates hexose (galactose) at the ends of the antennae, and the D and [D-18]$^-$ ions at m/z 688 and 670, respectively, show that the 6-antenna contains Gal-GlcNAc. The $^{0,3}A_3$ ion at m/z 424 is a cross-ring ion (Gal-GlcNAc-O-CH=CH-O$^-$). (d) The presence of the bisecting GlcNAc residue produces the abundant [M-221]$^-$ ion at m/z 670. (e) The branched 3-antenna gives rise to the abundant E ion at m/z 831. The D and [D-18]$^-$ ions remain at m/z 688 and 670, respectively. (f) The branching pattern is revealed by the absence of the E ion at m/z 831 and the presence of D and [D-18]$^-$ ions at m/z 1053 and 1035. A third diagnostic ion is present at m/z 1017 ([D-36]$^-$) [315].

Table 3.4 Ions defining structural features in the negative ion spectra of N-linked glycans.

Structural feature	Ion	Ionic composition	*m/z*
Antenna sequence	C	Gal, Man, Glc	179
		GlcNAc, GalNAc	220
		[Fuc]Gal	325
		Gal-[Fuc]GlcNAc	528
		αGal-Gal, Man-Man	341
		Man-[Man]Man	503
		GalNAc-GlcNAc	423
Antenna composition	F	Man	262
		GlcNAc	303
		Gal-GlcNAc	424
		Gal-[Fuc]GlcNAc	570
		αGal-Gal	586
		GalNAc-GlcNAc	627
		$(Gal-GlcNAc)_2$	789
		$(Gal-GlcNAc)_2Fuc$	935
		$(Gal-GlcNAc)_3$	1154
Fucose at 6-position of reducing terminus	$^{2,4}A_R$	$[M-Cl-307]^-$	$[M-342]^-$
		$[M-NO_3-307]^-$	$[M-369]^-$
		$[M-H_2PO_4-307]^-$	$[M-405]^-$
Absence of fucose at 6-position of reducing terminus	$^{2,4}A_R$	$[M-Cl-161]^-$	$[M-196]^-$
		$[M-NO_3-161]^-$	$[M-223]^-$
		$[M-H_2PO_4-161]^-$	$[M-259]^-$
Composition of 6-antenna	D and $[D-18]^-$ ($[D-36]^-$)	GlcNAc	526, 508
		Gal-GlcNAc	688, 670
		Gal-[Fuc]GlcNAc	834, 816
		$(Gal-GlcNAc)_2$	1053, 1035 (1017)
		$(GalGlcNAc)_2Fuc$	1199, 1181 (1163)
		Man_3	647, 629
		Man_4	808, 791
		Man_5	971, 953

Table 3.4 (Continued)

Structural feature	Ion	Ionic composition	m/z
	$^{0,3}A_{R-2}$ and $^{0,4}A_{R-2}$	GlcNAc	292, 262
		GalGlcNAc	454, 424
		Gal-[Fuc]GlcNAc	600, 570
		$(Gal\text{-}GlcNAc)_2$	819, 789
		$(GalGlcNAc)_2Fuc$	965, 935
		Man_3	251, 221
		Man_4	413, 383
		Man_5	575, 545
Composition of 3-antenna	$^{0,4}A_{R-3}$ (E) ion	Gal-GlcNAc	466
		$GlcNAc_2$	507
		$Gal\text{-}GlcNAc_2$	669
		$(Gal\text{-}GlcNAc)_2$	831
		Gal-[Fuc]GlcNAc	977
Presence of bisect	Abundant $[D-221]^-$ ion	GlcNAc	508
		Gal-GlcNAc	670
		Gal-[Fuc]GlcNAc	816
		$(Gal\text{-}GlcNAc)_2$	1035
		$(Gal\text{-}GlcNAc)_2Fuc$	1181
		Man_3	629
Presence of sialic acid	B_1	Neu5Ac	290
		Neu5Gc	306
Presence of $\alpha 2 \rightarrow 6$-linked sialic acid	$^{0,4}A_2\text{-}CO_2$	Neu5Ac	306
		Neu5Gc	322

Unfortunately, acidic carbohydrates, such as those containing sialic acid or carboxyl-containing derivatives such as 2-AA [316], ionize by loss of one or more protons from acidic groups to give $[M-nH]^{n-}$ ions with localized charges and, consequently, restricted fragmentation. 2-AA derivatives should, therefore, be avoided in this context. The spectra of sialylated glycans, however, do yield useful, although restricted, fragmentation [314, 317]. The type of sialic acid attached to the glycan is revealed by the mass of the prominent B_1 fragment (m/z 290 for Neu5Ac and 306 for Neu5Gc), and the presence of $\alpha 2 \rightarrow 6$-linked sialic acids can be determined by the presence of fragments at m/z 306 and 322 for these two sialic acids, respectively. Glycans containing sulfated GalNAcGlcNAc moieties fragment to give two very prominent ions at m/z 282 and 485.

Negative ion fragmentation spectra of O-linked glycans appear to offer similar specificity to that seen in the spectra of the N-linked glycans. Karlsson et al. [318] have investigated negative ion spectra of O-linked alditols from salivary mucin MUC5B and found major Z- and Y-type fragments. C- and A-type cleavages provided information on the structure of the reducing termini. However, cross-ring fragments were not as abundant as in the spectra of the N-linked glycans.

3.8.4   Ion Mobility

The recent introduction of ion mobility to analytical mass spectrometers had provided an additional method of separation before fragmentation [319, 320]. The technique involves the movement of ions through a buffer gas under the influence of an electric field and separates ions on the basis of charge and shape rather than m/z. This field can be fixed or in the form of a traveling wave (TWIMS). The physical parameter associated with the drift time of ions separated by ion mobility is the rotationally averaged collisional cross section, a parameter specific to the glycan that can be used to aid identification. These cross sections can be measured directly with drift tube instruments, but traveling wave-type instruments require suitable calibration. A table of positive ion cross sections has been published by Fenn and McLean [321], and Pagel et al. have published extensive tables of positive [322] and negative [323] ion cross sections of N-glycans measured in both helium and nitrogen. A third type of ion mobility is high-field asymmetric waveform ion mobility spectrometry (FAIMS), an atmospheric pressure method that separates ions according to differentially mobility in a high electric field. Applications of ion mobility to the analysis of carbohydrates have recently been reviewed [324].

Although resolutions on current instruments are still comparatively low, they are still sufficient to separate some glycan isomers [321, 325–327]. Two isomers of Man$_3$GlcNAc$_3$ could be separated to baseline as its [M+Na]$^+$ ion [326], and structures could be assigned by negative ion CID. The arrival time distribution (ATD) of the doubly charged [M+Na$_2$]$^{2+}$ ion from permethylated Hex$_5$GlcNAc$_2$ from chicken ovalbumin has shown evidence for three isomers [325]. Isailovic et al. [328] have noted differences in the ATD profiles of sialylated biantennary N-glycans from human serum, but specific structures were not identified.

Another use of ion mobility is for group separations of compounds such as carbohydrate, peptides, and lipids [329–331]. N-glycans from ribonuclease could be separated from peptides [331]. The method has proved to be invaluable for extracting N-glycan profiles from contaminated samples ionized by both ESI and MALDI where sometimes no evidence of glycan ions was

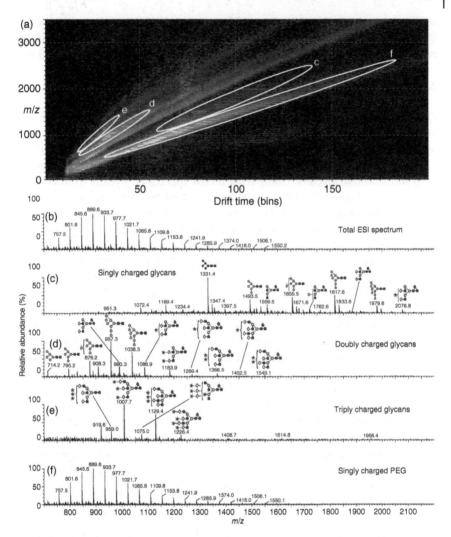

Figure 3.9 (a) Driftscope (*m/z*:drift time, log scale) display of the negative ions from a sample of released gp120 glycans contaminated with PEG. Circled regions are labeled with those of the panels below. (b) Total electrospray spectrum. (c) Extracted singly charged N-glycan ions ([M+H$_2$PO$_4$]$^-$ except *m/z* 2076, which is [M–H]$^-$). (d) Extracted doubly charged N-glycan ions (high-mannose glycans) are [M+(H$_2$PO$_4$)$_2$]$^{2-}$, glycans with one sialic acid are [M–H+H$_2$PO$_4$]$^{2-}$, and the disialylated glycans are [M–H$_2$]$^{2-}$). (e) Extracted triply charged N-glycan ions ([M–H$_3$]$^{3-}$). (f) Extracted singly charged PEG ions.

apparent in the original spectra [332–334]. Figure 3.9 shows an example from a sample of released N-glycans contaminated with polyethylene glycol (PEG) [334]. Singly and multiply charged glycan ions could also be separated from small amounts of material extracted from recombinant virus samples [335]. When combined with negative ion CID, this method has proved to one of the most powerful for N-linked glycan analysis and is described in detail in [336].

3.8.5 Quantitative Measurements

Quantitative aspects of MALDI spectra have been discussed in Section 3.8.1.3, but quantitation has also been achieved with a number of other ionization techniques. Recent reviews summarize quantitative methods for glycoproteins [337, 338]. A number of applications have incorporated stable isotopes into reagents used to derivatize glycans. Comparing mixed spectra of glycans from control and test samples where differential derivatization has been performed with normal and isotopically labeled reagents allows, for example, changes in glycosylation in disease states to be monitored. Thus, Kang et al. [339] have used CH_3 and C^2H_3 derivatization to monitor changes in breast cancer by MALDI-TOF MS. Atwood et al. [340] have used a mixture of $^{13}CH_3I$ and $^{12}C^2H_1H_2I$ for derivatization. Although these derivatives have the same nominal mass, derivatized glycans could be separated at 30,000 resolution with an FT-ICR instrument.

Other isotope-labeled derivatives have involved derivatization at the reducing terminus and have included $^{13}C_6$-labeled 2-AA [341] and aniline [342] and $^2H_4$-labeled pyridine [343]. Hahne has produced two reagents (Figure 3.10) with various stable isotope labels. Of these, the aminooxy reagent (B), which formed an oxime with the glycan, showed the greatest labeling efficiency. Four ^{13}C and one ^{15}N labels were incorporated in the positions shown. Thus, all three labeled derivatives introduced a five unit increase in mass and experiments were conducted using labeled and unlabeled reagents as mentioned earlier. However, on fragmentation, the fragment containing the dimethylpyridine group exhibited a different mass according to which reagent had been used. Consequently, the glycan profile from glycans derivatized with any of the three labeled derivatives looked the same, but, if mixed, the glycans labeled with each specific reagent could be identified by the mass of the fragment.

3.9 Computer Interpretation of MS Data

Although glycomics resources have not reached the level of proteomics with respect to databases, there are nevertheless many carbohydrate sites that provide databases and other tools for glycomics. Space does not permit a comprehensive coverage of all available resources, but the CFG site http://www. functionalglycomics.org/static/index.shtml contains a wealth of information on databases and tools for predicting glycan composition from masses and for interpreting fragmentation spectra. Links to most major resources can be found on this site. Other resources of interest include GlycoWorkbench

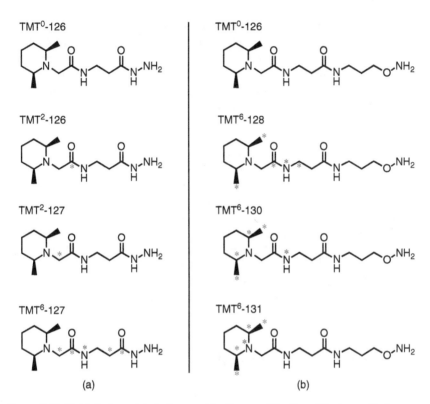

TMT⁰-126   TMT⁰-126

TMT²-126   TMT⁶-128

TMT²-127   TMT⁶-130

TMT⁶-127   TMT⁶-131

(a)   (b)

Figure 3.10 Stable isotope–labeled reagent for the quantification of N-glycans. Red asterisks indicate ^{13}C and the blue asterisks show ^{15}N.

(http://glycomics.ccrc.uga.edu/eurocarb/gwb/home.action) [344], which is a useful tool for drawing glycan structures (both the CFG and Oxford systems are supported), for interpreting fragmentation spectra and for annotating spectra.

SimGlycan (http://www.premierbiosoft.com/glycan/index.html) is a tool for predicting the structures of carbohydrates from MS^n data and accepts data directly from various mass spectrometers, and Proteome Systems has a tool, GlycoMod [345] (http://www.expasy.org/tools/glycomod/), for obtaining glycan composition from glycan masses. GlycoSpectrumScan, developed by Deshpande et al. [346], is a tool for assigning potential glycopeptide compositions based on peptide sequences and potential N- or O-glycan compositions, and GlycoPep DB (http://hexose.chem.ku.edu/sugar.php) is an application that uses "smart searching" to find glycopeptides in a database. Only the compositions of previously characterized glycans are used in order to avoid glycopeptides with implausible structures [347]. CASy (http://www.casy.org) is a site that has a database describing the families of structurally related catalytic and carbohydrate-binding modules of enzymes that degrade, modify, or create glycosidic bonds. UniCarbKB (http://www.unicarbkb.org/) is an online information storage and search platform for glycomics and glycobiology research. It contains UniCarb-DB (http://www.unicarb-db.org/), which is an online LC/MS/MS database that continues that initiated as EurocarbDB, and GlycoBase (http://glycobase.nibrt.ie/), a database of HPLC data. Cartoonist [348] is a tool for structurally annotating MALDI spectra.

Following the demise of the Complex Carbohydrate Structure Database (Carbbank) in 1997, a number of other initiatives featuring databases have been launched to aid glycomics. Among these is Kyoto Encyclopedia of Genes and Genomes (KEGG) ([349, 350] (http://www.genome.jp/kegg/glycan), which has its own database of carbohydrate structures, including those from the CarbBank database and a database of known biosynthetic pathways. GlycomeDB (http://www.glycome-db.org) is a database that attempts to combine data from earlier databases such as KEGG, CarbBank, and GlycoCT. During the construction of this database, a critical analysis of data in the CarbBank database found that about 35% of the records contain errors, and it was concluded that data contained in that database cannot, therefore, be used and referenced without verification [351]. Lists of databases and other tools for glycobiology have been published in several reviews [352–357].

3.10   Total Glycomics Methods

The last few years have seen the publication of many methods for the structural analysis of glycans from their parent glycoproteins incorporating several of the techniques discussed earlier. Most rely on preliminary purification of the glycoprotein, site analysis, and then removal of the glycans for further analysis by ESI or MALDI MS and/or HPLC. The nature of the monosaccharide constituents is further investigated by HPLC or GC/MS after suitable derivatization or by exoglycosidase digestions. As a representative example, Morelle and Michalski [358] have published full practical details involving PNGase F release of the glycans from SDS-PAGE gel-separated glycoproteins followed by cleanup with porous graphitized carbon. O-glycans were then released by reductive elimination and cleaned with Dowex beads. Exoglycosidase digestion and methylation analysis by GC/MS was performed to identify the monosaccharides, and glycan profiles and glycan compositions were obtained by MALDI-TOF MS. Finally, nano-MS/MS was performed to obtain sequence and linkage information.

In another method where practical details were reported [359], PNGase F was again used to release the glycans, which were then labeled with 2-AA. This procedure lends itself to analysis by HPLC and capillary electrophoresis (CE) as well as to analysis by MALDI-TOF. Glycans were examined from human plasma glycoproteins and glycan cleanup employed hydrophilic interaction chromatography, which allowed the separation of the labeled glycans from an excess of labeling reagent, proteins, lipids, and salts. High throughput was achieved by conducting the sample preparation in 96-well plate.

Full experimental details have also been published of a method for the analysis of human plasma glycoproteins and their attached glycans [360]. The target low-abundance proteins were concentrated by immunoprecipitation and removal of albumin, IgG, IgA, transferrin, haptoglobin, and α-1-antitrypsin and then separated by 1D anion-exchange chromatography with an eight-step salt elution. Fractions from each elution steps were transferred onto a 2D reversed-phase HPLC column, and the eluted compounds were digested with trypsin. MALDI MS was performed with a quadrupole ion trap (QIT)-TOF and LC/MS employed an Orbitrap. Finally, Wuhrer et al. [361, 362] have described a nanoscale LC/MS technique using NP LC/MS that is capable of producing fragmentation data from small N-linked glycans at the low femtomole level.

3.11 Conclusions

Many developments in the structural and quantitative analysis have been made in recent years, and progress shows no sign of slowing. The modern ionization techniques of MALDI and ESI have replaced older methods such as FAB and enabled most types of carbohydrate to be ionized and analyzed by MS. New instrumentation such as the Orbitrap^{TM} series of mass spectrometers provide the high resolution needed for resolution of many glycoforms, and ion mobility separations are showing promise for very fast separations of isomeric glycans and for handling very small quantities of material. The coming years should see further developments, particularly in the area of bioinformatics and the ability to examine even smaller and more complicated samples.

Abbreviations

2-AA	2-Aminobenzoic acid
AAL	*Aleuria aurantia* lectin
2-AB	2-aminobenzamide
2-AP	2-aminopyridine
ATD	arrival time distribution
3-AQ	3-aminoquinolibe
ATT	6-aza-2-thiothymine
BAD-lectins	boronate-decorated lectins
CE	capillary electrophoresis
CFG	Consortium for Functional Glycomics
CHCA	α-cyano-4-hydroxycinnamic acid

CID	collision-induced decomposition
CMBT	5-chloro-2-mercaptobenzothiazole
ConA	concanavalin A
Da	Dalton
DHB	2,5-dihydroxybenzoic acid
DMSO	dimethyl sulfoxide
ECD	electron capture dissociation
ECL	*Erythrina crista-galli* lectin
EI	electron impact
endoH (S)	endoglycosidase H (S)
ESI	electrospray ionization
ETD	electron transfer dissociation
FAB	fast atom bombardment
FAIMS	high-field asymmetric waveform ion mobility spectrometry
FT	Fourier transform
Gal	galactose
GalNAc	*N*-acetylgalactosamine
GC/MS	gas chromatography/mass spectrometry
Glc	glucose
GlcNAc	*N*-acetylglucosamine
HABA	2-(4′-hydroxyphenylazo)benzoic acid
Hex	hexose
HexNAc	*N*-acetylaminohexose
HILAC	hydrophilic interaction chromatography
HIQ	1-amino*iso*quinoline
HPA	hydroxypicolinic acid
HPLC	high-performance liquid chromatography
ICR	ion cyclotron resonance
IgG (AM)	immunoglobulin G (A or M)
IRMPD	infrared multiphoton dissociation
ISD	in-source decay
KEGG	Kyoto Encyclopedia of Genes and Genomes
LCA	*Lens culinaris* agglutinin
LC/MS	liquid chromatography/mass spectrometry
MALDI	matrix-assisted laser desorption/ionization
Man	mannose
MS	mass spectrometry
Neu5Ac	*N*-acetylneuraminic acid (sialic acid)
Neu5Gc	*N*-acetyl-glycolylneuraminic acid
NP	normal phase
PAGE	polyacrylamide gel electrophoresis
PEG	polyethylene glycol
PHA-L	*Phaseolus vulgaris* lectin

PMAA	partially methylated alditol acetates
PNA	peanut agglutinin
PNGase	protein *N*-glycosidase
PSD	postsource decay
Q	quadrupole
QIT	quadrupole ion trap
SA	sinapinic acid
SDS	sodium dodecyl sulfate
SNA-1	*Sambucus nigra* agglutinin
TFA	trifluoroacetic acid
THAP	2,4,6-trihydroxyacetophenone
TMAPA	trimethyl(4-aminophenyl)ammonium chloride
TOF	time-of-flight
TWIMS	traveling wave ion mobility spectrometry
UEA-1	*Ulex europaeus* agglutinin I
UV	ultraviolet
WGA	wheat germ agglutinin.

References

1 Varki A. Biological roles of oligosaccharides: All of the theories are correct. *Glycobiology* 1993;**3**:97–130.
2 Dwek RA. Glycobiology: Towards understanding the function of sugars. *Chem Rev* 1996;**96**:683–720.
3 Alley WR Jr, Mann BF, Novotny MV. High-sensitivity analytical approaches for the structural characterization of glycoproteins. *Chem Rev* 2013;**113**:2668–2732.
4 Alley WR Jr, Novotny MV. Structural glycomic analyses at high sensitivity: A decade of progress. *Annu Rev Anal Chem* 2013;**6**:237–265.
5 An HJ, Lebrilla CB. Structure elucidation of native N- and O-linked glycans by tandem mass spectrometry (Tutorial). *Mass Spectrom Rev* 2011;**30**:560–578.
6 Budnik BA, Lee RS, Steen JAJ. Global methods for protein glycosylation analysis by mass spectrometry. *Biochim Biophys Acta* 2006;**1764**:1870–1880.
7 Cortes DF, Kabulski JL, Lazar AC, Lazar IM. Recent advances in the MS analysis of glycoproteins: Capillary and microfluidic workflows. *Electrophoresis* 2011;**32**:14–29.
8 Dalpathado DS, Desaire H. Glycopeptide analysis by mass spectrometry. *Analyst* 2008;**133**:731–738.
9 Desaire H. Glycopeptide analysis, recent developments and applications. *Mol Cell Proteomics* 2013;**12**:893–901.
10 Han L, Costello CE. Mass spectrometry of glycans. *Biochemistry* 2013;**78**:710–720.

11 Harvey DJ. Matrix-assisted laser desorption/ionization mass spectrometry of carbohydrates and glycoconjugates. *Int J Mass Spectrom* 2003;**226**:1–35.

12 Harvey DJ. Proteomic analysis of glycosylation: structural determination of N- and O-linked glycans by mass spectrometry. *Expert Rev Proteomics* 2005;**2**:87–101.

13 Harvey DJ. Structural determination of N-linked glycans by matrix-assisted laser desorption/ionization and electrospray ionization mass spectrometry. *Proteomics* 2005;**5**:1774–1786.

14 Haslam SM, North SJ, Dell A. Mass spectrometric analysis of *N*- and *O*-glycosylation of tissues and cells. *Curr Opin Struct Biol* 2006;**16**:584–591.

15 Haslam SM, Khoo KH, Dell A. Sequencing of oligosaccharides and glycoproteins. In: Wong C-H, editor. *Carbohydrate-Based Drug Discovery.* Wiley VCH; 2006. p 461–482.

16 Morelle W, Faid V, Chirat F, Michalski JC. Analysis of N- and O-linked glycans from glycoproteins using MALDI-TOF mass spectrometry. *Methods Mol Biol* 2009;**534**:5–21.

17 Novotny MV, Alley WR Jr, Mann BF. Analytical glycobiology at high sensitivity: current approaches and directions. *Glycoconj J* 2013;**30**:89–117.

18 Novotny MV, Alley WR Jr. Recent trends in analytical and structural glycobiology. *Curr Opin Chem Biol* 2013;**17**:832–840.

19 Uçaktürk E. Analysis of glycoforms on the glycosylation site and the glycans in monoclonal antibody biopharmaceuticals. *J Sep Sci* 2012;**35**:341–350.

20 Zaia J. Mass spectrometry of oligosaccharides. *Mass Spectrom Rev* 2004;**23**:161–227.

21 Zaia J. Mass spectrometry and glycomics. *OMICS J Integrat Biol* 2010;**14**:401–418.

22 Zauner G, Kozak RP, Gardner RA, Fernandes DL, Deelder AM, Wuhrer M. Protein O-glycosylation analysis. *Biol Chem* 2012;**393**:687–708.

23 Laine RA. A calculation of all possible oligosaccharide isomers both branched and linear yields 105 x 1012 structures for a reducing hexasaccharide: the Isomer Barrier to development of single-method saccharide sequencing or synthetic systems. *Glycobiology* 1994;**4**:759–767.

24 Kennedy JF. *Carbohydrate Chemistry.* Oxford: Clarendon Press; 1988.

25 Hounsell EF, Davies MJ, Renouf DV. O-Linked protein glycosylation structure and function. *Glycoconj J* 1996;**13**:19–26.

26 Satomi Y, Shimonishi Y, Takao T. N-glycosylation at Asn in the Asn-Xaa-Cys motif of human transferrin. *FEBS Lett* 2004;**576**:51–56.

27 Varki A, Cummings RD, Esko JD, Freeze HH, Stanley P, Bertozzi CR, Hart GW, Etzler ME. *Essentials of Glycobiology.* 2nd ed. Cold Spring Harbor Laboratory Press; 2008.

28 Cooper CA, Wilkins MR, Williams KL, Packer NH. BOLD - A biological O-linked glycan database. *Electrophoresis* 1999;**20**:3589–3598.

29 Harvey DJ, Merry AH, Royle L, Campbell MP, Rudd PM. Symbol nomenclature for representing glycan structures: Extension to cover different carbohydrate types. *Proteomics* 2011;**11**:4291–4295.

30 Kornfeld R, Kornfeld S. Assembly of asparagine-linked oligosaccharides. *Annu Rev Biochem* 1985;**54**:631–664.

31 Chakel JA, Pungor E Jr, Hancock WS, Swedberg SA. Analysis of recombinant DNA-derived glycoproteins via high-performance capillary electrophoresis coupled with off-line matrix-assisted laser desorption ionization time-of-flight mass spectrometry. *J Chromatogr B* 1997;**689**:215–220.

32 Apffel A, Chakel JA, Hancock WS, Souders C, M'Timkulu T, Pungor E Jr. Application of high-performance liquid chromatography-electrospray ionization mass spectrometry and matrix-assisted laser desorption ionization time-of-flight mass spectrometry in combination with selective enzymatic modifications in the characterization of glycosylation patterns in single-chain plasminogen activator. *J Chromatogr A* 1996;**732**:27–42.

33 Fanayan S, Hincapie M, Hancock WS. Using lectins to harvest the plasma/serum glycoproteome. *Electrophoresis* 2012;**33**:1746–1754.

34 Kobayashi Y, Tateno H, Ogawa H, Yamamoto K, Hirabayashi J. Comprehensive list of lectins: Origins, natures, and carbohydrate specificities. *Methods Mol Biol* 2014;**1200**:555–577.

35 Sparbier K, Asperger A, Resemann A, Kessler I, Koch S, Wenzel T, Stein G, Vorwerg L, Suckau D, Kostrzewa M. Analysis of glycoproteins in human serum by means of glycospecific magnetic bead separation and LC-MALDI-TOF/TOF analysis with automated glycopeptide detection. *J Biomol Tech* 2007;**18**:252–258.

36 Sparbier K, Koch S, Kessler I, Wenzel T, Kostrzewa M. Selective isolation of glycoproteins and glycopeptides for MALDI-TOF MS detection supported by magnetic particles. *J Biomol Tech* 2005;**16**:407–413.

37 Tang J, Liu Y, Yin P, Yao G, Yan G, Deng C, Zhang X. Concanavalin A-immobilized magnetic nanoparticles for selective enrichment of glycoproteins and application to glycoproteomics in hepatocellular carcinoma cell line. *Proteomics* 2010;**10**:2000–2014.

38 Yang Z, Hancock WS. Approach to the comprehensive analysis of glycoproteins isolated from human serum using a multi-lectin affinity column. *J Chromatogr A* 2004;**1053**:79–88.

39 Qiu R, Regnier FE. Use of multidimensional lectin affinity chromatography in differential glycoproteomics. *Anal Chem* 2005;**77**:2802–2809.

40 Madera M, Mechref Y. I., I., Novotny, M. V.: High-sensitivity profiling of glycoproteins from human blood serum through multiple-lectin affinity chromatography and liquid chromatography/tandem mass spectrometry. *J Chromatogr B* 2007;**845**:121–137.

41 Zhang X, He X, Chen L, Zhang Y. Boronic acid modified magnetic nanoparticles for enrichment of glycoproteins *via* azide and alkyne click chemistry. *J Mater Chem* 2012;**22**:16520–16526.

42 Lin Z-A, Zheng J-N, Lin F, Zhang L, Cai Z, Chen G-N. Synthesis of magnetic nanoparticles with immobilized aminophenylboronic acid for selective capture of glycoproteins. *J Mater Chem* 2011;**21**:518–524.

43 Pan M, Sun Y, Zheng J, Yang W. Boronic acid-functionalized core-shell-shell magnetic composite microspheres for the selective enrichment of glycoprotein. *ACS Appl Mater Interfaces* 2013;**5**:8351–8358.

44 Lin ZA, Pang JL, Lin Y, Huang H, Cai ZW, Zhang L, Chen GN. Preparation and evaluation of a phenylboronate affinity monolith for selective capture of glycoproteins by capillary liquid chromatography. *Analyst* 2011;**136**:3281–3288.

45 Lin Z, Pang J, Yang H, Cai Z, Zhang L, Chen G. One-pot synthesis of an organic-inorganic hybrid affinity monolithic column for specific capture of glycoproteins. *Chem Commun* 2011;**47**:9675–9677.

46 Lu Y-W, Chien C-W, Lin P-C, Huang L-D, Chen C-Y, Wu S-W, Han C-L, Khoo K-H, Lin C-C, Chen Y-J. BAD-lectins: Boronic acid-decorated lectins with enhanced binding affinity for the selective enrichment of glycoproteins. *Anal Chem* 2013;**85**:8268–8276.

47 Zhang H, Li X, Martin DB, Aebersold R. Identification and quantification of N-linked glycoproteins using hydrazide chemistry, stable isotope labeling and mass spectrometry. *Nat Biotechnol* 2003;**21**:660–666.

48 Sun B, Ranish JA, Utleg AG, White JT, Yan X, Lin B, Hood L. Shotgun glycopeptide capture approach coupled with mass spectrometry for comprehensive glycoproteomics. *Mol Cell Proteomics* 2007;**6**:141–149.

49 Tian Y, Zhou Y, Elliott S, Aebersold R, Zhang H. Solid-phase extraction of N-linked glycopeptides. *Nat Protoc* 2007;**2**:334–339.

50 Larsen MR, Jensen SS, Jakobsen LA, Heegaard NHH. Exploring the sialiome using titanium dioxide chromatography and mass spectrometry. *Mol Cell Proteomics* 2007;**6**:1778–1787.

51 Wohlgemuth J, Karas M, Eichhorn T, Hendriks R, Andrecht S. Quantitative site-specific analysis of protein glycosylation by LC-MS using different glycopeptide-enrichment strategies. *Anal Biochem* 2009;**395**:178–188.

52 Thompson NJ, Rosati S, Rose RJ, Heck AJR. The impact of mass spectrometry on the study of intact antibodies: from post-translational modifications to structural analysis. *Chem Commun* 2013;**49**:538–548.

53 Yang Y, Barendregt A, Kamerling JP, Heck AJR. Analyzing protein micro-heterogeneity in chicken ovalbumin by high-resolution native mass spectrometry exposes qualitatively and semi-quantitatively 59 proteoforms. *Anal Chem* 2013;**85**:12037–12045.

54 Imre T, Schlosser G, Pocsfalvi G, Siciliano R, Molnár-Szöllsi É, Kremmer T, Malorni A, Vékey K. Glycosylation site analysis of human alpha-1-acid glycoprotein (AGP) by capillary liquid chromatography - electrospray mass spectrometry. *J Mass Spectrom* 2005;**40**:1472–1483.

55 Lin S, Yao G, Qi D, Li Y, Deng C, Yang P, Zhang X. Fast and efficient proteolysis by microwave-assisted protein digestion using trypsin-immobilized magnetic silica microspheres. *Anal Chem* 2008;**80**:3655–3665.

56 Annesley TM. Ion suppression in mass spectrometry. *Clin Chem* 2003;**49**:1041–1044.

57 Carr SA, Huddleston MJ, Bean MF. Selective identification and differentiation of N- and O-linked oligosaccharides in glycoproteins by liquid chromatography-mass spectrometry. *Protein Sci* 1993;2:183–196.

58 Mirgorodskaya E, Roepstorff P, Zubarev RA. Localization of O-glycosylation sites in peptides by electron capture dissociation in a Fourier transform mass spectrometer. *Anal Chem* 1999;71:4431–4436.

59 Syka JEP, Coon JJ, Schroeder MJ, Shabanowitz J, Hunt DF. Peptide and protein sequence analysis by electron transfer dissociation mass spectrometry. *Proc Natl Acad Sci USA* 2004;101:9528–9533.

60 Håkansson K, Cooper HJ, Emmett MR, Costello CE, Marshall AG, Nilsson CL. Electron capture dissociation and infrared multiphoton dissociation MS/MS of an N-glycosylated tryptic peptic to yield complementary sequence information. *Anal Chem* 2001;73:4530–4536.

61 Gonzalez J, Takao T, Hori H, Besada V, Rodriguez R, Padron G, Shimonishi Y. A method for determination of N-glycosylation sites in glycoproteins by collision-induced dissociation analysis in fast atom bombardment mass spectrometry: Identification of the positions of carbohydrate-linked asparagine in recombinant α-amylase by treatment with peptide-N-glycosidase F in ^{18}O-labelled water. *Anal Biochem* 1992;205:151–158.

62 Lee B-S, Krishnanchettiar S, Lateef SS, Gupta S. Characterization of oligosaccharide moieties of glycopeptides by microwave-assisted partial acid hydrolysis and mass spectrometry. *Rapid Commun Mass Spectrom* 2005;19:1545–1550.

63 Müller S, Goletz S, Packer N, Gooley A, Lawson AM, Hanisch F-G. Localization of O-glycosylation sites on glycopeptide fragments from lactation-associated MUC1. All putative sites within the tandem repeat are glycosylation targets in vivo. *J Biol Chem* 1997;272:24780–24793.

64 Juhasz P, Martin SA. The utility of nonspecific proteases in the characterization of glycoproteins by high-resolution time-of-flight mass spectrometry. *Int J Mass Spectrom Ion Processes* 1997;169/170:217–230.

65 Coddeville B, Girardet J-M, Plancke Y, Campagna S, Linden G, Spik G. Structure of the O-glycopeptides isolated from bovine milk component PP3. *Glycoconj J* 1998;15:371–378.

66 An HJ, Peavy TR, Hedrick JL, Lebrilla CB. Determination of N-glycosylation sites and site heterogeneity in glycoproteins. *Anal Chem* 2003;75:5628–5637.

67 Schiel JE, Smith NJ, Phinney KW. Universal proteolysis and MS^n for N- and O-glycan branching analysis. *J Mass Spectrom* 2013;48:533–538.

68 Hua S, Hu CY, Kim BJ, Totten SM, Oh MJ, Yun N, Nwosu CC, Yoo JS, Lebrilla CB, An HJ. Glyco-analytical multispecific proteolysis (Glyco-AMP): A simple method for detailed and quantitative glycoproteomic characterization. *J Proteome Res* 2013;12:4414–4423.

69 Dodds ED, Seipert RR, Clowers BH, German JB, Lebrilla CB. Analytical performance of immobilized pronase for glycopeptide footprinting and

implications for surpassing reductionist glycoproteomics. *J Proteome Res* 2009;**8**:502–512.

70 Alving K, Paulsen H, Peter-Katalinic J. Characterization of O-glycosylation sites in MUC2 glycopeptides by nanoelectrospray QTOF mass spectrometry. *J Mass Spectrom* 1999;**34**:395–407.

71 Czeszak X, Morelle W, Ricart G, Tétaert D, Lemoine J. Localization of the O-glycosylated sites in peptides by fixed-charge derivatization with a phosphonium group. *Anal Chem* 2004;**76**:4320–4324.

72 Zhou H, Froehlich JW, Briscoe AC, Lee RS. The GlycoFilter: A simple and comprehensive sample preparation platform for proteomics, *N*-glycomics and glycosylation site assignment. *Mol Cell Proteomics* 2013;**12**:2981–2991.

73 Takasaki S, Misuochi T, Kobata A. Hydrazinolysis of asparagine-linked sugar chains to produce free oligosaccharides. *Methods Enzymol* 1982;**83**:263–268.

74 Patel T, Bruce J, Merry A, Bigge C, Wormald M, Jaques A, Parekh R. Use of hydrazine to release in intact and unreduced form both N- and O-linked oligosaccharides from glycoproteins. *Biochemistry* 1993;**32**:679–693.

75 Merry AH, Neville DCA, Royle L, Matthews B, Harvey DJ, Dwek RA, Rudd PM. Recovery of intact 2-aminobenzamide-labeled O-glycans released from glycoproteins by hydrazinolysis. *Anal Biochem* 2002;**304**:91–99.

76 Tanabe K, Ikenaka K. In-column removal of hydrazine and N-acetylation of oligosaccharides released by hydrazinolysis. *Anal Biochem* 2006;**348**:324–326.

77 Bendiac B, Cumming DA. Hydrazinolysis-N-reacetylation of glycopeptides and glycoproteins. Model studies using 2-acetamido-1-N-(L-aspart-4-oyl)-2-deoxy-α-d-glucopyranosylamine. *Carbohydr Res* 1985;**144**:1–12.

78 Carlson DM. Oligosaccharides isolated from pig submaxillary mucin. *J Biol Chem* 1966;**241**:2984–2986.

79 Almeida A, Ferreira JA, Teixeira F, Gomes C, Cordeiro MNDS, Osório H, Santos LL, Reis CA, Vitorino R, Amado F. Challenging the limits of detection of sialylated Thomsen-Friedenreich antigens by in-gel deglycosylation and nano-LC-MALDI-TOF-MS. *Electrophoresis* 2013;**34**:2337–2341.

80 Kumagai T, Katoh T, Nix DB, Tiemeyer M, Aoki K. In-gel β-elimination and aqueous-organic partition for improved O- and sulfoglycomics. *Anal Chem* 2013;**85**:8692–8699.

81 Rademaker GJ, Pergantis SA, Blok-Tip L, Langridge JI, Kleen A, Thomas-Oates JE. Mass spectrometric determination of the sites of O-glycan attachment with low picomolar sensitivity. *Anal Biochem* 1998;**257**:149–160.

82 Huang Y, Mechref Y, Novotny MV. Microscale nonreductive release of O-linked glycans for subsequent analysis through MALDI mass spectrometry and capillary electrophoresis. *Anal Chem* 2001;**73**:6063–6069.

83 Huang Y, Konse T, Mechref Y, Novotny MV. Matrix-assisted laser desorption/ionization mass spectrometry compatible β-elimination of O-linked oligosaccharides. *Rapid Commun Mass Spectrom* 2002;**16**:1199–1204.

84 Miura Y, Kato K, Takegawa Y, Kurogochi M, Furukawa J-i, Shinohara Y, Nagahori N, Amano M, Hinou H, Nishimura S-I. Glycoblotting-assisted O-glycomics: Ammonium carbamate allows for highly efficient O-glycan release from glycoproteins. *Anal Chem* 2010;**82**:10021–10029.

85 Tarelli E. Resistance to deglycosylation by ammonia of IgA1 O-glycopeptides: implications for the β-elimination of O-glycans linked to serine and threonine. *Carbohydr Res* 2007;**342**:2322–2325.

86 Yu G, Zhang Y, Zhang Z, Song L, Wang P, Chai W. Effect and limitation of excess ammonium on the release of O-glycans in reducing forms from glycoproteins under mild alkaline conditions for glycomic and functional analysis. *Anal Chem* 2010;**82**:9534–9542.

87 Zheng Y, Guo Z, Cai Z. Combination of β-elimination and liquid chromatography/quadrupole time-of-flight mass spectrometry for the determination of O-glycosylation sites. *Talanta* 2009;**78**:358–363.

88 Tarentino AL, Gómez CM, Plummer TH Jr. Deglycosylation of asparagine-linked glycans by peptide:N-glycosidase F. *Biochemistry* 1985;**24**:4665–5671.

89 Küster B, Harvey DJ. Ammonium-containing buffers should be avoided during enzymatic release of glycans from glycoproteins when followed by reducing terminal derivatization. *Glycobiology* 1997;**7**:vii–ix.

90 Harvey DJ, Rudd PM. Identification of by-products formed during the release of N-glycans with protein N-glycosidase F in the presence of dithiothreitol. *J Mass Spectrom* 2010;**45**:815–819.

91 Tretter V, Altmann F, März L. Peptide-N^4-(N-acetyl-glucosaminyl)asparagine amidase F cannot release glycans with fucose attached α-1-3 to the asparagine-linked N-acetylglucosamine residue. *Eur J Biochem* 1991;**199**:647–652.

92 Sandoval WN, Arellano F, Arnott D, Raab H, Vandlen R, Lill JR. Rapid removal of N-linked oligosaccharides using microwave assisted enzyme catalyzed deglycosylation. *Int J Mass Spectrom* 2007;**259**:117–123.

93 Tzeng Y-K, Chang C-C, Huang C-N, Wu C-C, Han C-C, Chang H-C. Facile MALDI-MS analysis of neutral glycans in NaOH-doped matrixes: Microwave-assisted deglycosylation and one-step purification with diamond nanoparticles. *Anal Chem* 2008;**80**:6809–6814.

94 Szabo Z, Guttman A, Karger BL. Rapid release of N-linked glycans from glycoproteins by pressure-cycling technology. *Anal Chem* 2010;**82**:2588–2593.

95 Zhou Q, Park S-H, Boucher S, Higgins E, Lee K, Edmunds T. N-linked oligosaccharide analysis of glycoprotein bands from isoelectric focusing gels. *Anal Biochem* 2004;**335**:10–16.

96 Küster B, Wheeler SF, Hunter AP, Dwek RA, Harvey DJ. Sequencing of N-linked oligosaccharides directly from protein gels: In-gel deglycosylation followed by matrix-assisted laser desorption/ionization mass spectrometry and normal-phase high performance liquid chromatography. *Anal Biochem* 1997;**250**:82–101.

97 Wheeler SF, Harvey DJ. Extension of the in-gel release method for structural analysis of neutral and sialylated N-linked glycans to the analysis of sulphated glycans. *Anal Biochem* 2001;**296**:92–100.

98 Charlwood J, Skehel JM, Camilleri P. Immobilisation of antibodies in gels allows the improved release and identification of glycans. *Proteomics* 2001;**1**:275–284.

99 Mills PB, Mills K, Johnson AW, Clayton PT, Winchester BG. Analysis by matrix assisted laser desorption/ionisation-time of flight mass spectrometry of the post-translational modifications of α₁-antitrypsin isoforms separated by two-dimensional polyacrylamide gel electrophoresis. *Proteomics* 2001;**1**:778–786.

100 Charlwood J, Skehel JM, Camilleri P. Analysis of N-linked oligosaccharides released from glycoproteins separated by two-dimensional gel electrophoresis. *Anal Biochem* 2000;**284**:49–59.

101 Charlwood J, Bryant D, Skehel JM, Camilleri P. Analysis of N-linked oligosaccharides: progress towards the characterization of glycoprotein-linked carbohydrates. *Biomol Eng* 2001;**18**:229–240.

102 Callewaert N, Vervecken W, Van Hecke A, Contreras R. Use of a meltable polyacrylamide matrix for sodium dodecyl sulfate-polyacrylamide gel electrophoresis in a procedure for N-glycan analysis on picomole amounts of glycoproteins. *Anal Biochem* 2002;**303**:93–95.

103 Kolarich D, Altmann F. N-glycan analysis by matrix-assisted laser desorption/ionization mass spectrometry of electrophoretically separated nonmammalian proteins: Application to peanut allergen Ara h 1 and olive pollen allergen Ole e 1. *Anal Biochem* 2000;**285**:64–75.

104 Altmann F, Paschinger K, Dalik T, Vorauer K. Characterisation of peptide-N^4-(N-acetyl-β-glucosaminyl)asparagine amidase A and its N-glycans. *Eur J Biochem* 1998;**252**:118–123.

105 Goodfellow JJ, Baruah K, Yamamoto K, Bonomelli C, Krishna B, Harvey DJ, Crispin M, Scanlan CN, Davis BG. An endoglycosidase with alternative glycan specificity allows broadened glycoprotein remodelling. *J Am Chem Soc* 2012;**134**:8030–8033.

106 Yu YQ, Gilar M, Kaska J, Gebler JC. A rapid sample preparation method for mass spectrometric characterization of N-linked glycans. *Rapid Commun Mass Spectrom* 2005;**19**:2331–2336.

107 Packer NH, Lawson MA, Jardine DR, Redmond JW. A general approach to desalting oligosaccharides released from glycoproteins. *Glycoconj J* 1998;**15**:737–747.

108 Rouse JC, Vath JE. On-the-probe sample cleanup strategies for glycoprotein-released carbohydrates prior to matrix-assisted laser desorption-ionization time-of-flight mass spectrometry. *Anal Biochem* 1996;**238**:82–92.

109 Börnsen KO, Mohr MD, Widmer HM. Ion exchange and purification of carbohydrates on a Nafion$^{(R)}$ membrane as a new sample pretreatment for

matrix-assisted laser desorption-ionization mass spectrometry. *Rapid Commun Mass Spectrom* 1995;**9**:1031–1034.

110 Dedvisitsakul P, Jacobsen S, Svensson B, Bunkenborg J, Finnie C, Hägglund P. Glycopeptide enrichment using a combination of ZIC-HILIC and cotton wool for exploring the glycoproteome of wheat flour albumins. *J Proteome Res* 2014;**13**:2696–2703.

111 Nishimura S-I, Niikura K, Kurogochi M, Matsushita T, Fumoto M, Hinou H, Kamitani R, Nakagawa H, Deguchi K, Miura N, Monde K, Kondo H. High-throughput protein glycomics: combined use of chemoselective glycoblotting and MALDI-TOF/TOF mass spectrometry. *Angew Chem Int Ed Engl* 2005;**44**:91–96.

112 Bigge JC, Patel TP, Bruce JA, Goulding PN, Charles SM, Parekh RB. Nonselective and efficient fluorescent labeling of glycans using 2-aminobenzamide and anthranilic acid. *Anal Biochem* 1995;**230**:229–238.

113 Anumula KR. Quantitative monosaccharide analysis of glycoproteins as anthranilyl derivatives by reverse phase HPLC. *Glycobiology* 1993;**3**:511.

114 Anumula KR. Unique anthranilic acid chemistry facilitates profiling and characterization of Ser/Thr-linked sugar chains following hydrazinolysis. *Anal Biochem* 2008;**373**:104–111.

115 Anumula KR. Single tag for total carbohydrate analysis. *Anal Biochem* 2014;**457**:31–37.

116 Hase S, Ibuki T, Ikenaka T. Reexamination of the pyridylamination used for fluorescence labelling of oligosaccharides and its application to glycoproteins. *J Biochem* 1984;**95**:197–203.

117 Hase S. Analysis of sugar chains by pyridylamination. *Methods Mol Biol* 1993;**14**:69–80.

118 Okamoto M, Takahashi K, Doi T, Takimoto Y. High-sensitivity detection and postsource decay of 2-aminopyridine-derivatized oligosaccharides with matrix-assisted laser desorption-ionization mass spectrometry. *Anal Chem* 1997;**69**:2919–2926.

119 Harvey DJ. *N*-[2-Diethylamino)ethyl-4-aminobenzamide derivatives for high sensitivity mass spectrometric detection and structure determination of N-linked carbohydrates. *Rapid Commun Mass Spectrom* 2000;**14**:862–871.

120 Poulter L, Burlingame AL. Desorption mass spectrometry of oligosaccharides coupled with hydrophobic chromophores. *Methods Enzymol* 1990;**193**:661–689.

121 Takao T, Tambara Y, Nakamura A, Yoshino K-I, Fukuda H, Fukuda M, Shimonishi Y. Sensitive analysis of oligosaccharides derivatised with 4-aminobenzoic acid 2-(diethylamino)ethyl ester by matrix-assisted laser desorption/ionization mass spectrometry. *Rapid Commun Mass Spectrom* 1996;**10**:637–640.

122 Naven TJP, Harvey DJ. Cationic derivatization of oligosaccharides with Girard's T reagent for improved performance in matrix-assisted laser

desorption/ionization and electrospray mass spectrometry. *Rapid Commun Mass Spectrom* 1996;**10**:829–834.

123 Li M, Kinzer JA. Structural analysis of oligosaccharides by a combination of electrospray mass spectrometry and bromine isotope tagging of reducing-end sugars with 2-amino-5-bromopyridine. *Rapid Commun Mass Spectrom* 2003;**17**:1462–1466.

124 Harvey DJ. Halogeno-substituted 2-aminobenzoic acid derivatives for negative ion fragmentation studies of N-linked carbohydrates. *Rapid Commun Mass Spectrom* 2005;**19**:397–400.

125 Suzuki S, Fujimori T, Yodoshi M. Recovery of free oligosaccharides from derivatives labeled by reductive amination. *Anal Biochem* 2006;**354**:94–103.

126 Küster B, Naven TJP, Harvey DJ. Effect of the reducing-terminal substituents on the high energy collision-induced dissociation matrix-assisted laser desorption/ionization mass spectra of oligosaccharides. *Rapid Commun Mass Spectrom* 1996;**10**:1645–1651.

127 Lattova E, Perreault H. Labelling saccharides with phenylhydrazine for electrospray and matrix-assisted laser desorption-ionization mass spectrometry. *J Chromatogr B* 2003;**793**:167–179.

128 Lattova E, Perreault H. Profiling of N-linked oligosaccharides using phenylhydrazine derivatization and mass spectrometry. *J Chromatogr A* 2003;**1016**:71–87.

129 Lattová E, Perreault H. The usefulness of hydrazine derivatives for mass spectrometric analysis of carbohydrates. *Mass Spectrom Rev* 2013;**32**:366–385.

130 Lattová E, Kapková P, Krokhin O, Perreault H. Method for investigation of oligosaccharides from glycopeptides: Direct determination of glycosylation sites in proteins. *Anal Chem* 2006;**78**:2977–2984.

131 Shinohara Y, Furukawa J, Niikura K, Miura N, Nishimura S-I. Direct N-glycan profiling in the presence of tryptic peptides on MALDI-TOF by controlled ion enhancement and suppression upon glycan-selective derivatization. *Anal Chem* 2004;**76**:6989–6997.

132 Gouw JW, Burgers PC, Trikoupis MA, Terlouw JK. Derivatization of small oligosaccharides prior to analysis by matrix-assisted laser desorption/ionization using glycidyltrimethylammonium chloride and Girard's reagent T. *Rapid Commun Mass Spectrom* 2002;**16**:905–912.

133 Gil G-C, Kim Y-G, Kim B-G. A relative and absolute quantification of neutral N-linked oligosaccharides using modification with carboxymethyl trimethylammonium hydrazide and matrix-assisted laser desorption/ionization time-of-flight mass spectrometry. *Anal Biochem* 2008;**379**:45–59.

134 Rajesh T, Jeon J-M, Song E, Park H-M, Seo HM, Kim H-J, Yi D-H, Kim Y-H, Choi K-Y, Kim Y-G, Park H-Y, Lee YK, Yang Y-H. Putative role of a *Streptomyces coelicolor*-derived α-mannosidase in deglycosylation and antibiotic production. *Appl Biochem Biotechnol* 2014;**172**:1639–1651.

135 Chen P, Novotny MV. 2-Methyl-3-oxo-4-phenyl-2,3-dihydrofuran-2-yl acetate: a fluorogenic reagent for detection and analysis of primary amines. *Anal Chem* 1997;**69**:2806–2811.

136 Harvey DJ. Derivatization of carbohydrates for analysis by chromatography, electrophoresis and mass spectrometry. *J Chromatogr B* 2011;**879**:1196–1225.

137 Lamari FN, Kuhn R, Karamanos NK. Derivatization of carbohydrates for chromatographic, electrophoretic and mass spectrometric structure analysis. *J Chromatogr B* 2003;**793**:15–36.

138 Ruiz-Matute AI, Hernández-Hernández O, Rodríguez-Sánchez S, Sanz ML, Martínez-Castro I. Derivatization of carbohydrates for GC and GC-MS analyses. *J Chromatogr B* 2011;**879**:1226–1240.

139 Bern M, Brito AE, Pang P-C, Rekhi A, Dell A, Haslam SM. Polylactosaminoglycan glycomics: Enhancing the detection of high-molecular-weight N-glycans in matrix-assisted laser desorption ionization time-of-flight profiles by matched filtering. *Mol Cell Proteomics* 2013;**12**:996–1004.

140 Hakomori S. A rapid permethylation of glycolipid, and polysaccharide catalysed by methylsulfinyl carbanion in dimethyl sulfoxide. *J Biochem* 1964;**55**:205–208.

141 Levery SB. Use of permethylation with GC/MS for linkage and sequence analysis of oligosaccharides: Historical perspectives and recent developments. In: Large DG, Warren CD, editors. *Glycopeptides and Related Compounds: Synthesis, Analysis and Applications*. New York: Marcel Dekker Inc.; 1997. p 541–592.

142 Ciucanu I, Kerek F. A simple and rapid method for the permethylation of carbohydrates. *Carbohydr Res* 1984;**131**:209–217.

143 Ciucanu I, Costello CE. Elimination of oxidative degradation during the per-O-methylation of carbohydrates. *J Am Chem Soc* 2003;**125**:16213–16219.

144 Weiskopf AS, Vouros P, Harvey DJ. Characterization of oligosaccharide composition and structure by quadrupole ion trap mass spectrometry. *Rapid Commun Mass Spectrom* 1997;**11**:1493–1504.

145 Kang P, Mechref Y, Klouckova I, Novotny MV. Solid-phase permethylation of glycans for mass spectrometric analysis. *Rapid Commun Mass Spectrom* 2005;**19**:3421–3428.

146 Mechref Y, Kang P, Novotny MV. Solid-phase permethylation for glycomic analysis. *Methods Mol Biol* 2009;**534**:53–64.

147 Kang P, Mechref Y, Novotny MV. High-throughput solid-phase permethylation of glycans prior to mass spectrometry. *Rapid Commun Mass Spectrom* 2008;**22**:721–734.

148 Robinson S, Routledge A, Thomas-Oates J. Characterisation and proposed origin of mass spectrometric ions observed 30 Th above the ionised molecules of per-O-methylated carbohydrates. *Rapid Commun Mass Spectrom* 2005;**19**:3681–3688.

149 Jay A. The methylation reaction in carbohydrate analysis. *J Carbohydr Chem* 1996;**15**:897–923.

150 Reinhold VN, Reinhold BB, Costello CE. Carbohydrate molecular weight profiling, sequence, linkage and branching data: ES-MS and CID. *Anal Chem* 1995;**67**:1772–1784.

151 Morelle W, Slomianny MC, Diemer H, Schaeffer C, Dorsselaer AV, Michalski JC. Fragmentation characteristics of permethylated oligosaccharides using a matrix-assisted laser desorption/ionization two-stage time-of-flight (TOF/ TOF) tandem mass spectrometer. *Rapid Commun Mass Spectrom* 2004;**18**:2637–2649.

152 Yu SY, Wu SW, Khoo KH. Distinctive characteristics of MALDI-Q/TOF and TOF/TOF tandem mass spectrometry for sequencing of permethylated complex type N-glycans. *Glycoconj J* 2006;**23**:355–369.

153 Björndal H, Lindberg B, Svensson S. Mass spectrometry of partially methylated alditol acetates. *Carbohydr Res* 1967;**5**:433–440.

154 Hellerqvist CG, Lindberg B, Svensson S, Holme T, Lindberg AA. Structural studies on the O-specific side-chains of the cell wall lipopolysaccharide from *Salmonella typhimurium* 395 ms. *Carbohydr Res* 1968;**8**:43–55.

155 Lönngren J, Svensson S. Mass spectrometry in structural analysis of natural carbohydrates. *Adv Carbohydr Chem Biochem* 1974;**29**:41–106.

156 Björndal H, Hellerqvist CG, Lindberg B, Svensson S. Gas-liquid chromatography and mass spectrometry in methylation analysis of polysaccharides. *Angew Chem Int Ed Engl* 1970;**9**:610–619.

157 Lindberg B. Methylation analysis of polysaccharides. *Methods Enzymol* 1972;**28**:178–195.

158 Lindberg B, Lönngren J. Methylation analysis of complex carbohydrates: General procedure and application for sequence analysis. *Methods Enzymol* 1978;**50**:3–33.

159 Carpita NC, Shea EM. Linkage structure of carbohydrates by gas chromatography-mass spectrometry (GC-MS) of partially methylated alditol acetates. In: Biermann CJ, McGinnis GD, editors. *Analysis of Carbohydrates by GLC and MS*. Boca Raton: CRC Press; 1989. p 157–216.

160 Hellerqvist CG. Linkage analysis using Lindberg method. *Methods Enzymol* 1990;**193**:554–573.

161 Hanisch FG. Methylation analysis of complex carbohydrates: overview and critical comments. *Biol Mass Spectrom* 1994;**23**:309–312.

162 Mizuochi T, Yonemasu K, Yamashita K, Kobata A. The asparagine-linked sugar chains of subcomponent Clq of the first component of human complement. *J Biol Chem* 1978;**253**:7404–7409.

163 Sutton CW, O'Neill JA, Cottrell JS. Site-specific characterization of glycoprotein carbohydrates by exoglycosidase digestion and laser desorption mass spectrometry. *Anal Biochem* 1994;**218**:34–46.

164 Royle L, Campbell MP, Radcliffe CM, White DM, Harvey DJ, Abrahams JL, Kim Y-G, Henry GW, Shadick NA, Weinblatt ME, Lee DM, Rudd PM, Dwek RA. HPLC-based analysis of serum N-glycans on a 96-well plate platform with dedicated database software. *Anal Biochem* 2008;**376**:1–12.

165 Dwek RA, Edge CJ, Harvey DJ, Wormald MR, Parekh RB. Analysis of glycoprotein-associated oligosaccharides. *Annu Rev Biochem* 1993;**62**:65–100.

166 Mechref Y, Novotny MV. Mass spectrometric mapping and sequencing of N-linked oligosaccharides derived from submicrogram amounts of glycoproteins. *Anal Chem* 1998;**70**:455–463.

167 Küster B, Naven TJP, Harvey DJ. Rapid approach for sequencing neutral oligosaccharides by exoglycosidase digestion and matrix-assisted laser desorption/ionization time-of-flight mass spectrometry. *J Mass Spectrom* 1996;**31**:1131–1140.

168 Colangelo J, Orlando R. On-target exoglycosidase digestions, MALDI-MS for determining the primary structures of carbohydrate chains. *Anal Chem* 1999;**71**:1479–1482.

169 Schmitt S, Glebe D, Alving K, Tolle TK, Linder M, Geyer H, Linder D, Peter-Katalinic J, Gerlich WH, Geyer R. Analysis of the pre-S2 N- and O-linked glycans of the M surface protein from human hepatitis B virus. *J Biol Chem* 1999;**274**:11945–11957.

170 Geyer H, Schmitt S, Wuhrer M, Geyer R. Structural analysis of glycoconjugates by on-target enzymatic digestion and MALDI-TOF-MS. *Anal Chem* 1999;**71**:476–482.

171 Wuhrer M, Deelder AM, Hokke CH. Protein glycosylation analysis by liquid chromatography–mass spectrometry. *J Chromatogr B* 2005;**825**:124–133.

172 Novotny MV, Mechref Y. New hyphenated methodologies in high-sensitivity glycoprotein analysis. *J Sep Sci* 2005;**28**:1956–1968.

173 Mechref Y, Novotny MV. Miniaturized separation techniques in glycomic investigations. *J Chromatogr B* 2006;**841**:65–78.

174 Royle L, Mattu TS, Hart E, Langridge JI, Merry AH, Murphy N, Harvey DJ, Dwek RA, Rudd PM. An analytical and structural database provides a strategy for sequencing O-glycans from microgram quantities of glycoproteins. *Anal Biochem* 2002;**304**:70–90.

175 Morelle W, Page A, Michalski J-C. Electrospray ionization ion trap mass spectrometry for structural characterization of oligosaccharides derivatized with 2-aminobenzamide. *Rapid Commun Mass Spectrom* 2005;**19**:1145–1158.

176 Karlsson J, Momcilovic D, Wittgren B, Schülein M, Tjerneld F, Brinkmalm G. Enzymatic degradation of carboxymethyl cellulose hydrolyzed by the endoglucanases Cel5A, Cel7B, and Cel45A from *Humicola insolens* and Cel7B, Cel12A and Cel45A core from *Trichoderma reesei*. *Biopolymers* 2002;**63**:32–40.

177 Friedl CH, Lochnit G, Zähringer U, Bahr U, Geyer R. Structural elucidation of zwitterionic carbohydrates derived from glycosphingolipids of the porcine parasitic nematode *Ascaris suum. Biochem J* 2003;**369**:89–102.

178 Karlsson NG, Wilson NL, Wirth H-J, Dawes P, Joshi H, Packer NH. Negative ion graphitised carbon nano-liquid chromatography/mass spectrometry increases sensitivity for glycoprotein oligosaccharide analysis. *Rapid Commun Mass Spectrom* 2004;**18**:2282–2292.

179 Liu Y, Urgaonkar S, Verkade JG, Armstrong DW. Separation and characterization of underivatized oligosaccharides using liquid chromatography and liquid chromatography-electrospray ionization mass spectrometry. *J Chromatogr A* 2005;**1079**:146–152.

180 Alpert AJ, Shukla M, Shukla AK, Zieske LR, Yuen SW, Ferguson MA, Mehlert A, Pauly M, Orlando R. Hydrophilic-interaction chromatography of complex carbohydrates. *J Chromatogr A* 1994;**676**:191–202.

181 Zauner G, Deelder AM, Wuhrer M. Recent advances in hydrophilic interaction liquid chromatography (HILIC) for structural glycomics. *Electrophoresis* 2011;**32**:3456–3466.

182 Froesch M, Bindila LM, Baykut G, Allen M, Peter-Katalinic J, Zamfir AD. Coupling of fully automated chip electrospray to Fourier transform ion cyclotron resonance mass spectrometry for high-performance glycoscreening and sequencing. *Rapid Commun Mass Spectrom* 2004;**18**:3084–3092.

183 Zamfir A, Vakhrushev S, Sterling A, Niebel HJ, Allen M, Peter-Katalinic J. Fully automated chip-based mass spectrometry for complex carbohydrate system analysis. *Anal Chem* 2004;**76**:2046–2054.

184 Zhang S, Chelius D. Characterization of protein glycosylation using chip-based infusion nanoelectrospray linear ion trap. *J Biomol Tech* 2004;**15**:120–133.

185 Karlsson H, Hansson GC. Gas chromatography and gas chromatography/ mass spectrometry for the characterization of complex mixtures of large oligosaccharides. *J High Resolut Chromatogr* 1988;**11**:820–824.

186 Hansson GC, Li YT, Karlsson H. Characterization of glycosphingolipid mixtures with up to ten sugars by gas chromatography and gas chromatography-mass spectrometry as permethylated oligosaccharides and ceramides released by ceramide glycanase. *Biochemistry* 1989;**28**:6672–6678.

187 Karlsson H, Carlstedt I, Hansson GC. The use of gas chromatography and gas chromatography - mass spectrometry for the characterisation of permethylated oligosaccharides with molecular mass up to 2300. *Anal Biochem* 1989;**182**:438–446.

188 Hansson GC, Karlsson H. High-mass gas chromatography-mass spectrometry of permethylated oligosaccharides. *Methods Enzymol* 1990;**193**:733–738.

189 Hansson GC, Karlsson H. Gas chromatography and gas chromatography-mass spectrometry of glycoprotein oligosaccharides. In: Hounsell EF, editor. *Methods Molec. Biol.* Totowa: Humana Press; 1993. p 47–54.

190 Karlsson H, Karlsson N, Hansson GC. High-temperature gas chromatography-mass spectrometry of glycoprotein and glycosphingolipid oligosaccharides. *Mol Biotechnol* 1994;**1**:165–180.

191 Barber M, Bordoli RS, Sedgwick RD, Tyler AN. Fast atom bombardment of solids (FAB): A new ion source for mass spectrometry. *Chem Commun* 1981:325–327.

192 Dell A. FAB Mass spectrometry of carbohydrates. *Adv Carbohydr Chem Biochem* 1987;**45**:19–72.

193 Dell A, Carman NH, Tiller PR, Thomas-Oates JE. Fast atom bombardment mass spectrometric strategies for characterising carbohydrate-containing biopolymers. *Biomed Environ Mass Spectrom* 1987;**16**:19–24.

194 Dell A, Thomas-Oates JE. Fast atom bombardment-mass spectrometry (FAB-MS): Sample preparation and analytical strategies. In: Biermann CJ, McGinnis GD, editors. *Analysis of Carbohydrates by GLC and MS*. Boca Raton: CRC Press; 1989. p 217–235.

195 Dell A, Morris HR. Glycoprotein structure determination by mass spectrometry. *Science* 2001;**291**:2351–2356.

196 Karas M, Bachmann D, Bahr U, Hillenkamp F. Matrix-assisted ultraviolet laser desorption of non-volatile compounds. *Int J Mass Spectrom Ion Processes* 1987;**78**:53–68.

197 Mock KK, Davy M, Cottrell JS. The analysis of underivatised oligosaccharides by matrix-assisted laser desorption mass spectrometry. *Biochem Biophys Res Commun* 1991;**177**:644–651.

198 Harvey DJ. Matrix-assisted laser desorption/ionization mass spectrometry of carbohydrates. *Mass Spectrom Rev* 1999;**18**:349–451.

199 Harvey DJ. Analysis of carbohydrates and glycoconjugates by matrix-assisted laser desorption/ionization mass spectrometry: An update covering the period 1999-2000. *Mass Spectrom Rev* 2006;**25**:595–662.

200 Harvey DJ. Analysis of carbohydrates and glycoconjugates by matrix-assisted laser desorption/ionization mass spectrometry: An update covering the period 2001-2002. *Mass Spectrom Rev* 2008;**27**:125–201.

201 Harvey DJ. Analysis of carbohydrates and glycoconjugates by matrix-assisted laser desorption/ionization mass spectrometry: An update for 2003-2004. *Mass Spectrom Rev* 2009;**28**:273–361.

202 Harvey DJ. Analysis of carbohydrates and glycoconjugates by matrix-assisted laser desorption/ionization mass spectrometry: An update for the period 2005-2006. *Mass Spectrom Rev* 2011;**30**:1–100.

203 Harvey DJ. Analysis of carbohydrates and glycoconjugates by matrix-assisted laser desorption/ionization mass spectrometry: An update for 2007-2008. *Mass Spectrom Rev* 2012;**31**:183–311.

204 Harvey DJ. Analysis of carbohydrates and glycoconjugates by matrix-assisted laser desorption/ionization mass spectrometry: An update for 2009-2010. *Mass Spectrom Rev* 2015;**34**:268–422.

205 Strupat K, Karas M, Hillenkamp F. 2,5-Dihydroxybenzoic acid: a new matrix for laser desorption-ionization mass spectrometry. *Int J Mass Spectrom Ion Processes* 1991;**111**:89–102.

206 Kussmann M, Nordhoff E, Rehbek-Nielsen H, Haebel S, Rossel-Larsen M, Jakobsen L, Gobom J, Mirgorodskaya E, Kroll-Kristensen A, Palm L, Roepstorff P. Matrix-assisted laser desorption/ionization mass spectrometry sample preparation techniques designed for various peptide and protein analytes. *J Mass Spectrom* 1997;**32**:593–601.

207 Karas M, Ehring H, Nordhoff E, Stahl B, Strupat K, Hillenkamp F, Grehl M, Krebs B. Matrix-assisted laser desorption/ionization mass spectrometry with additives to 2,5-dihydroxybenzoic acid. *Org Mass Spectrom* 1993;**28**:1476–1481.

208 Mohr MD, Börnsen KO, Widmer HM. Matrix-assisted laser desorption/ ionization mass spectrometry: Improved matrix for oligosaccharides. *Rapid Commun Mass Spectrom* 1995;**9**:809–814.

209 Gusev AI, Wilkinson WR, Proctor A, Hercules DM. Improvement of signal reproducibility and matrix/comatrix effects in MALDI analysis. *Anal Chem* 1995;**67**:1034–1041.

210 Mechref Y, Novotny MV. Matrix-assisted laser desorption/ionization mass spectrometry of acidic glycoconjugates facilitated by the use of spermine as a co-matrix. *J Am Soc Mass Spectrom* 1998;**9**:1292–1302.

211 Harvey DJ. Quantitative aspects of the matrix-assisted laser desorption mass spectrometry of complex oligosaccharides. *Rapid Commun Mass Spectrom* 1993;**7**:614–619.

212 Tsarbopoulos A, Bahr U, Pramanik BN, Karas M. Glycoprotein analysis by delayed extraction and post-source decay MALDI-TOF-MS. *Int J Mass Spectrom Ion Processes* 1997;**169/170**:251–261.

213 Harvey DJ, Hunter AP, Bateman RH, Brown J, Critchley G. The relationship between in-source and post-source fragment ions in the MALDI mass spectra of carbohydrates recorded with reflectron-TOF mass spectrometers. *Int J Mass Spectrom Ion Processes* 1999;**188**:131–146.

214 Powell AK, Harvey DJ. Stabilisation of sialic acids in N-linked oligosaccharides and gangliosides for analysis by positive ion matrix-assisted laser desorption-ionization mass spectrometry. *Rapid Commun Mass Spectrom* 1996;**10**:1027–1032.

215 Mechref Y, Kang P, Novotny MV. Differentiating structural isomers of sialylated glycans by matrix-assisted laser desorption/ionization time-of-flight/time-of-flight tandem mass spectrometry. *Rapid Commun Mass Spectrom* 2006;**20**:1381–1389.

216 Sekiya S, Wada Y, Tanaka K. Derivatization for stabilizing sialic acids in MALDI-MS. *Anal Chem* 2005;**77**:4962–4968.

217 Zhang Q, Feng X, Li H, Liu B-F, Lin Y, Liu X. Methylamidation for isomeric profiling of sialylated glycans by nanoLC-MS. *Anal Chem* 2014;**86**:7913–7919.

218 Ueki M, Yamaguchi M. Analysis of acidic carbohydrates as their quaternary ammonium or phosphonium salts by matrix-assisted laser desorption/ionization mass spectrometry. *Carbohydr Res* 2005;**340**:1722–1731.

219 Wheeler SF, Domann P, Harvey DJ. Derivatization of sialic acids for stabilization in matrix-assisted laser desorption/ionization mass spectrometry and concomitant differentiation of α(2-3) and α(2-6) isomers. *Rapid Commun Mass Spectrom* 2009;**23**:303–312.

220 Papac DI, Wong A, Jones AJS. Analysis of acidic oligosaccharides and glycopeptides by matrix assisted laser desorption/ionization time-of-flight mass spectrometry. *Anal Chem* 1996;**68**:3215–3223.

221 Schulz E, Karas M, Rosu F, Gabelica V. Influence of the matrix on analyte fragmentation in atmospheric pressure MALDI. *J Am Soc Mass Spectrom* 2006;**17**:1005–1013.

222 Von Seggern CE, Moyer SC, Cotter RJ. Liquid infrared atmospheric pressure matrix-assisted laser desorption/ionization ion trap mass spectrometry of sialylated carbohydrates. *Anal Chem* 2003;**75**:3212–3218.

223 Von Seggern CE, Zarek PE, Cotter RJ. Fragmentation of sialylated carbohydrates using infrared atmospheric pressure MALDI ion trap mass spectrometry from cation-doped liquid matrixes. *Anal Chem* 2003;**75**:6523–6530.

224 Tan PV, Taranenko NI, Laiko VV, Yakshin MA, Prasad CR, Doroshenko VM. Mass spectrometry of N-linked oligosaccharides using atmospheric pressure infrared laser ionization from solution. *J Mass Spectrom* 2004;**39**:913–921.

225 Von Seggern CE, Gardner BD, Cotter RJ. Infrared atmospheric pressure MALDI ion trap mass spectrometry of frozen samples using a Peltier-cooled sample stage. *Anal Chem* 2004;**76**:5887–5893.

226 Moyer SC, Marzilli LA, Woods AS, Laiko VV, Doroshenko VM, Cotter RJ. Atmospheric pressure matrix-assisted laser desorption/ionization (AP MALDI) on a quadrupole ion trap mass spectrometer. *Int J Mass Spectrom* 2003;**226**:133–150.

227 Zhang J, LaMotte L, Dodds ED, Lebrilla CB. Atmospheric pressure MALDI Fourier transform mass spectrometry of labile oligosaccharides. *Anal Chem* 2005;**77**:4429–4438.

228 Dell A, Morris HR, Greer F, Redfern JM, Rogers ME, Wisshaar G, Hiyama J, Renwick AGC. Fast-atom-bombardment mass spectrometry of sulphated oligosaccharides from ovine lutropin. *Carbohydr Res* 1991;**209**:33–50.

229 Harvey DJ, Bousfield GR. Differentiation between sulphated and phosphated carbohydrates in low-resolution matrix-assisted laser desorption/ionization mass spectra. *Rapid Commun Mass Spectrom* 2005;**19**:287–288.

230 Irungu J, Dalpathado DS, Go EP, Jiang H, Ha H-V, Bousfield GR, Desaire H. Method for characterizing sulfated glycoproteins in a glycosylation site-specific fashion, using ion pairing and tandem mass spectrometry. *Anal Chem* 2006;**78**:1181–1190.

231 Zhang Y, Jiang H, Go EP, Desaire H. Distinguishing phosphorylation and sulfation in carbohydrates and glycoproteins using ion-pairing and mass spectrometry. *J Am Soc Mass Spectrom* 2006;**17**:1282–1288.

232 Takashiba M, Chiba Y, Jigami Y. Identification of phosphorylation sites in N-linked glycans by matrix-assisted laser desorption/ionization time-of-flight mass spectrometry. *Anal Chem* 2006;**78**:5208–5213.

233 Lei M, Mechref Y, Novotny MV. Structural analysis of sulfated glycans by sequential double-permethylation using methyl iodide and deuteromethyl iodide. *J Am Soc Mass Spectrom* 2009;**20**:1660–1671.

234 Stahl B, Steup M, Karas M, Hillenkamp F. Analysis of neutral oligosaccharides by matrix-assisted laser desorption/ionization mass spectrometry. *Anal Chem* 1991;**63**:1463–1466.

235 Nonami H, Tanaka K, Fukuyama Y, Erra-Balsells R. β-Carboline alkaloids as matrices for UV-Matrix-assisted laser desorption/ionization time-of-flight mass spectrometry in positive and negative ion modes. Analysis of proteins of high molecular mass, and of cyclic and acyclic oligosaccharides. *Rapid Commun Mass Spectrom* 1998;**12**:285–296.

236 Yamagaki T, Suzuki H, Tachibana K. In-source and postsource decay in negative-ion matrix-assisted laser desorption/ionization time-of-flight mass spectrometry of neutral oligosaccharides. *Anal Chem* 2005;**77**:1701–1707.

237 Domann P, Spencer DIR, Harvey DJ. Production and fragmentation of negative ions from neutral N-linked carbohydrates ionized by matrix-assisted laser desorption/ionization. *Rapid Commun Mass Spectrom* 2012;**26**:469–479.

238 Cai Y, Jiang Y, Cole RB. Anionic adducts of oligosaccharides by matrix-assisted laser desorption/ionization time-of-flight mass spectrometry. *Anal Chem* 2003;**75**:1638–1644.

239 Becher J, Muck A, Mithöfer A, Svatoš A, Boland W. Negative ion mode matrix-assisted laser desorption/ionisation time-of-flight mass spectrometric analysis of oligosaccharides using halide adducts and 9-aminoacridine matrix. *Rapid Commun Mass Spectrom* 2008;**22**:1153–1158.

240 Tholey A, Heinzle E. Ionic (liquid) matrices for matrix-assisted laser desorption/ionization mass spectrometry-applications and perspectives. *Anal Bioanal Chem* 2006;**386**:24–37.

241 Mank M, Stahl B, Boehm G. 2,5-Dihydroxybenzoic acid butylamine and other ionic liquid matrixes for enhanced MALDI-MS analysis of biomolecules. *Anal Chem* 2004;**76**:2938–2950.

242 Naven TJP, Harvey DJ. Effect of structure on the signal strength of oligosaccharides in matrix-assisted laser desorption/ionization mass spectrometry on time-of-flight and magnetic sector instruments. *Rapid Commun Mass Spectrom* 1996;**10**:1361–1366.

243 Siemiatkoski J, Lyubarskaya Y, Houde D, Tep S, Mhatre R. A comparison of three techniques for quantitative carbohydrate analysis used in characterization of therapeutic antibodies. *Carbohydr Res* 2006;**341**:410–419.

244 Harvey DJ. Collision-induced fragmentation of underivatised N-linked carbohydrates ionized by electrospray. *J Mass Spectrom* 2000;**35**:1178–1190.

245 Harvey DJ, Bateman RH, Green MR. High-energy collision-induced fragmentation of complex oligosaccharides ionized by matrix-assisted laser desorption/ionization mass spectrometry. *J Mass Spectrom* 1997;**32**:167–187.

246 Harvey DJ, Bateman RH, Bordoli RS, Tyldesley R. Ionization and fragmentation of complex glycans with a Q-TOF mass spectrometer fitted with a MALDI ion source. *Rapid Commun Mass Spectrom* 2000;**14**:2135–2142.

247 Harvey DJ. Electrospray mass spectrometry and collision-induced fragmentation of 2-aminobenzamide-labelled neutral N-linked glycans. *Analyst* 2000;**125**:609–617.

248 Harvey DJ. Electrospray mass spectrometry and fragmentation of N-linked carbohydrates derivatised at the reducing terminus. *J Am Soc Mass Spectrom* 2000;**11**:900–915.

249 Wong AW, Cancilla MT, Voss LR, Lebrilla CB. Anion dopant for oligosaccharides in matrix-assisted laser desorption/ionization mass spectrometry. *Anal Chem* 1999;**71**:205–211.

250 Cole RB, Zhu J. Chloride ion attachment in negative ion electrospray ionization mass spectrometry. *Rapid Commun Mass Spectrom* 1999;**13**:607–611.

251 Wong AW, Wang H, Lebrilla CB. Selection of anionic dopant for quantifying desialylation reactions with MALDI-FTMS. *Anal Chem* 2000;**72**:1419–1425.

252 Zhu J, Cole RB. Formation and decomposition of chloride adduct ions, $[M + Cl]^-$, in negative ion electrospray ionization mass spectrometry. *J Am Soc Mass Spectrom* 2000;**11**:932–941.

253 Cai Y, Concha MC, Murray JS, Cole RB. Evaluation of the role of multiple hydrogen bonding in offering stability to negative ion adducts in electrospray mass spectrometry. *J Am Soc Mass Spectrom* 2002;**13**:1360–1369.

254 Harvey DJ. Fragmentation of negative ions from carbohydrates: Part 1; Use of nitrate and other anionic adducts for the production of negative ion electrospray spectra from N-linked carbohydrates. *J Am Soc Mass Spectrom* 2005;**16**:622–630.

255 Bahr U, Pfenninger A, Karas M, Stahl B. High sensitivity analysis of neutral underivatized oligosaccharides by nanoelectrospray mass spectrometry. *Anal Chem* 1997;**69**:4530–4535.

256 Spengler B, Kirsch D, Kaufmann R, Lemoine J. Structure analysis of branched oligosaccharides using post-source decay in matrix-assisted laser desorption/ionization mass spectrometry. *J Mass Spectrom* 1995;**30**:782–787.

257 Shevchenko A, Loboda A, Shevchenko A, Ens W, Standing KG. MALDI Quadrupole time-of-flight mass spectrometry: a powerful tool for proteomic research. *Anal Chem* 2000;**72**:2132–2141.

258 Verhaert P, Uttenweiler-Joseph S, de Vries M, Loboda A, Ens W, Standing KG. Matrix-assisted laser desorption/ionization quadrupole time-of-flight mass spectrometry: An elegant tool for peptidomics. *Proteomics* 2001;**1**:118–131.

259 Loboda AV, Krutchinsky AN, Bromirski M, Ens W, Standing KG. A quadrupole/time-of-flight mass spectrometer with a matrix-assisted laser desorption/ionization source: design and performance. *Rapid Commun Mass Spectrom* 2000;**14**:1047–1057.

260 Spina E, Sturiale L, Romeo D, Impallomeni G, Garozzo D, Waidelich D, Glueckmann M. New fragmentation mechanisms in matrix-assisted laser desorption/ionization time-of-flight/time-of-flight tandem mass spectrometry of carbohydrates. *Rapid Commun Mass Spectrom* 2004;**18**:392–398.

261 Domon B, Costello CE. A systematic nomenclature for carbohydrate fragmentations in FAB-MS/MS spectra of glycoconjugates. *Glycoconj J* 1988;**5**:397–409.

262 Harvey DJ, Martin RL, Jackson KA, Sutton CW. Fragmentation of N-linked glycans with a MALDI-ion trap time-of-flight mass spectrometer. *Rapid Commun Mass Spectrom* 2004;**18**:2997–3007.

263 Huberty MC, Vath JE, Yu W, Martin SA. Site-specific carbohydrate identification in recombinant proteins using MALD-TOF MS. *Anal Chem* 1993;**65**:2791–2800.

264 Kaufmann R, Chaurand P, Kirsch D, Spengler B. Post-source decay and delayed extraction in matrix-assisted laser desorption/ionization-reflectron time-of-flight mass spectrometry. Are there trade-offs? *Rapid Commun Mass Spectrom* 1996;**10**:1199–1208.

265 Clayton E, Bateman RH. Time-of-flight mass analysis of high-energy collision-induced dissociation fragment ions. *Rapid Commun Mass Spectrom* 1992;**6**:719–720.

266 Morelle W, Slomianny M-C, Diemer H, Schaeffer C, van Dorsselaer A, Michalski J-C. Structural characterization of 2-aminobenzamide-derivatized oligosaccharides using a matrix-assisted laser desorption/ionization two-stage time-of-flight tandem mass spectrometer. *Rapid Commun Mass Spectrom* 2005;**19**:2075–2084.

267 Conboy JJ, Henion JD. The determination of glycopeptides by liquid chromatography/mass spectrometry with collision-induced dissociation. *JxAm Soc Mass Spectrom* 1992;**3**:804–814.

268 Huddleston MJ, Bean MF, Carr SA. Collisional fragmentation of glycopeptides by electrospray ionization LC/MS and LC/MS/MS - Methods for selective detection of glycopeptides in protein digests. *Anal Chem* 1993;**65**:877–884.

269 Wuhrer M, Catalina MI, Deelder AM, Hokke CH. Glycoproteomics based on tandem mass spectrometry of glycopeptides. *J Chromatogr B* 2007;**849**:115–128.

270 Kolli V, Dodds ED. Energy-resolved collision-induced dissociation pathways of model N-linked glycopeptides: implications for capturing glycan

connectivity and peptide sequence in a single experiment. *Analyst* 2014;**139**:2144–2153.

271 Han L, Costello CE. Electron transfer dissociation of milk oligosaccharides. *J Am Soc Mass Spectrom* 2011;**22**:997–1013.

272 Xie Y, Lebrilla CB. Infrared multiphoton dissociation of alkali metal-coordinated oligosaccharides. *Anal Chem* 2003;**75**:1590–1598.

273 Lancaster KS, An HJ, Li B, Lebrilla CB. Interrogation of N-linked oligosaccharides using infrared multiphoton dissociation in FT-ICR mass spectrometry. *Anal Chem* 2006;**78**:4990–4997.

274 Adamson JT, Håkansson K. Infrared multiphoton dissociation and electron capture dissociation of high-mannose type glycopeptides. *J Proteome Res* 2006;**5**:493–501.

275 Zhang J, Schubothe K, Li B, Russell S, Lebrilla CB. Infrared multiphoton dissociation of O-linked mucin-type oligosaccharides. *Anal Chem* 2005;**77**:208–214.

276 Weiskopf AS, Vouros P, Harvey DJ. Electrospray ionization-ion trap mass spectrometry for structural analysis of complex N-linked glycoprotein oligosaccharides. *Anal Chem* 1998;**70**:4441–4447.

277 Ashline D, Singh S, Hanneman A, Reinhold V. Congruent strategies for carbohydrate sequencing. 1. Mining structural details by MS^n. *Anal Chem* 2005;**77**:6250–6262.

278 Ashline DJ, Lapadula AJ, Liu Y-H, Lin ML, Grace M, Pramanik B, Reinhold VN. Carbohydrate structural isomers analyzed by sequential mass spectrometry. *Anal Chem* 2007;**79**:3830–3842.

279 Ojima N, Masuda K, Tanaka K, Nishimura O. Analysis of neutral oligosaccharides for structural characterization by matrix-assisted laser desorption/ionization quadrupole ion trap time-of-flight mass spectrometry. *J Mass Spectrom* 2005;**40**:380–388.

280 Kurimoto A, Daikoku S, Mutsuga S, Kanie O. Analysis of energy-resolved mass spectra at MS^n in a pursuit to characterize structural isomers of oligosaccharides. *Anal Chem* 2006;**78**:3461–3466.

281 Deguchi K, Ito H, Takegawa Y, Shinji N, Nakagawa H, Nishimura SI. Complementary structural information of positive- and negative-ion MSn spectra of glycopeptides with neutral and sialylated N-glycans. *Rapid Commun Mass Spectrom* 2006;**20**:741–746.

282 Takemori N, Komori N, Matsumoto H. Highly sensitive multistage mass spectrometry enables small-scale analysis of protein glycosylation from two-dimensional polyacrylamide gels. *Electrophoresis* 2006;**27**:1394–1406.

283 Wuhrer M, Deelder AM. Matrix-assisted laser desorption/ionization in-source decay combined with tandem time-of-flight mass spectrometry of permethylated oligosaccharides: targeted characterization of specific parts of the glycan structure. *Rapid Commun Mass Spectrom* 2006;**20**:943–951.

284 Kovácik V, Hirsch J, Kovác P, Heerma W, Thomas-Oates J, Haverkamp J. Oligosaccharide characterization using collision-induced dissociation fast atom bombardment mass spectrometry: Evidence for internal monosaccharide residue loss. *J Mass Spectrom* 1995;**30**:949–958.

285 Brüll LP, Heerma W, Thomas-Oates J, Haverkamp J, Kovácik V, Kovác P. Loss of internal 1-6 substituted monosaccharide residues from underivatized and per-O-methylated trisaccharides. *J Am Soc Mass Spectrom* 1997;**8**:43–49.

286 Warrack BM, Hail ME, Triolo A, Animati F, Seraglia R, Traldi P. Observation of internal monosaccharide losses in the collisionally activated dissociation mass spectra of anthracycline aminodisaccharides. *J Am Soc Mass Spectrom* 1998;**9**:710–715.

287 Mattu TS, Royle L, Langridge J, Wormald MR, Van den Steen PE, Van Damme J, Opdenakker G, Harvey DJ, Dwek RA, Rudd PM. O-glycan analysis of natural human neutrophil gelatinase B using a combination of normal phase- HPLC and online tandem mass spectrometry: Implications for the domain organization of the enzyme. *Biochemistry* 2000;**39**:15695–15704.

288 Wuhrer M, Koeleman CA, Hokke CH, Deelder AM. Mass spectrometry of proton adducts of fucosylated N-glycans: fucose transfer between antennae gives rise to misleading fragments. *Rapid Commun Mass Spectrom* 2006;**20**:1747–1754.

289 Franz AH, Lebrilla CB. Evidence for long-range glycosyl transfer reactions in the gas phase. *J Am Soc Mass Spectrom* 2002;**13**:325–337.

290 Harvey DJ, Mattu TS, Wormald MR, Royle L, Dwek RA, Rudd PM. "Internal residue loss": rearrangements occurring during the fragmentation of carbohydrates derivatized at the reducing terminus. *Anal Chem* 2002;**74**:734–740.

291 Brüll LP, Kovácik V, Thomas-Oates JE, Heerma W, Haverkamp J. Sodium-cationized oligosaccharides do not appear to undergo 'internal residue loss' rearrangement processes on tandem mass spectrometry. *Rapid Commun Mass Spectrom* 1998;**12**:1520–1532.

292 Ngoka LC, Gal J-F, Lebrilla CB. Effects of cations and charge types on the metastable decay rates of oligosaccharides. *Anal Chem* 1994;**66**:692–698.

293 Cancilla MT, Penn SG, Carroll JA, Lebrilla CB. Coordination of alkali metals to oligosaccharides dictates fragmentation behavior in matrix assisted laser desorption ionization/Fourier transform mass spectrometry. *J Am Chem Soc* 1996;**118**:6736–6745.

294 Orlando R, Bush CA, Fenselau C. Structural analysis of oligosaccharides by tandem mass spectrometry: Collisional activation of sodium adduct ions. *Biomed Environ Mass Spectrom* 1990;**19**:747–754.

295 Mechref Y, Novotny MV, Krishnan C. Structural characterization of oligosaccharides using MALDI-TOF/TOF tandem mass spectrometry. *Anal Chem* 2003;**75**:4895–4903.

296 Stephens E, Maslen SL, Green LG, Williams DH. Fragmentation characteristics of neutral N-linked glycans using a MALDI-TOF/TOF tandem mass spectrometer. *Anal Chem* 2004;**76**:2343–2354.

297 Wuhrer M, Deelder AM. Negative-mode MALDI-TOF/TOF-MS of oligosaccharides labeled with 2-aminobenzamide. *Anal Chem* 2005;**77**:6954–6959.

298 Lewandrowski U, Resemann A, Sickmann A. Laser-induced dissociation/high-energy collision-induced dissociation fragmentation using MALDI-TOF/TOF-MS instrumentation for the analysis of neutral and acidic oligosaccharides. *Anal Chem* 2005;**77**:3274–3283.

299 Devakumar A, Thompson MS, Reilly JP. Fragmentation of oligosaccharide ions with 157 nm vacuum ultraviolet light. *Rapid Commun Mass Spectrom* 2005;**19**:2313–2320.

300 Kurogochi M, Nishimura S-I. Structural characterization of *N*-glycopeptides by matrix-dependent selective fragmentation of MALDI-TOF/TOF tandem mass spectrometry. *Anal Chem* 2004;**76**:6097–6101.

301 Boutreau L, Léon E, Salpin J-Y, Amekraz B, Moulin C, Tortajada J. Gas-phase reactivity of silver and copper coordinated monosaccharide cations studied by electrospray ionization and tandem mass spectrometry. *Eur J Mass Spectrom* 2003;**9**:377–390.

302 Harvey DJ. Ionization and fragmentation of N-linked glycans as silver adducts by electrospray mass spectrometry. *Rapid Commun Mass Spectrom* 2005;**19**:484–492.

303 Harvey DJ. Ionization and collision-induced fragmentation of N-linked and related carbohydrates using divalent cations. *J Am Soc Mass Spectrom* 2001;**12**:926–937.

304 Harvey DJ. Fragmentation of negative ions from carbohydrates: Part 2, Fragmentation of high-mannose N-linked glycans. *J Am Soc Mass Spectrom* 2005;**16**:631–646.

305 Harvey DJ. Fragmentation of negative ions from carbohydrates: Part 3, Fragmentation of hybrid and complex N-linked glycans. *J Am Soc Mass Spectrom* 2005;**16**:647–659.

306 Harvey DJ, Royle L, Radcliffe CM, Rudd PM, Dwek RA. Structural and quantitative analysis of N-linked glycans by MALDI and negative ion nanospray mass spectrometry. *Anal Biochem* 2008;**376**:44–60.

307 Cai Y, Cole RB. Stabilization of anionic adducts in negative ion electrospray mass spectrometry. *Anal Chem* 2002;**74**:985–991.

308 Jiang Y, Cole RB. Oligosaccharide analysis using anion attachment in negative mode electrospray mass spectrometry. *J Am Soc Mass Spectrom* 2005;**16**:60–70.

309 Yamagaki T, Suzuki H, Tachibana K. Semiquantitative analysis of isomeric oligosaccharides by negative-ion mode UV-MALDI TOF postsource decay mass spectrometry and their fragmentation mechanism study at *N*-acetyl hexosamine moiety. *J Mass Spectrom* 2006;**41**:454–462.

310 Takegawa Y, Deguchi K, Ito S, Yoshioka S, Nakagawa H, Nishimura S-I. Structural assignment of isomeric 2-aminopyridine-derivatized oligosaccharides using negative-ion MS^n spectral matching. *Rapid Commun Mass Spectrom* 2005;**19**:937–946.

311 Chai W, Piskarev V, Lawson AM. Negative-ion electrospray mass spectrometry of neutral underivatized oligosaccharides. *Anal Chem* 2001;**73**:651–657.

312 Chai W, Piskarev V, Lawson AM. Branching pattern and sequence analysis of underivatized oligosaccharides by combined MS/MS of singly and doubly charged molecular ions in negative-ion electrospray mass spectrometry. *J Am Soc Mass Spectrom* 2002;**13**:670–679.

313 Sagi D, Peter-Katalinic J, Conradt HS, Nimtz M. Sequencing of tri- and tetraantennary N-glycans containing sialic acid by negative mode ESI QTOF tandem MS. *J Am Soc Mass Spectrom* 2002;**13**:1138–1148.

314 Wheeler SF, Harvey DJ. Negative ion mass spectrometry of sialylated carbohydrates: Discrimination of *N*-acetylneuraminic acid linkages by matrix-assisted laser desorption/ionization-time-of-flight and electrospray-time-of-flight mass spectrometry. *Anal Chem* 2000;**72**:5027–5039.

315 Harvey DJ, Crispin M, Scanlan C, Singer BB, Lucka L, Chang VT, Radcliffe CM, Thobhani S, Yuen C-T, Rudd PM. Differentiation between isomeric triantennary N-linked glycans by negative ion tandem mass spectrometry and confirmation of glycans containing galactose attached to the bisecting (β1-4-GlcNAc) residue in N-glycans from IgG. *Rapid Commun Mass Spectrom* 2008;**22**:1047–1052.

316 Harvey DJ. Collision-induced fragmentation of negative ions from N-linked glycans derivatized with 2-aminobenzoic acid. *J Mass Spectrom* 2005;**40**:642–653.

317 Harvey DJ, Rudd PM. Fragmentation of negative ions from N-linked carbohydrates. Part 5: Anionic N-linked glycans. *Int J Mass Spectrom* 2011;**305**:120–130.

318 Karlsson NG, Schulz BL, Packer NH. Structural determination of neutral O-linked oligosaccharide alditols by negative ion LC-electrospray-MS^n. *J Am Soc Mass Spectrom* 2004;**15**:659–672.

319 Bohrer BC, Merenbloom SI, Koeniger SL, Hilderbrand AE, Clemmer DE. Biomolecule analysis by ion mobility spectrometry. *Annu Rev Anal Chem* 2008;**1**:293–327.

320 Kanu AB, Dwivedi P, Tam M, Matz L, Hill HHJ. Ion mobility-mass spectrometry. *J Mass Spectrom* 2008;**43**:1–22.

321 Fenn LS, McLean JA. Structural resolution of carbohydrate positional and structural isomers based on gas-phase ion mobility-mass spectrometry. *Phys Chem Chem Phys* 2011;**13**:2196–2205.

322 Pagel K, Harvey DJ. Ion mobility mass spectrometry of complex carbohydrates - collision cross sections of sodiated N-linked glycans. *Anal Chem* 2013;**85**:5138–5145.

323 Hofmann J, Struwe WB, Scarff CA, Scrivens JH, Harvey DJ, Pagel K. Estimating collision cross sections of negatively charged N-glycans using travelling wave ion mobility-mass spectrometry. *Anal Chem* 2014;**86**:10789–10795.

324 Huang Y, Gelb SA, Dodds ED. Carbohydrate and glycoconjugate analysis by ion mobility mass spectrometry: Opportunities and challenges. *Curr Metabolomics* 2013;**1**:291–305.

325 Plasencia MD, Isailovic D, Merenbloom SI, Mechref Y, Clemmer DE. Resolving and assigning N-linked glycan structural isomers from ovalbumin by IMS-MS. *J Am Soc Mass Spectrom* 2008;**19**:1706–1715.

326 Williams JP, Grabenauer M, Carpenter CJ, Holland RJ, Wormald MR, Giles K, Harvey DJ, Bateman RH, Scrivens JH, Bowers MT. Characterization of simple isomeric oligosaccharides and the rapid separation of glycan mixtures by ion mobility mass spectrometry. *Int J Mass Spectrom* 2010;**298**:119–127.

327 Harvey DJ, Edgeworth M, Krishna BA, Bonomelli C, Allman S, Crispin M, Scrivens JH. Fragmentation of negative ions from N-linked carbohydrates: Part 6: Glycans containing one *N*-acetylglucosamine in the core. *Rapid Commun Mass Spectrom* 2014;**28**:2008–2018.

328 Isailovic D, Plasencia MD, Gaye MM, Stokes ST, Kurulugama RT, Pungpapong V, Zhang M, Kyselova Z, Goldman R, Mechref Y, Novotny MV, Clemmer DE. Delineating diseases by IMS-MS profiling of serum N-linked glycans. *J Proteome Res* 2012;**11**:576–585.

329 Fenn LS, Kliman M, Mahsut A, Zhao SR, McLean JA. Characterizing ion mobility-mass spectrometry conformation space for the analysis of complex biological samples. *Anal Bioanal Chem* 2009;**394**:235–244.

330 Fenn LS, McLean JA. Biomolecular structural separations by ion mobility–mass spectrometry. *Anal Bioanal Chem* 2008;**391**:905–909.

331 Fenn LS, McLean JA. Simultaneous glycoproteomics on the basis of structure using ion mobility-mass spectrometry. *Mol Biosyst* 2009;**5**:1298–1302.

332 Harvey DJ, Sobott F, Crispin M, Wrobel A, Bonomelli C, Vasiljevic S, Scanlan CN, Scarff C, Thalassinos K, Scrivens JH. Ion mobility mass spectrometry for extracting spectra of N-glycans directly from incubation mixtures following glycan release: Application to glycans from engineered glycoforms of intact, folded HIV gp120. *J Am Soc Mass Spectrom* 2011;**22**:568–581.

333 Harvey DJ, Scarff CA, Crispin M, Scanlan CN, Bonomelli C, Scrivens JH. MALDI-MS/MS with traveling wave ion mobility for the structural analysis of N-linked glycans. *J Am Soc Mass Spectrom* 2012;**23**:1955–1966.

334 Harvey DJ, Scarff CA, Edgeworth M, Crispin M, Scanlan CN, Sobott F, Allman S, Baruah K, Pritchard L, Scrivens JH. Travelling wave ion mobility and negative ion fragmentation for the structural determination of N-linked glycans. *Electrophoresis* 2013;**34**:2368–2378.

335 Crispin M, Harvey DJ, Bitto D, Halldorsson S, Bonomelli C, Edgeworth M, Scrivens JH, Huiskonen JT, Bowden TA. Uukuniemi phlebovirus assembly

and secretion leave a functional imprint on the virion glycome. *J Virol* 2014;**88**:10244–10251.

336 Bitto D, Harvey DJ, Halldorsson S, Doores KJ, Huiskonen JT, Bowden TA, Crispin M. Determination of N-linked glycosylation in UUKV glycoproteins by negative ion mass spectrometry and ion mobility. *Methods Mol Biol* 2015;**1331**:93–121.

337 Ahn YH, Kim JY, Yoo JS. Quantitative mass spectrometric analysis of glycoproteins combined with enrichment methods. *Mass Spectrom Rev* 2015;**34**:148–165.

338 Goldman R, Sanda M. Targeted methods for quantitative analysis of protein glycosylation. *Proteomics Clin Appl* 2015;**9**:17–32.

339 Kang P, Mechref Y, Kyselova Z, Goetz JA, Novotny MV. Comparative glycomic mapping through quantitative permethylation and stable-isotope labeling. *Anal Chem* 2007;**79**:6064–6073.

340 Atwood JA III, Cheng L, Alvarez-Manilla G, Warren NL, York WS, Orlando R. Quantitation by isobaric labeling: Applications to glycomics. *J Proteome Res* 2008;**7**:367–374.

341 Prien JM, Prater BD, Qin Q, Cockrill SL. Mass spectrometric-based stable isotopic 2-aminobenzoic acid glycan mapping for rapid glycan screening of biotherapeutics. *Anal Chem* 2010;**82**:1498–1508.

342 Ridlova G, Mortimer JC, Maslen SL, Dupree P, Stephens E. Oligosaccharide relative quantitation using isotope tagging and normal-phase liquid chromatography/mass spectrometry. *Rapid Commun Mass Spectrom* 2008;**22**:2723–2730.

343 Hashii N, Kawasaki N, Itoh S, Nakajima Y, Kawanishi T, Yamaguchi T. Alteration of N-glycosylation in the kidney in a mouse model of systemic lupus erythematosus: relative quantification of N-glycans using an isotope-tagging method. *Immunology* 2009;**126**:336–345.

344 Ceroni A, Maass K, Geyer H, Geyer R, Dell A, Haslam SM. GlycoWorkbench: A tool for the computer-assisted annotation of mass spectra of glycans. *J Proteome Res* 2008;**7**:1650–1659.

345 Cooper CA, Gasteiger E, Packer NH. GlycoMod - A software tool for determining glycosylation compositions from mass spectrometric data. *Proteomics* 2001;**1**:340–349.

346 Deshpande N, Jensen PH, Packer NH, Kolarich D. GlycoSpectrumScan: Fishing glycopeptides from MS spectra of protease digests of human colostrum sIgA. *J Proteome Res* 2010;**9**:1063–1075.

347 Go EP, Rebecchi KR, Dalpathado DS, Bandu ML, Zhang Y, Desaire H. GlycoPep DB: A tool for glycopeptide analysis using a "Smart Search". *Anal Chem* 2007;**79**:1708–1713.

348 Goldberg D, Sutton-Smith M, Paulson J, Dell A. Automatic annotation of matrix-assisted laser desorption/ionization N-glycan spectra. *Proteomics* 2005;**5**:865–875.

349 Aoki K, Yamaguchi A, Ueda N, Akutsu T, Mamitsuka H, Goto S, Kanehisa M. KCaM (KEGG Carbohydrate Matcher): a software tool for analyzing the structures of carbohydrate sugar chains. *Nucleic Acids Res* 2004;**32**:W267–W272.

350 Hashimoto K, Goto S, Kawano S, Aoki-Kinoshita KF, Ueda N, Hamajima M, Kawasaki T, Kanehisa M. KEGG as a glycome informatics resource. *Glycobiology* 2006;**16**:63R–70R.

351 Egorova KS, Toukach PV. Critical analysis of CCSD data quality. *J Chem Inf Model* 2012;**52**:2812–2814.

352 Yamada K, Kakehi K. Recent advances in the analysis of carbohydrates for biomedical use. *J Pharm Biomed Anal* 2011;**55**:702–727.

353 Artemenko NV, McDonald AG, Davey GP, Rudd PM. Databases and tools in glycobiology. *Methods Mol Biol* 2012;**899**:325–350.

354 Lütteke T. The use of glycoinformatics in glycochemistry. *Beilstein J Org Chem* 2012;**8**:915–929.

355 Mazola Y, Chinea G, Musacchio A. Integrating bioinformatics tools to handle glycosylation. *PLoS Comput Biol* 2011;**7**:e1002285.

356 Aoki-Kinoshita KF. Using databases and web resources for glycomics research. *Mol Cell Proteomics* 2013;**12**:1036–1045.

357 Hizal DB, Wolozny D, Colao J, Jacobson E, Tian Y, Krag SS, Betenbaugh MJ, Zhang H. Glycoproteomic and glycomic databases. *Clin Proteomics* 2014;**11**:Article 15.

358 Morelle W, Michalski JC. Analysis of protein glycosylation by mass spectrometry. *Nat Protoc* 2007;**2**:1585–1602.

359 Ruhaak LR, Huhn C, Koeleman CAM, Deelder AM, Wuhrer M. Robust and high-throughput sample preparation for (semi-)quantitative analysis of N-glycosylation profiles from plasma samples. *Methods Mol Biol* 2012;**893**:371–385.

360 Wang H, Wong C-H, Chin A, Taguchi A, Taylor A, Hanash S, Sekiya S, Takahashi H, Murase M, Kajihara S, Iwamoto S, Tanaka K. Integrated mass spectrometry-based analysis of plasma glycoproteins and their glycan modifications. *Nat Protoc* 2011;**6**:253–269.

361 Wuhrer M, Koeleman CAM, Deelder AM, Hokke CH. Normal-phase nanoscale liquid chromatography-mass spectrometry of underivatized oligosaccharides at low-femtomole sensitivity. *Anal Chem* 2004;**76**:833–838.

362 Wuhrer M, Koeleman CAM, Hokke CH, Deelder AM. Nano-scale liquid chromatography-mass spectrometry of 2-aminobenzamide-labeled oligosaccharides at low femtomole sensitivity. *Int J Mass Spectrom* 2004;**232**:51–57.

4

Protein Acetylation and Methylation

Caroline Evans

Department of Chemical and Biological Engineering, University of Sheffield, Sheffield, UK

4.1   Overview of Protein Acetylation and Methylation

Protein acetylation and methylation are post-translational modifications (PTMs) occurring predominantly on arginine (methylation) and lysine (acetylation and methylation) residues. Acetylation can also occur at the N-terminus and acts as multifunctional regulator including regulation of protein stability, via a specific degradation signal, termed the Ac/N-degron [1]. Originally considered epigenetic modifications that occurred mainly or exclusively on histones, it has become clear that acetylation and methylation also occur on a range of nonhistone proteins [2, 3] with modification rates that are subject to metabolic control [4]. As such these PTMs can be considered key regulatory events in cell function. Acetylation and methylation, as for other PTMs, can occur singly or in combination providing mechanisms for regulating protein activity through generating functional variants and regulatory sites. MS analysis of acetylation and methylation is directed not only to site identification and localization but also characterization of their combinations [5, 6].

4.1.1   Protein Acetylation

Acetylated lysine (Kac) is generated via the addition of an acetyl group to the free ε amine on the lysine side chain. Lysine acetyltransferase enzymes mediate the transfer of the acetyl group, utilizing acetyl-CoA as the donor molecule. Lysine deacetylase enzymes catalyze removal of the acetyl group. These enzymes were originally termed histone deacetylases and histone acetyltransferases but are generally abbreviated to HDAC/KDAC or HAT/KAT since they also catalyze the addition and removal of acetyl groups from nonhistone proteins. The presence of such enzymatic regulation has led researchers to pose

Analysis of Protein Post-Translational Modifications by Mass Spectrometry,
First Edition. Edited by John R. Griffiths and Richard D. Unwin.
© 2017 John Wiley & Sons, Inc. Published 2017 by John Wiley & Sons, Inc.

the question of "Acetylation: a regulatory modification to rival phosphorylation?" discussing the functional significance of lysine acetylation [7]. This was due to the identification of lysine acetylation as an enzyme-mediated dynamic process (analogous to kinase/phosphatase) with acetylation sites providing specific recruitment sites via conserved interaction domains. Importantly too, the question was sparked by the recent discovery of novel, nonhistone acetylated proteins that represented a wide range of cellular proteins, including transcription factors, nuclear import factors, and the cytoskeletal protein, α-tubulin. Prior to this, acetylation had been considered to be a histone modification, co-occurring with protein methylation at multiple sites and with other PTMs to provide "marks" for epigenetic (heritable) regulation. Termed the histone code, PTM marks provide a mechanism for transcriptional regulation by determining dynamic transitions between transcriptionally active or transcriptionally silent chromatin states [8]. Today, lysine acetylation is established as a conserved mechanism across species, central to cellular regulation, with acetyl-CoA, the donor of the acetyl group, proposed as a key node or indicator of the cell metabolic state via alterations in protein acetylation patterns, including of metabolic enzymes mediating their activity [2, 4]. Lysine acetylation can also occur via a nonenzymatic acetyl phosphate-dependent mechanism and is proposed to occur when carbon flux exceeds the capacity of the central metabolic pathways, providing a means to regulate carbon flow through central metabolic pathways [4, 9].

Protein acetylation additionally occurs on the N-terminus of proteins. N-terminal acetylation occurs after the removal of initiator methionine residues, with the acetyl moiety transferred from acetyl-CoA to the α-amino group by the action of N-terminal transferase enzyme. Unlike lysine acetylation, N-terminal acetylation is an irreversible modification. It occurs at predominantly the cotranslational level but can also occur as a PTM [10].

4.1.2 Protein Methylation

Protein methylation was identified as a PTM in the 1980s; conversion of *S*-adenosyl-L-methionine (SAM) to *S*-adenosyl-L-homocysteine mediates transfer of methyl groups to a protein, a reaction catalyzed by a family of protein methyltransferases and subject to metabolic control [4]. Methylated lysine exists in mono-, di-, and trimethylated forms. Methylated arginine exists in mono- and dimethylated forms [11, 12]. Trimethyl-arginine has not been identified to date. Methylation of arginine results in the addition of methyl groups to the guanidine nitrogens to form monomethyl-arginine and dimethyl-arginine. Symmetric and asymmetric dimethyl-arginines are both naturally occurring forms, resulting from the addition of two methyl groups to one or two guanidino nitrogen atoms, respectively. Although the extent to which demethylation occurs has been debated, it is clearly accepted that lysine methylation status is enzyme mediated with the addition and removal of methylation catalyzed by lysine methyltransferases (KMTs) and demethylases (KDMs), two enzyme superfamilies. For arginine methylation, a key finding in this area is the identification of reversible arginine methylation of TRAF6 as a novel mechanism for dynamic regulation of innate immune pathways [13]. Nonhistone lysine methylation also regulates signaling pathways including p53- and NFkB-mediated signaling [12].

4.1.3 Functional Aspects

In general, proteins associated with signals from acetylated and methylated proteins can be broadly classified into the writer–reader–eraser model. In this scheme, enzymes mediating the addition and removal of the acetyl or methyl groups are "writers" and "erasers." Proteins interacting with the post-translationally modified proteins are "readers" [2, 14]. The primary "readers" of methylation and acetylation marks are Tudor domain- and bromodomain-containing proteins [15, 16]. Acetyl and methyl groups do not themselves confer charge but can modulate charge by removing the positive charge from the ε-amino group of lysine. Addition of these groups also introduces steric hindrance, with the net result of modulating protein–protein and protein–nucleic acid interactions. The biological significance of both protein acetylation and methylation is underlined by the development of specific inhibitors against methyltransferase (lysine and arginine) and deacetylases as "druggable" therapeutic targets for clinical application [17, 18]. Beyond clinical application, these inhibitors also provide biochemical reagents for probing of functional analysis [19]. Acetylation and methylation have combinatorial activity with other PTMs; thus a number of proteoforms/protein can exist. The term proteoform describes biological variability and complexity at the level of protein primary structure, including isoform and PTM variants for the products of a single gene [20].

4.1.4 Mass Spectrometry Analysis

In principle, MS can be applied to the study of acetylation and methylation since these PTMs are chemical modifications resulting in specific elemental mass additions to peptides. Mono-, di-, and trimethylation add +14.0156, +28.0312, and +42.0469 Da, respectively. In terms of data interpretation, monomethylation is isobaric with several amino acid substitutions such as Asp–Asn, Gly–Ala, and Val–Leu/Ile. Acetylation results in the addition of +42.0105 Da, a small mass difference of 0.0363 Da relative to trimethylation but distinguishable by the use of high-resolution MS [21]. Acetylation and methylation are stable modifications relative to the more labile modifications such as phosphorylation and glycosylation [22]. As such they are retained intact during sample preparation and MS analyses, typically operated in positive mode.

N-terminal tails of histone proteins possess a number of acetylation and methylation sites; as such, these are often exemplified for method development/proof of concept of novel PTM approaches for the analysis of post-translationally modified and isomeric peptides, which are discussed in this chapter. Analysis of acetylation and methylation is achieved by using three MS-complementary approaches: bottom up, top down, and middle down [23, 24]. Bottom-up strategies are directed to peptides to provide information to identify and localize sites. This is achieved by fragmentation to generate MS2 (or MS/MS) product ion spectra. PTM assignment is aided by the generation of modification-specific diagnostic fragments. Acetyl- and methyl-specific enrichment strategies are applied for improved sensitivity of detection. Quantification strategies to profile alterations of acetylated and methylated proteins, including analysis of modification dynamics and stoichiometry are also typically, but not exclusively, applied at the peptide level. Intact proteins or larger fragments are analyzed in top-down and middle-down approaches for characterization of complex PTM patterns in individual proteins.

MS analysis benefits from coupling liquid chromatography (LC) instrumentation or other prefractionation methods that reduce peptide complexity. Such methods also provide additional parameters such as LC retention time information to aid analysis. For example, lysine trimethylation and acetylation can also be distinguished by their distinct chromatographic retention times [25]. In general, the co-occurrence of acetylation and methylation sites makes their analysis challenging. In particular, isomeric peptides can result, which possess the same modification(s) at different sites, and thus share physicochemical properties that result in coelution and cofragmentation during LC-MS analysis. Chimeric product ion spectra make PTM site assignment problematic, particularly when no unique fragment ions are detected.

4.2 Mass Spectrometry Behavior of Modified Peptides

4.2.1 MS Fragmentation Modes

Fragmentation of peptides detected in the MS survey scan (MS1) to generate product ion spectra plays an important role in the identification and localization of PTM. The key requirement is to generate sufficient fragment ion information to accurately assign acetylation and/or methylation status and to determine the amino acid sequence of the precursor for site assignment.

Collision-induced dissociation (CID) and increasingly higher-energy collisional activation/C-trap dissociation (HCD) are both used in the large majority of proteomics experiments for peptide identification following proteolytic digestion. CID and HCD fragmentations occur along the peptide backbone at the amide bonds to generate b and y ions, which contain the N- and C-termini, respectively. Amino-immonium (IM) ions, which are unique for the majority of naturally occurring amino acids, except lysine/glutamine or leucine/isoleucine, confirm the presence of specific amino acids. In general, CID and HCD are most effective for low-charged tryptic peptides (typically 9–12 amino acids), which are doubly and triply charged. Trypsin cleavage is highly specific and reproducible, cleaving C-terminal to lysine and arginine residues, except when lysine and arginine have proline C-terminal to their position or are N-linked to aspartic acid. Typically, CID or HCD are used either alone or in combination with electron transfer dissociation (ETD) for PTM-directed analytical workflows [22]. ETD provides an alternative fragmentation strategy to CID and generates complementary sequence information to y- and b-type ions resulting from CID and HCD [26]. ETD cleavage occurs at the Cα–N bond producing c- and z-type ions, with a higher tendency for retention of PTM side-chain groups relative to CID/HCD, a particular benefit to PTM site localization [27]. ETD also has compatibility with higher charge state peptides such as those characteristic of both miscleaved tryptic peptides (which can result from acetylation or methylation) and nontryptic peptides, for example, Arg-C, Glu-C, Asp-N, and chymotrypsin-generated peptides [26]. The use of alternating CID/ETD has been shown to be beneficial for the detection of acetylation- and methylation-modified peptides [28, 29]. The fragmentation of longer peptides and proteins by ETD has also found application for top-down and middle-down proteomics studies (see Section 4.3).

4.2.2 Acetylation- and Methylation-Specific Diagnostic Ions in MS Analysis

The generation of PTM-specific and thus diagnostic ions upon peptide fragmentation is a key feature used for the assignment of acetylated and methylated peptides. PTM diagnostic ions include immonium ions, side-chain fragment ions, and those resulting from neutral loss (NL). Details of acetyl and methyl PTM diagnostic ions, their mass, and fragment type are listed in Table 4.1, together with specific references in which their application was described. These ions can be used to discriminate between acetylation and trimethylation [21, 30] and also between methyl-arginine and methyl-lysine forms [11]. IM ions indicate the presence of acetylation or methylation in a peptide but do not provide information on site localization. Many of the diagnostic masses were first determined from CID spectra in lower resolution instruments in the early 2000s, typically cited to first or no decimal place of mass value. These low-mass diagnostic ions can also be observed in HCD spectra that are measured using high-resolution MS instrumentation [31]. HCD offers an advantage over CID since the beam-type energy deposited during fragmentation improves the generation of both IM and other sequence-related ions [31]. Including information

Table 4.1 PTM diagnostic ions for the assignment of acetylation and methylation status.

PTM	Mass of ion or neutral loss	Type	Mode of fragmentation	References
Acetyl-ε-lysine side chain, *N*-Acetyl	126.091	IM-NH3	CID	[34, 48, 51]
Monomethyl-lysine	98.0964	IM-NH$_3$	CID	[30, 37]
Dimethyl-lysine	112.1	IM	CID	[30, 37, 44]
	45.0578	NL		
Trimethyl-lysine	59.0735	NL	CID	[30, 37, 44]
Monomethyl-arginine	73.064	NL monomethyl-guanidine	ETD	[28, 37, 39, 40, 42, 44]
	31.0422	NL monomethylamine	ETD	
Dimethyl-arginine	46.0651	Side-chain dimethylammonium	CID	[39, 41]
Symmetric > asymmetric dimethyl-arginine	71.0604	Side-chain dimethyl-carbodiimidium	CID	[39, 41]
Symmetric dimethyl-arginine	31.0417	Neutral loss of monomethylamine	CID	[38–40, 42]
	70.0525	Neutral loss of dimethyl-carbodiimidium	CID or ETD	
Asymmetric dimethyl-arginine	45.0573	Neutral loss of dimethylamine	CID or ETD	[38–40, 42]

Notes: Hung et al., 2007 also provide additional resources by listing the accurate masses for modified y1 and dipeptide a and b ions to aid sequence assignment of acetylated and methylated peptides [43]. Zhang et al. provide information on the mechanisms of IM and NL fragment generated from acetylated and methylated peptides [30]. Gehrig et al. also provide mechanistic information on fragmentation of methylated peptides [38].

on the presence and absence of diagnostic neutral losses and IM ions aids software-based PTM assignment and localization [32, 33]. High mass accuracy measurement (<2 ppm) aids sequence and PTM analysis assignment since the chemical composition for the majority of detected MS2 fragment ions can be unambiguously assigned [33]. For lower resolution instruments, the use of heavy isotope–labeled synthetic peptide standards and retention time information enables confident assignment of acetylation and methylation sites [34].

For acetylation, it is the IM–NH$_3$ (m/z, 126.091), rather than the IM (m/z, 143.1179), that is diagnostic [35]. The 143.1179 ion is not unique: isobaric masses can also result from GlyLeu-, LeuGly-, GlyIsoleu-, and IsoleuGly-containing peptides by the formation of either a2 or internal fragment. A specific feature of the Kac diagnostic ion is that the ion intensity is higher for peptides with N-terminal lysine relative to internal lysine ions [35]. In terms of fragmentation, the presence of Kac promotes selective cleavage of Lys–Xxx amide bond to generate more information of the peptide backbone sequence relative to the nonacetylated form. This is due to the presence of b(n)+ ions as the most abundant primary product ions [36]. The site of acetylation can be additionally inferred by the 170 Da mass differences between y and y + 1 ion or corresponding b ions (Figure 4.1). Note that high mass accuracy and resolution discriminate lysine acetylation (Δ170.1056) from lysine trimethylation (Δ170.1420) [21].

Peptides containing different types of methylations produce specific, diagnostic ions and neutral loss fragments [37–41, 44]. The diagnostic mass for monomethyl-lysine immonium-specific ion at m/z 98.1 can be used to discriminate

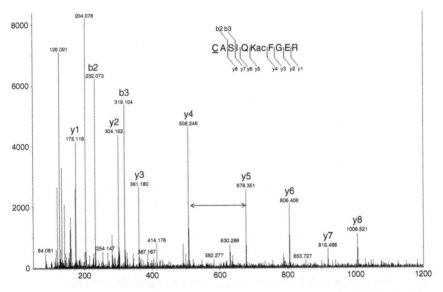

Figure 4.1 Representative spectrum of an acetylated peptide analyzed by TOF MS. Peptide sequence CASIQKacFGER, precursor m/z 619.3, with the acetyl lysine diagnostic IM-NH3 (126.1) and mass difference of 170 Da, which together confirm the presence of acetyl lysine.

monomethyl-lysine from monomethyl-arginine [30, 37]. The symmetric and asymmetric forms of dimethyl-arginine can be discriminated by different neutral loss ions following CID [38, 40, 41] and ETD [28, 42]. Dimethyl- and trimethyl-lysine-containing peptides undergo specific neutral losses from the precursor, MH^+-45 and MH^+-59, respectively [30, 37, 44]. Neutral loss is only observed in ion series containing the modified lysines, information which enables site assignment of the PTM. The m/z values of monopeptide and dipeptide ions of y, a, and b types provide confirmatory data for N- and C-terminal end amino acids [43]. For example, trimethylation can be assigned to the N-terminus of the peptide when the b2-59 ion is detected [30].

A comparative analysis of CID and ETD revealed that ETD was superior for the analysis of protein methylation, particularly since CID spectra are complex and neutral loss-derived product ions can be at low abundance relative to ions resulting from $-H_2O$ losses [45]. Lysine-methylated peptides do not produce significant losses during ETD fragmentation, but despite this ETD has proved particularly useful for the analysis of monomethyl- and dimethyl-lysine, since they form multiply charged peptides [42]. Methylated peptides are often present at substoichiometric levels, and the sensitivity of detection can be enhanced by the use of targeted inclusion lists of theoretical m/z values for methylated peptides for selection and fragmentation – an approach of potential value to other PTMs [45].

Stable isotope labeling of methylated peptides *in vivo* with ($^{13}CD_3$)-methionine enables high-confidence detection of protein methylation sites by MS [39]. This technique, heavy methyl SILAC is a variant of Stable isotope labeling with amino acids in cell culture (SILAC); resulting in a mass difference of 4Da/methyl group, relative to the unlabeled (light) form of the peptide. Following mixing of heavy and light samples, the presence of a 1:1 methyl heavy:light precursor pair in the MS1 scan corroborates the assignment of the fragmentation spectrum to a methylated peptide. The mass difference within the SILAC pairs enables assignment of the number of methyl groups per identified peptide. While of proven utility in the identification of methylated peptides, the method has the drawback that heavy methyl SILAC also generates L-methionine or L-methionine-$^{13}CD_3$ SILAC pairs. A refinement of the method overcomes this limitation by the substitution of L-methionine with isomethionine (L-methionine-$^{13}C_4$). This "iMethyl-SILAC" approach results in near-isobaric methionine peptide pairs, thus iMethyl-SILAC pairs are specific to methylated peptides for improved confidence in the identification of methylated peptides: iMethyl SILAC led to a sixfold reduction in false discovery rate when compared with label-free identification of methylation sites [46].

4.2.3 Application of MS Methodologies for the Analysis of PTM Status

Knowledge of the precursor mass and the diagnostic ion facilitates analytical approaches, both discovery and targeted. The specific pair of m/z values associated with a precursor and product ion (including the PTM diagnostic ion) is termed a "transition," which provides the basis for "reaction monitoring." Selected reaction monitoring (SRM)/multiple reaction monitoring (MRM) are based on iterative precursor isolation and scanning of MS2 transition ions, typically using triple quadrupole MS, where the Q1 and Q3 mass analyzers are used as static mass filters for the selected m/z values of precursor–product transition ions. CID occurs in Q2. MRM-initiated detection and sequencing (MIDAS) is a technique for PTM analysis [47] with application to protein acetylation based on precursor–126.1 transition [48]. This approach is more specific and 10-fold more sensitive than precursor ion scanning for 126.1 ion, where precursor masses are scanned in Q1, with Q3 static for the product ion mass [49, 50]. MIDAS employs a hybrid triple quadrupole/linear ion trap to enable full product ion scan information – triggered by the detection of PTM diagnostic. The MIDAS approach requires *a priori* knowledge of the protein sequence for targeted analysis of the predicted, theoretical precursor–PTM diagnostic ion transitions.

An alternative approach, collision-induced release of acetyl diagnostic (CIRAD) MS method, utilizes the acetylation diagnostic ion for discovery of acetylation in a knowledge-independent manner to assign acetylated precursor m/z [51]. This is achieved by the use of broadband CID, rapid cycling between low and elevated collision energies to generate intact and fragment high-resolution mass spectra (<2 ppm accuracy) from all precursors simultaneously. The presence of the 126.091 m/z fragment ion is monitored and aligned to the precursor m/z using an accurate mass-retention time information-type approach. Processing of the LC-MS raw data with a specific "dissect" extraction algorithm links precursors and product ions (pseudo tandem MS information for database searching) by retention time, traced across the LC elution peak using both the high- and low-energy data. The CIRAD approach complements and informs other MS analysis methods by providing precursor m/z, precursor–product transitions, and retention time information independent of primary sequence or information on PTM status.

4.2.4 Quantification Strategies

In general, quantification can be achieved by two main strategies – use of stable isotope labeling or without, so-called "label-free." The principle of labeling-based quantitative proteomics is the introduction of heavy stable isotopes such as ^{2}H, ^{13}C, and ^{15}N to proteins, which can be discriminated from unlabeled proteins for comparative analysis. The introduction of an isotope label results in a predictable mass difference between a labeled peptide and the unlabeled counterpart, and thus stable isotope labeling provides a means for sample comparison in the same analytical LC-MS run. Stable isotope quantification can be achieved via metabolic or chemical labeling. Stable isotope labeling of amino acids in cell culture (SILAC and related variants) is a widely used form of metabolic labeling, typically employing heavy lysine and arginine for protein labeling. SILAC achieves MS1 level quantification: peptides to be compared exist in heavy (labeled) and light (unlabeled) forms, for the simplest implementation in duplex format. As previously discussed (Section 4.2.2), heavy methyl SILAC and isomethionine (i) methyl SILAC can be used for analysis of protein methylation [39, 46]. In contrast, chemical labeling with isobaric tags provide MS2 level relative quantification. Examples include the isobaric tag for relative and absolute quantification (iTRAQ) and tandem mass tag (TMT) reagents. iTRAQ (4 plex, 8 plex) and TMT (2, 6, and 10 plex) labeled peptide sets coelute to form a single MS1 peak, and quantification of the component peptides is achieved by measurement of distinct reporter ions released upon peptide fragmentation. SILAC, iTRAQ, TMT, and label-free methods are typically used to quantify proteome and protein subsets [52]. Specific enrichment strategies increase the sensitivity of analysis of methylated and acetylated peptide subsets of the proteome (see Section 4.4). Isotope labeling can be applied to determine PTM stoichiometry as described later (see Section 4.2.5.2).

Quantitative MS can also be performed in a label-free manner by comparing peptide intensities in a series of m/z and retention time-aligned LC-MS runs. The approach requires high interrun reproducibility, since each sample is run individually. Label-free methods are based on MS2 or MS1 level intensity information. For MS1-based quantification methods, the area under the curve from the extracted ion chromatogram of the peptide precursor provides a quantitative measure, with MS2 data used for the identification of peptides. MS1 quantification assumes that each precursor is a single species and cannot discriminate isobaric peptides; the availability of MS2 data is very useful where unique products provide a means to discriminate. Spectral counting and intensity-based measurements of peptide fragments at the MS2 level provide a measure of protein abundance, but this MS2 level quantification is not generally applied to PTM peptides [53]. Label-free methods enable sample comparison based on relative or absolute quantification. Label-free methods can be used for discovery and targeted proteomics, while absolute quantification is widely applied to targeted proteomics.

Absolute quantification can be achieved by the inclusion of a synthetic reference (AQUA) peptide, that is, same sequence as peptide of interest with heavy label, for example, ^{13}C, spiked at a known concentration into the sample to be analyzed. The synthetic AQUA peptide and corresponding peptide of interest have the same physicochemical properties and thus coelute and share ionization and fragmentation characteristics but can be distinguished on the basis of (known) mass difference [54].

4.2.4.1 Single Reaction Monitoring/Multiple Reaction Monitoring

SRM is typically performed using triple quadrupole MS, where specificity is achieved by monitoring of precursor–product transition [55]. SRM/MRM utilizes peptides unique to the protein or a post-translationally modified peptide. Best practice assay design principles and implementation have been described [56]. Optimal transitions and retention time information are typically determined from laboratory-acquired data: prior data-dependent acquisition (DDA) or analysis of synthetic peptides can also be designed computationally [57]. SRM/MRM plots typically contain a single peak, from which integrated peak area provides a quantitative measure. Internal reference unmodified peptides from the target protein can be used for signal normalization for label-free quantification. Pseudo-SRM is a variant of this technique, where a target inclusion list directs acquisition of full-scan MS2 spectra for selected precursors. The resulting extracted ion chromatograms are used to generate peak areas for transitions of interest and are used for quantitation [58]. The MRM method has a wide linear dynamic range of up to five orders of magnitude and high precision, making it well suited for the analysis of low-abundance peptides, which typically include post-translationally modified peptides.

For absolute quantification, a selected analyte–AQUA peptide pair is analyzed using two alternating SRM experiments performed during a single LC-SRM analysis. The SRM transition and LC retention time information are determined using synthetic peptides. This method is applicable to post-translationally modified and unmodified peptides [54], including acetylation and methylation as exemplified by a library of 93 reference peptides, which include lysine mono-, di-, and trimethylation; arginine monomethylation; serine/threonine phosphorylation; lysine acetylation; and N-terminal acetylation [59]. Analysis of this reference set demonstrated that detection efficiency did not correlate with molecular weight, hydrophobicity, basicity, or modification type. However, differences in detection efficiencies suggested a positional effect of lysine acetylation within a given peptide sequence [59].

4.2.4.2 Parallel Reaction Monitoring

Parallel reaction monitoring (PRM) represents a key improvement over lower resolution triple quad SRM methods, the specificity of which is limited by coanalysis of near-isobaric ions [55, 60]. Substitution of Q3 of a triple quad with a high resolution and accurate mass (HRAM) analyzer enables the detection of all target product ions in a single scan to monitor all possible transitions in parallel. PRM utilizes extracted ion chromatograms with narrow m/z windows, which increases specificity by resolving interferences to enable confident assignment of precursor to products [61]. Proof of principle of the utility of PRM to map acetylation and methylation sites was demonstrated for histones H3 and H4; novel methyl (mono-, di-, tri-), acetyl, and monomethyl/acetyl dual modifications were assigned based on the presence of diagnostic immonium ions and neutral losses of methylamine following HCD fragmentation. The data were generated with the purpose of establishing novel precursor–product pairs for SRM quantification, particularly for discriminating peptide isoforms since PRM product ion scan information enables the assignment of distinct b and y sequence ions to specific peptide PTM variants [62]. A quantification approach coupling PRM with the use of stable isotope–labeled internal standards enabled dynamic analysis and determination of alterations in specific PTM sites, including downregulation of H3 K_{36} trimethylation in response to vitamin C [63]. This exemplifies two types of PRM applications: screening mode, directed to defining specific peptides of interest [62], and quantification mode, which focuses on the accurate quantification of selected peptides by reference to internal standards [63].

4.2.4.3 Data-Independent Acquisition MS

Data-independent acquisition (DIA) offers the combined benefits of SRM (targeted) and DDA (discovery) MS modes. In this approach, all precursors within a given retention time and m/z window MS1 scan are subjected to a "fragment all" MS2 fragmentation. DDA results in the fragmentation of individual peptides triggered based on an intensity threshold, which occurs on a stochastic basis and thus does not acquire data for all peptides. DIA is distinct from DDA in acquiring all precursor and fragment signals from all peptides across a sample set. As such, DIA provides a comprehensive peptide profile. DIA data can be mined for targeted extraction of both qualitative (PTM diagnostic, isomer-specific ions) and quantitative (intensity) information using DDA-derived spectral libraries containing retention time and fragmentation information for the desired peptide species. DIA spectral libraries can be derived by the alignment of precursor and fragment chromatographic features to form pseudo tandem mass spectra [64]. A major benefit of DIA techniques, in general, is that the data can be interrogated at a later stage, for example, for additional/new PTMs. Quantitative information can be derived at the level of both precursor and product ions, from the MS1 and MS2 scans [65].

DIA mode of acquisition is available on quadrupole-orbitrap (all-ion fragmentation) and quadrupole TOF MS (SWATH, MS^E). All-ion HCD fragmentation is performed on Q Exactive MS instrumentation alternating MS and "all-ion fragmentation" MS2 scans [66]. In MS^E, all ions are transferred to the collision cell for CID, intact peptides are measured in the low-energy scans, whereas fragment ions are measured in the high-energy scans, which cycle throughout the run [67]. In SWATH acquisition, the MS scans through the entire mass range of precursors in user-defined window sizes; all precursor ions in each SWATH window are simultaneously fragmented. The MS1 survey scan for the full mass range (shorter acquisition time relative to DDA) is followed by a SWATH MS2 acquisition series. The mass window is typically ≤25 Da, but these scan accumulation times can be adjusted based on m/z distribution through the LC run to enhance the sensitivity of the analysis, based on MS1 intensity [65]. In general, DIA MS1 precursor ions can be subject to interference, which can be minimized by the reduction of the MS1 window size. MS2 ions are more selective and thus subject to less interference and thus have superior linear response and dynamic range relative to MS1 [65]. A specific benefit of DIA methodology is the ability to resolve and quantify isobaric H3 peptides, for example, the Kac18 and Kac23 peptides [68].

DIA is of particular value for acetylated and/or methylated peptides lacking unique b and y type transition ions. These are refractory to SRM/MRM/PRM since they cannot be discriminated. In DIA the intensity values of precursor and fragments (MS1 and MS2 signals, respectively) can be analyzed to calculate the contributions of individual peptides to the overall abundance of a given set of coeluting isobaric peptides [69]. To do this, the average areas of the chimeric precursor intensities are extracted, and where possible, each peptide of the subset is quantified on the basis of unique sets of b and y ions. Where no unique b or y ions exist, the relative contribution of those peptides is calculated as the proportion of the total remaining. This makes the assumption that ionization efficiencies are similar for the peptide set, which was demonstrated to be appropriate in this study but might not generally translate to other protein PTM peptides. Interestingly, fold change values were strongly correlated between MS1 and MS2 indicating that these values can be used either alone or combined for accurate analysis. Profiling of 62 histone H3 and H4 protein PTMs monitored changes in methylation and acetylation in response to the HDAC inhibitor, suberoylanilide hydroxamic acid (SAHA). DIA acquisition was performed on a Q Exactive MS with MS1 scan in the range 390–910 m/z, and DIA was performed by dividing the m/z range into smaller m/z ranges for sequential isolation (10 m/z units) and fragmentation, using a defined list of target m/z values for specificity [69].

4.2.4.4 Ion Mobility MS

Ion mobility separation (IM/IMS) is of value to the analysis of post-translationally modified peptides, particularly of positional isomers. So-called "conventional IMS" measures the absolute mobility at moderate electric field strength. Field asymmetric waveform IMS (FAIMS, also known as differential mobility separation) is based on the difference between absolute mobility values at high and low electric field strengths. UltraFAIMS and SelexION devices are variants of the FAIMS device. A range of mass analyzers can be coupled to IMS including TOF, linear quadrupole, ion trap, and orbitrap instrumentation.

The introduction of differential IMS, based on gas-phase collisional cross-section (CCS), reflects the ion topology of ions and provides an orthogonal measurement to m/z. The use of IMS provides an additional ion characteristic, termed drift time (order of milliseconds), to combine with retention time, m/z, and intensity information, providing another dimension of separation to extend the resolving power of HPLC by increasing selectivity and peak capacity [70, 71]. The combination of retention and drift time information improves the accuracy of assignment of CID fragments to precursors as demonstrated for TWIMS, by resolving chimeric MS2 spectra of coeluting peptides and reduction of chemical noise. [72, 73]. Additional benefits accrue for improved quantitative performance of isobaric tag-based quantitation by reducing coisolation of mixed precursors of same m/z value relative to DDA [74]. IMS-DDA has been applied to protein acetylation research using TMT quantitative chemoproteomic analysis to identify cellular proteins binding immobilized HDAC inhibitors [75].

The combination of TWIMS with DIA provides high-definition MS^E [72]. The introduction of ETD capability offers complementary data to IMS-CID data for PTM assignment and quantification [76]. IMS post-ETD enables the assignment of fragments to a specific charge-reduced state of the precursor, a particular benefit for top-down analysis of proteins [77]. Furthermore, combination of IMS-CID and ETD-MS generates multistage MS (MS_n) data, which can aid PTM assignment [76]. IMS is an appropriate resolving tool for PTMs where the PTM induces a change in the CCS of the analyte under investigation. FAIMS spectra have been used to resolve methylation and acetylation site localization variants at the bottom-up and middle-down levels [78].

4.2.5   Use of Stable Isotope–Labeled Precursors

4.2.5.1   Dynamics of Acetylation and Methylation

Acetylation and methylation dynamics can be monitored by the use of heavy-labeled precursors, typically ^{13}C labeled for *in vivo* metabolic labeling. For acetylation, the precursor can either be incorporated into acetyl-CoA (^{13}C-labeled glucose or ^{13}C-labeled glutamine) or the substrate, acetate, provided in ^{13}C-labeled form [79]. High-resolution LC-MS analysis discriminates new versus existing "old" acetylation on the basis of ^{13}C incorporation, assessed by isotope distribution patterns of coeluting labeled and unlabeled peptides. An optimization-based model simultaneously considers MS isotopic distribution; MS/MS fragment ions and relative peptide hydrophobicity relationships identify and quantify PTM isoforms for individual peptides. Raw abundance is normalized across all observable labeled states within a specific modified state to control for variable ionization efficiencies. ^{13}C glucose is advantageous as the label source, since it is metabolized to alanine and thus is incorporated into proteins, enabling "new" and "old" assignment of protein pools. The method has been applied to defining the turnover of histone acetylation [79] but in principle can be applied at a cellular level and combined with immunoaffinity Kac enrichment strategies for higher proteome coverage.

Dynamic analysis of protein methylation has been achieved by combining dual stable isotope labeling of protein and methylation sites with heavy arginine $[^{13}C_6]$ and heavy methionine $[^{13}C_{12}H_3]$ [80]. Switching cells into *light medium* enabled the discrimination of "old" from "new," based on the analysis of the amounts of heavy or light components by analysis of SILAC intermediates. This approach was applied to the analysis of histone H3 $K_{27}-K_{36}$ peptide isomers. Monitoring the H3 $K_{27}-K_{36}$ peptide methylation states across time enabled MS-based measurement and modeling of histone methylation kinetics (M4K) to detail bidirectional antagonism between H3K27 and H3K36 as a control for writing and erasing of histone marks [80].

4.2.5.2 Stoichiometry of Acetylation and Methylation

Beyond identification of a peptide or protein as acetylated or methylated and relative quantification across different biological states, assessment of functional significance requires determination of the PTM occupancy of specific sites by the PTM. This requires calculation of the amount of the post-translationally modified peptide relative to the corresponding unmodified peptide. When measuring stoichiometry, it can be assumed that PTM abundance changes are inversely proportional to their unmodified counterpart [81]. This approach, in its simplest form, requires the unmodified and modified peptides to have similar ionization efficiencies for direct comparison based on MS intensity measurements. This in inherently difficult for acetylated and methylated peptides, since these PTMs result in missed tryptic cleavage peptides; thus an unmodified peptide will be shorter relative to the modified form. Either proteolytic digestion with Arg-C or chemical derivatization to promote Arg-C type cleavage generates lysine acetylated and nonacetylated counterpart peptides that are of the same length to provide a useful tool for direct comparison to assess stoichiometry [82, 83] with application at the level of the entire proteome without requirement for specific enrichment strategies [84]

A mass spectrometry method using a combination of isotope labeling, Arg-C digestion, and detection of Kac diagnostic ion at m/z 126.09 to determine the stoichiometry of protein lysine acetylation, applicable to proteome-*wide profiling, has also* been described [82]. This technique complements relative quantification of lysine-acetylated peptides postimmunoaffinity enrichment. Sites of lysine acetylation are discriminated from the unmodified form by chemical acetylation using a stable isotope, ^{13}C form of acetic anhydride [1,1'-[13]C$_2$-acetic anhydride], to label unmodified lysines. The ^{13}C form and preexisting "light" lysine acetylated form coelute during HPLC analysis; cofragmentation enables the calculation of acetylation stoichiometry using the intensities of the heavy and light PTM forms of the diagnostic ions of $\Delta 1.0033$ Da. It should be noted that peptide-level calculations do not provide individual site stoichiometry data for peptides containing multiple lysine residues [82]. This can be mitigated by prior fractionation of peptides but represents an issue for isobaric peptides.

A related technique has applicability to acetylation and (mono)methylated peptides [83]. Chemical derivatization with deuterated (d_6) acetic anhydride prior to tryptic digestion restricts digestion to arginine residues (Arg-C like) and produces a chemically identical set of peptides, which can be discriminated based on isotope distribution of mixed heavy and light peptides for the calculation of stoichiometry. Methylation stoichiometry is determined from relative intensities of the deuteroacetylated and monomethylated peptide compared with the deuteroacetylated but unmethylated analog. High-resolution accurate mass MS2 measurements enable the discrimination of lysine acetylation from lysine trimethylation. Monomethylated but not di- or trimethylated residues are derivatized in this technique. The ability to determine the stoichiometries of trimethylation and acetylation in the same peptide enables the analysis of PTM cross talk.

Methylation stoichiometry can be determined by iMethyl-SILAC labeling using L-Methionine-$^{13}C_4$ for heavy labeling of methylated peptides and comparing relative differences with cells grown in a "light" methionine-containing media. Normalization of changes in levels of methylated peptides to changes in protein expression is used to derive fold changes in site occupancy. This was exemplified for a discovery workflow of arginine methylation peptides identifying 365 unique arginine-methylated peptides before and after stimulation of T cells, corresponding to 319 distinct arginine methylation sites (from a total of 1411) in 202 proteins [45].

4.3 Global Analysis

4.3.1 Top-Down Proteomics

Analysis of intact proteins, termed top-down proteomics, has the potential to provide comprehensive profiling of proteoforms particular since the connectivity of multiple PTM is retained. While bottom-up analysis for identification and relative quantification provides a catalog of post-translationally modified sites, information on PTM combinations at the proteoform level is lost, particularly for sites in different regions of the protein. Intact proteins are amenable to MS, particularly since the occurrence of PTMs has a much smaller effect on the ionization/detection efficiency of intact proteins relative to peptides [85].

A typical top-down workflow employs accurate mass determination of the intact protein at the MS level, followed by primary fragmentation and fragment mass MS2 analysis. Isolation of individual fragments for further MS3 fragmentation builds up amino acid sequence and PTM information [86]. The technique is typically applied to a single protein and requires mass resolution and software for deconvolution of the isotopic peaks to determine the charge states and calculate precursor and product ion masses from complex spectra.

A major technical breakthrough has been the ability to resolve intact proteins to reduce sample complexity for analysis by a top-down approach [87]. Proteins were fractionated by in solution isoelectric focusing (sIEF) followed by gel-eluted liquid fraction electrophoresis GELFrEE. Isoelectric focusing separates post-translationally modified proteoforms since acetylation and methylation act to neutralize charge relative to the unmodified protein. Combining sIEF with molecular weight separation resolves proteoforms, in a manner analogous to 2D gel separation, but instead in liquid phase enabling recovery of intact proteins for resolution by LC-MS analysis. Protein identification was achieved from m/z values of intact mass and matching MS2 fragments from the N- and C-termini [88]. The presence of a PTM on intact proteins can be deduced by observing an increase in the precursor mass and a shift in the masses of the fragment ions containing the PTM. This study generated proof of concept of being able to identify pairs of protein intact mass values with mass differences consistent with mono-, di-, and trimethylation, acetylation in addition to mono- and diphosphorylated proteins in a complex mixture. A total of 3093 proteoforms corresponding to 1043 proteins were identified, for which 538 lysine acetylations and 158 methylations were assigned. The use of SDS in the GELFrEE system-enabled analysis of membrane proteins and proteins up to 80 kDa were evaluated. The method is time consuming, requiring two rounds of offline fractionation and multiple fractions/sample from which SDS must be removed for compatibility with LC-MS analysis, but achieves significant coverage of the proteome for proteome wide monitoring using 0.5–1 mg starting material of cell lysates or mitochondrial fractions [87].

A technically simpler, but high-resolution protein separation employing two-dimensions of HPLC separation, couples hydrophobic interaction chromatography (HIC) to orthogonal RP-HPLC to provide an alternative separation method for top-down proteomic analysis [89]. The use of MS compatible ammonium tartrate buffer for the HIC separation results in separation orthogonal to that of RP-HPLC. The method was evaluated by analysis of mixes of 4 or 10 protein standards and an *Escherichia coli* lysate providing proof of principle of applicability to top-down proteomics. The study was not PTM directed but has potential applicability to the analysis of acetylated and methylated proteoforms.

4.3.2 Middle Down

The middle-down approach analyzes longer peptides (3000–10,000 Da), which are typically generated using nontryptic methods, for example, alternative enzymes such as Glu-C, Asp-N, OmpT, or chemical cleavage. The technique combines the benefits of being a relatively simple workflow compared with top-down analysis with a greater ability to map combinatorial modifications than bottom-up peptide-centric approaches. In terms of fragmentation methods, middle-down approaches have been used with both CID [90] and ETD [91]. ETD is widely employed due to applicability to higher charge states and generation of fragments with sequencing and acetylation and methylation site assignment potential [91]. Step change in method utility comes from implementation of combined ETD-PRM for targeted analysis of Histone H3 acetylation and methylation variants [92].

4.4 Enrichment

Acetylated and methylated peptides, in common with other PTM peptides, are typically present at substoichiometric levels relative to nonmodified peptides. As such, they are generally not detected by MS without specific enrichment. The classical biochemical method of immunoprecipitation has been widely employed. Naturally occurring, PTM-specific binding "reader" domains provide an alternative capture reagent with potential for customization via protein engineering for optimal pan recognition of acetylated peptides or for those within specific amino acid or PTM context [93]. Chemistry-based approaches, such as variant of the biotin switch technique, are unbiased with respect to consensus site motifs for acetylation and as such provide a useful addition to the analytical toolkit [94].

These are described later and can be used in isolation or combined with prior biochemical separations based on alterations in chemical and physical characteristics of the post-translationally modified protein, relative to unmodified forms. These properties include charge, pI, and hydrophobicity. Examples of fractionation techniques include isoelectric focusing [95] and different chromatography separations: basic reversed-phase HPLC [19, 96], HILIC and strong cation exchange (SCX) [97], and WCX/HILIC [25].

4.4.1 Immunoaffinity Enrichment

Immunoaffinity enrichment is based on specific interaction of post-translationally modified site with an antibody (polyclonal or monoclonal) directed to the PTM epitope. Antibodies are immobilized, typically in bead format, to provide bait for capture of the PTM site of interest. Postenrichment, methylated, or acetylated peptides can be identified and characterized by LC-MS/MS. It has become routine to incorporate isotopic labels (e.g., SILAC, iTRAQ) or label-free quantitative approaches to compare PTM profiles across multiple biological conditions. Heavy methyl-SILAC and iMethyl-SILAC labeling strategies aid identification and reduce false positives [46, 98].

Antibodies recognizing methylated arginine in the context of an Arg-Gly motif have been proved effective for enriching modified proteins and peptides [39, 99, 100]. Pan-specific immunoprecipitation using panels of antibodies against lysine methylation [101] and specific antibodies against monomethyl-arginine, asymmetric dimethyl-arginine; monomethyl-, dimethyl-, and trimethyl-lysine motifs have been generated for comprehensive profiling of the "methylome" [46, 100]. Prior to this, there were concerns over the effectiveness of methylation-specific antibodies for pan enrichment as reviewed by Carlson and Gozani [102]. Similarly, a panel of seven antibodies of complementary specificities has been raised for Kac affinity enrichment of peptides [19], which follows a series of landmark studies directed to cataloging of Kac sites. Other studies have focused on immunoprecipitation at the protein level using antibodies against methylated (arginine, lysine) [102] or lysine-acetylated proteins [103].

There is increasing recognition of PTM cross talk, which can be addressed experimentally via a serial enrichment strategy for quantitative analysis of alterations in protein abundance and PTM profile [96]. In this workflow, termed SEPTM, the samples to be compared are labeled with either heavy or light SILAC reagents and trypsin digested to generate peptides. Sequential enrichment of phosphorylated, ubiquitylated, and acetylated peptides is achieved using a series of immobilized capture reagents, with the flow-through from each enrichment step used as the input for the subsequent step. Proof of concept to cataloging of multiple PTM has been generated for human leukemia cells cultured in the presence and absence of the proteasome inhibitor, bortezomib. Fractionation by basic pH reversed-phase HPLC was evaluated and provides partitioning of PTM and unmodified peptides to significantly improve the depth of coverage for Kac peptides relative to unfractionated samples [19, 96]. Interestingly, while the ubiquitinome altered in response to proteasomal inhibition, the acetylome was unaltered, but 414 of 1554 Kac sites occurred on lysine residues that were also identified as sites of ubiquitination – indicating the power of the method for parallel analysis of PTM in the same sample set to provide functional information on mechanisms of PTM interaction [96].

4.4.2 Reader Domain-Based Capture

4.4.2.1 Kac-Specific Capture Reagents

Kacs are recognized and bound *in vivo* by bromodomains, via four α-helices linked by loop regions, which provide a specific binding site. A systematic evaluation of bromodomain efficacy as affinity capture reagents has been undertaken for *Saccharomyces cerevisiae* bromodomains [93]. The different bromodomains were demonstrated to have variable binding specificity, indicating pan lysine capture potential in a manner reflecting the natural diversity of acetylation sites. Proof of principle was demonstrated using immobilized GST-His6-tagged BDF1-B fusion protein for affinity capture followed by MS analysis and found to have enrichment efficiency similar to that of pan-acetyl antibodies. Furthermore, coupling of pairs of bromodomains enhanced capture efficiency as demonstrated for capture of histone lysine-acetylated peptides using both bromodomains of BDF-1 protein [93]. This strategy represents a potential step change in analysis of lysine acetylation, providing reagents that can be engineered to increase their selectivity and affinity toward acetylated sequences [104].

4.4.2.2 Methyl-Specific Capture Reagents

Naturally occurring methyl-lysine-binding domains provide alternative reagents to antibodies for the enrichment of methylated peptides or proteins. This approach is based on the observation that some methyl-lysine-binding domains, such as the triple MBT domains [3×MBT] of the protein L3MBTL1 bind to mono- and dimethylated lysine [105]. The 3×MBT domain exhibits minimal sequence specificity and thus has potential for pan methyl-lysine capture. A protocol for lysine methylome analysis has been developed using a recombinant 3×MBT-glutathione *S*-transferase (GST) anchored to beads functionalized with reduced glutathione [106]. Proteins enriched by 3×MBT can be separated by SDS-PAGE and analyzed by either western blotting or in-gel digestion with trypsin and LC-MS/MS. In this technique, a binding-null, inactive point mutant (D355N) of 3×MBT provides a negative control. SILAC labeling enables quantitative comparison of proteins captured by native 3×MBT and D355N mutant to discriminate specific from nonspecific binding and aid identifications by LC-MS/MS [106].

Immobilization of novel heterochromatin protein 1 β-chromodomain also provides a bait to capture methylated proteins, identifying dimethyl-lysine and lysine trimethylation sites [107]. It is interesting to note that there is no overlap among the methylation sites identified by this study and the 3×MBT domain capture study, which may be related to the use of different proteases prior to enrichment: trypsin [106] and Arg-C, Glu-C, chymotrypsin, and elastase [107].

4.4.3 Biotin Switch-Based Capture

Biotin switch methodology was first described for the study of *S*-nitrosothiols, whereby nitrosylated cysteines are converted to biotinylated cysteines using *N*-hydroxysuccinimide (NHS) chemistry. The biotinylated peptides are purified by avidin affinity chromatography and analyzed by LC-MS [108]. The procedure has been modified for application to detection of endogenously lysine-acetylated peptides, termed the acetyl-biotin switch technique [94]. In this technique, NHS forms a stable amide bond with free (N-terminal, lysine-ε) amines, resulting in a mixture of peptides with either NHS-blocked lysines or preexisting Kac. Lysine acetylation is removed *in vitro* by treatment with deacetylase enzyme. The resulting free lysines are derivatized with NHS-SS-biotin followed by affinity capture on streptavidin resin. The identity of the biotin-captured peptides is determined by LC-MS/MS. This approach has been applied to the identification of a range of potential substrates, including the 14-3-3ζ, as a substrate for Sirt1 deacetylase, a finding that links Sirt1 and 14-3-3ζ acetylation to control of caspase-2-dependent apoptotic cell death mediated by chemotherapeutic agents [94]. In general, biotin switch techniques can be compromised by incomplete reaction and side reactions. Technical aspects of the acetyl-biotin switch technique, including protocol details, advice on how to reduce false-positive protein identification, and coupling to quantitative MS approaches, including label-free analysis, are available [109]. In principle, the biotin technique could be applied to methylation by blocking of unmodified lysine and arginine residues, *in vitro* demethylation, and capture followed by MS analysis.

4.4.4 Enrichment of N-Terminally Acetylated Peptides

N-terminally acetylated peptides can be separated from internal tryptic peptides and unmodified N-terminal peptides due to differences in charge, resulting from blockage of the N-terminal amine. SCX at low pH (<3) can resolve these species based on charge state, but it is not completely selective since peptides of similar or lower positive charge such as phosphopeptides, C-terminal peptides, or Glu/Asp-containing peptides elute in a similar retention window. This can be improved by the use of LysN, which cleaves before lysine residues, to produce uncharged N-terminally acetylated peptides and singly charged internal peptides [110].

Step change for this strategy was achieved by introduction of a dimethylation step posttryptic digest to block free (N-terminal, lysine-ε) amines thus increasing the difference in basicity between N-acetylated and the rest of peptides in the sample. One-step purification of N-acetylated peptides is achieved by solid-phase extraction, using SCX in batch mode. A key benefit of this approach is that it can be utilized with stable isotope-labeled dimethylation reagents for relative quantification, an approach which has found application to distinguish protein isoforms that are N-terminally acetylated but differ in N-terminal amino acid sequence, for example, β-actin/γ-actin isoforms [111].

4.5 Bioinformatics

Data on sites of acetylation and methylation have been accrued and are continuing to be generated. Key challenges in this area are provision of database search tools for confident site and PTM assignment from mass spectrometry and accurate *in silico* prediction of PTM status. In terms of biological research, a key requirement for taking this forward is to use this information to gain understanding of the functions of these PTM sites, both individually and in combination. The challenges are to effectively mine proteomic data, catalog, and interrogate these multispecies PTM catalogs, including analysis of quantitative changes over time or in different biological states where these data are available. These will inform (i) follow-up experiments on selected proteins and (ii) computational prediction models to complement MS-based approaches for full definition of the acetylomes and methylomes.

4.5.1 Assigning Acetylation and Methylation Status

Database search engines process product ion spectra for peptide identification and were originally designed for data acquired via DDA, where individual tandem mass spectra are interpreted. Peptide identification is achieved by comparing the tandem mass spectra derived experimentally with theoretical tandem mass spectra generated by *in silico* digestion of a protein database using a number of different search algorithms (as reviewed by Noble and MacCoss [112]). Protein inference is accomplished by assigning peptide sequences to proteins, grouped on the basis of the assigned peptides being unique or shared with other proteins. Example search engines are Mascot, SEQUEST, PEAKS DB, ProteinPilot, pFind, Andromeda in MaxQuant, OMSSA in COMPASS, and X!Tandem. Assigning PTM status and localization, particularly when associated with missed cleavage peptides, as is the case for acetylated and methylated peptides, poses specific challenges [113], including a combinatorial increase in the search space, which impacts on the false discovery rates for site assignment (FDR), an issue discussed in detail by Fu and Wong using phosphorylation, carbamylation, and acetylation as reference PTMs [114].

Error-tolerant searching applies a "two-round-search" and is utilized in Mascot and X!Tandem software [115, 116] to identify proteins from unmodified peptides, and then identify post-translationally modified peptides. PTMTreeSearch (a plug-in to X!Tandem) employs a two-round peptide identification strategy analogous to X!Tandem and Mascot, where the first round is used to reduce the search space to likely solutions followed by an error-tolerant, more exhaustive search in the second round. A computational tree is created for each peptide, whereby the path from the root to the leaves is labeled with the amino acids of the peptide and branches represent PTMs [117]. The error-tolerant approach is limited by the requirement that the post-translationally modified protein must be assigned at least one high-scoring peptide in the first pass. A pragmatic approach, using iterative searching for unmodified, individually modified, and multiple modifications has been exemplified for histone tails. In this study, was achieved by accepting only identifications obtained in common from multiple search engines and applying <1% FDR [118].

MS analysis of spectra acquired via DIA is inherently limited without direct assignment of MS/MS fragments to precursor ions. Fragments can be grouped and assigned to a specific precursor based on LC retention time alignment. This poses a challenge for coeluting peptides, which can be ameliorated by the use of enhanced-resolution IMS in high-definition MS analysis (HDMS^E) or by using a SWATH-like approach to reduce complexity by using narrow mass windows. Software developments in this area include not only new algorithms but also modifications to existing packages (see review by Szabo and Janaky [6]). Of note, the DIA-Umpire software package processes DIA data using traditional database search tools [64]. Taking a peptide-centric approach to DIA data enables a conservative test to determine the presence and absence of specific query peptides using a library search strategy [119].

4.5.2 PTM Repositories and Data Mining Tools

There are a number of online PTM repositories including PhosphoSitePlus [120], SysPTM 2.0 [121], PTMCode [122], and ProteomeScout [123]. PTMCode indicates functional links between individual and multiple PTMs including acetylation and methylation [122]. Another software, ProteomeScout, provides a compendium of public PTM data, including quantitative proteomics and details on functional annotation [123]. These repositories aid design of follow-on experimental work and provide data for the development of *in silico* prediction tools.

4.5.3 Computational Prediction Tools for Acetylation and Methylation Sites

There has been a focus on the development and application of tools to predict acetylation and methylation sites to complement and extend analysis performed *in vitro* in a cost-effective manner. The challenges around site prediction have been outlined by review and evaluation of PMeS, PLMLA, MeMo, MASA, BPB-PPMS, MethK, and iMethyl-PseAAC using a common dataset of methylation sites [124]. It was concluded that performance needs to be optimized and that it would be of benefit to adopt different feature information for methyl-arginine and methyl-lysine predictors [124]. The PLMLA tool also predicts acetylation sites and has been compared with the LAceP, EnsemblePail, PHOSIDA, and PSKAcePred acetylation site prediction software [125]. The comparison utilized information on 13,810 acetylation sites from 6388 proteins, which was obtained from SysPTM 2.0 and PhosphoSitePlus databases. LAceP performed best, employing a logistic regression method to integrate information on amino acid sequence adjacent to acetylated sites, physicochemical properties, and the transition probability of adjacent amino acids [125].

A comprehensive and comparative analysis of acetylation, methylation, ubiquitinylation, and SUMOylation has identified conserved amino acid sequence association with secondary structure [126]. Information, combined with site specificity data for enzymes mediating addition and removal of acetyl and methyl groups, can benefit site prediction as has been demonstrated for lysine acetyltransferases [127]. Chemoselective reactions provide an MS method for the experimental determination of acetyl and methyl transferase activities. Examples include alkynyl-acetyl-CoA and SAM analogs for bioorthogonal click chemistry analysis of proteins that are substrates for acetylation and methylation, respectively [128–130]. These complement other MS-based analyses (Table 4.2).

Table 4.2 Biochemical methods coupled to MS for analysis of protein acetylation and methylation.

Method	Specificity	Example references
Immunoaffinity capture (peptides)	Acetylation (lysine)	[19]
	Methylation (arginine)	[39, 99, 100]
	Methylation (lysine)	[46, 100, 101]
Immunoaffinity capture (proteins)	Methylation (arginine, lysine)	[102]
	Acetylation (lysine)	[103]
Reader domain–based capture	Acetylation	[93]
	Methylation (mono-, dimethyl-lysine)	[105, 106]
	Methylation (di-, trimethyl-lysine)	[106, 107]
Biotin switch capture	Acetylation	[94, 108, 109]
Solid-phase extraction	N-terminal acetyl (including lysine)	[110, 111]
Chemical reporters (bioorthogonal reagents for click chemistry)	Acetylation (N-terminal, lysine)	[128, 129]
	Methylation (arginine, lysine)	[130]

Notes: Lysine acylations other than acetylation can be analyzed [129].

4.5.4 Information for Design of Follow-Up Experiments

For follow-up experiments, candidate proteins and specific PTM sites are selected on the basis of confidence of sequence and PTM assignments, stoichiometry data, and their potential role in the biological context [131]. Analysis of proteomic data to filter them based on protein function can be achieved using Gene Ontology terms as reviewed by Schmidt et al. [132]. Evolutionary conservation of PTMs across members of a domain family can be used to predict functionally important regulatory regions: PTMs that are known or predicted to be regulated *in vivo* are more likely to be conserved across species than average sites, particularly those involved in PTM cross talk [133]. *In silico* data demonstrate that the majority of lysine acetylation sites have the potential to impact protein phosphorylation, methylation, and ubiquitination status [134]. Thus, analysis of protein acetylation and methylation should be considered in the context of other PTMs in terms of determining regulatory pathways and mechanisms.

4.6 Summary

Acetylation and methylation PTMs are key regulatory events subject to research using mass spectrometry and related biochemical techniques. This chapter aims to capture the key findings and developments in the field in terms of identification, characterization, and monitoring of quantitative changes in different biological states.

References

1 Shemorry A, Hwang CS, Varshavsky A. Control of protein quality and stoichiometries by N-terminal acetylation and the N-end rule pathway. *Mol Cell* 2013;**50**:540–551.
2 Choudhary C, Weinert BT, Nishida Y, Verdin E, Mann M. The growing landscape of lysine acetylation links metabolism and cell signalling. *Nat Rev Mol Cell Biol* 2014;**15**:536–550.
3 Biggar KK, Li SS. Non-histone protein methylation as a regulator of cellular signalling and function. *Nat Rev Mol Cell Biol* 2015;**16**:5–17.
4 Su X, Wellen KE, Rabinowitz JD. Metabolic control of methylation and acetylation. *Curr Opin Chem Biol* 2016;**30**:52–60.
5 Venne AS, Kollipara L, Zahedi RP. The next level of complexity: crosstalk of post-translational modifications. *Proteomics* 2014;**14**:513–524.

6 Szabo Z, Janaky T. Challenges and developments in protein identification using mass spectrometry. *Trends Anal Chem* 2015;**69**:76–87.

7 Kouzarides T. Acetylation: a regulatory modification to rival phosphorylation? *EMBO J* 2000;**19**:1176–1179.

8 Ng MK, Cheung P. A brief histone in time: understanding the combinatorial functions of histone PTMs in the nucleosome context. *Biochem Cell Biol* 2015;**94**:33–42.

9 Schilling B, Christensen D, Davis R, Sahu AK, Hu LI, Walker-Peddakotla A, Sorensen DJ, Zemaitaitis B, Gibson BW, Wolfe AJ. Protein acetylation dynamics in response to carbon overflow in *Escherichia coli*. *Mol Microbiol* 2015;**98**:847–863.

10 Varland S, Osberg C, Arnesen T. N-terminal modifications of cellular proteins: the enzymes involved, their substrate specificities and biological effects. *Proteomics* 2015;**15**:2385–2401.

11 Afjehi-Sadat L, Garcia BA. Comprehending dynamic protein methylation with mass spectrometry. *Curr Opin Chem Biol* 2013;**17**:12–19.

12 Whetstine JR. Methylation: a multifaceted modification – looking at transcription and beyond. *Biochim Biophys Acta* 2014;**1839**:1351–1352.

13 Tikhanovich I, Kuravi S, Artigues A, Villar MT, Dorko K, Nawabi A, Roberts B, Weinman SA. Dynamic arginine methylation of tumor necrosis factor (TNF) receptor-associated factor 6 regulates toll-like receptor signalling. *J Biol Chem* 2015;**290**:22236–22249.

14 Declerck K, Vel Szic KS, Palagani A, Heyninck K, Haegeman G, Morand C, Milenkovic D, Berghe WV. Epigenetic control of cardiovascular health by nutritional polyphenols involves multiple chromatin-modifying writer-reader-eraser proteins. *Curr Top Med Chem* 2016;**16**:788–806.

15 Gayatri S, Bedford MT. Readers of histone methylarginine marks. *Biochim Biophys Acta* 2014;**1839**:702–10.

16 Sanchez R, Meslamani J, Zhou MM. The bromodomain: from epigenome reader to druggable target. *Biochim Biophys Acta* 2014;**1839**:676–85.

17 Kaniskan HÜ, Konze KD, Jin J. Selective inhibitors of protein methyltransferases. *J Med Chem* 2015;**58**:1596–629.

18 Van Dyke MW. Lysine deacetylase (KDAC) regulatory pathways: an alternative approach to selective modulation. *ChemMedChem* 2014;**9**:511–22.

19 Svinkina T, Gu H, Silva JC, Mertins P, Qiao J, Fereshetian S, Jaffe JD, Kuhn E, Udeshi ND, Carr SA. Deep, quantitative coverage of the lysine acetylome using novel anti-acetyl-lysine antibodies and an optimized proteomic workflow. *Mol Cell Proteomics* 2015;**14**:2429–2440.

20 Smith LM, Kelleher NL. Proteoform: a single term describing protein complexity. *Nat Methods* 2013;**10**:186–187.

21 Xiong L, Adhvaryu KK, Selker EU, Wang Y. Mapping of lysine methylation and acetylation in core histones of *Neurospora crassa*. *Biochemistry* 2010;**49**:5236–5243.

22 Quan, L., Liu, M.: CID, ETD and HCD fragmentation to study protein post-translational modifications. *Mod Chem Appl* **1**, e102 (2013).

23 Doll S, Burlingame AL. Mass spectrometry-based detection and assignment of protein post-translational modifications. *ACS Chem Biol* 2014;**10**:63–71.

24 Moradian A, Kalli A, Sweredoski MJ, Hess S. The top-down, middle-down, and bottom-up mass spectrometry approaches for characterization of histone variants and their post-translational modifications. *Proteomics* 2014;**14**:489–497.

25 Young NL, DiMaggio PA, Plazas-Mayorca MD, Baliban RC, Floudas CA, Garcia BA. High throughput characterization of combinatorial histone codes. *Mol Cell Proteomics* 2009;**8**:2266–2284.

26 Mikesh LM, Ueberheide B, Chi A, Coon JJ, Syka JE, Shabanowitz J, Hunt DF. The utility of ETD mass spectrometry in proteomic analysis. *Biochim Biophys Acta* 2006;**1764**:1811–1822.

27 Frese CK, Altelaar AM, Hennrich ML, Nolting D, Zeller M, Griep-Raming J, Heck AJ, Mohammed S. Improved peptide identification by targeted fragmentation using CID, HCD and ETD on an LTQ-Orbitrap Velos. *J Proteome Res* 2011;**10**:2377–2388.

28 Wang H, Straubinger RM, Aletta JM, Cao J, Duan X, Yu H, Qu J. Accurate localization and relative quantification of arginine methylation using nanoflow liquid chromatography coupled to electron transfer dissociation and orbitrap mass spectrometry. *J Am Soc Mass Spectrom* 2009;**20**:507–519.

29 Jufvas, A., Stralfors, P., Vener, A.V.: Histone variants and their post-translational modifications in primary human fat cells. *PLoS One* **6**, e15960 (2011).

30 Zhang K, Yau PM, Chandrasekhar B, New R, Kondrat R, Imai BS, Bradbury ME. Differentiation between peptides containing acetylated or trimethylated lysines by mass spectrometry: an application for determining lysine 9 acetylation and methylation of histone H3. *Proteomics* 2004;**4**:1–10.

31 Olsen JV, Macek B, Lange O, Makarov A, Horning S, Mann M. Higher-energy C-trap dissociation for peptide modification analysis. *Nat Methods* 2007;**4**:709–712.

32 Matthiesen R, Trelle MB, Højrup P, Bunkenborg J, Jensen ON. VEMS 3.0: algorithms and computational tools for tandem mass spectrometry based identification of post-translational modifications in proteins. *J Proteome Res* 2005;**4**:2338–2347.

33 Kelstrup CD, Frese C, Heck AJ, Olsen JV, Nielsen ML. Analytical utility of mass spectral binning in proteomic experiments by SPectral Immonium Ion Detection (SPIID). *Mol Cell Proteomics* 2014;**13**:1914–1924.

34 Karch KR, Zee BM, Garcia BA. High resolution is not a strict requirement for characterization and quantification of histone post-translational modifications. *J Proteome Res* 2014;**13**:6152–6159.

35 Trelle MB, Jensen ON. Utility of immonium ions for assignment of epsilon-*N*-acetyllysine-containing peptides by tandem mass spectrometry. *Anal Chem* 2008;**80**:3422–3430.

36 Fu L, Chen T, Xue G, Zu L, Fang W. Selective cleavage enhanced by acetylating the side chain of lysine. *J Am Soc Mass Spectrom* 2013;**48**:128–134.

37 Couttas TA, Raftery MJ, Bernardini G, Wilkins MR. Immonium ion scanning for the discovery of post-translational modifications and its application to histones. *J Proteome Res* 2008;**7**:2632–2641.

38 Gehrig PM, Hunziker PE, Zahariev S, Pongor S. Fragmentation pathways of N(G)-methylated and unmodified arginine residues in peptides studied by ESI-MS/MS and MALDI-MS. *J Am Soc Mass Spectrom* 2004;**15**:142–149.

39 Ong SE, Mittler G, Mann M. Identifying and quantifying *in vivo* methylation sites by heavy methyl SILAC. *Nat Methods* 2004;**1**:119–126.

40 Brame CJ, Moran MF, McBroom-Cerajewski LD. A mass spectrometry based method for distinguishing between symmetrically and asymmetrically dimethylated arginine residues. *Rapid Commun Mass Spectrom* 2004;**18**:877–881.

41 Rappsilber J, Friesen WJ, Paushkin S, Dreyfuss G, Mann M. Detection of arginine dimethylated peptides by parallel precursor ion scanning mass spectrometry in positive ion mode. *Anal Chem* 2003;**75**:3107–3114.

42 Snijders AP, Hung ML, Wilson SA, Dickman MJ. Analysis of arginine and lysine methylation utilizing peptide separations at neutral pH and electron transfer dissociation mass spectrometry. *J Am Soc Mass Spectrom* 2010;**21**:88–96.

43 Hung CW, Schlosser A, Wei J, Lehmann WD. Collision-induced reporter fragmentations for identification of covalently modified peptides. *Anal Bioanal Chem* 2007;**389**:1003–1016.

44 Hirota J, Satomi Y, Yoshikawa K, Takao T. ε-N,N,N-Trimethyllysine-specific ions in matrix-assisted laser desorption/ionization-tandem mass spectrometry. *Rapid Commun Mass Spectrom* 2003;**17**:371–376.

45 Hart-Smith G, Low JK, Erce MA, Wilkins MR. Enhanced methylarginine characterization by post-translational modification-specific targeted data acquisition and electron-transfer dissociation mass spectrometry. *J Am Soc Mass Spectrom* 2012;**23**:1376–1389.

46 Geoghegan V, Guo A, Trudgian D, Thomas B, Acuto O. Comprehensive identification of arginine methylation in primary T cells reveals regulatory roles in cell signalling. *Nat Commun* 2015;**7**:6758–6766.

47 Unwin RD, Griffiths JR, Whetton AD. A sensitive mass spectrometric method for hypothesis-driven detection of peptide post-translational modifications: multiple reaction monitoring-initiated detection and sequencing (MIDAS). *Nat Protoc* 2009;**4**:870–877.

48 Griffiths JR, Unwin RD, Evans CA, Leech SH, Corfe BM, Whetton AD. The application of a hypothesis-driven strategy to the sensitive detection and location of acetylated lysine residues. *J Am Soc Mass Spectrom* 2007;**18**:1423–1428.

49 Borchers C, Parker CE, Deterding LJ, Tomer KB. Preliminary comparison of precursor scans and liquid chromatography-tandem mass spectrometry on a hybrid quadrupole time-of-flight mass spectrometer. *J Chromatogr A* 1999;**854**:119–130.

50 Kim JY, Kim KW, Kwon HJ, Lee DW, Yoo JS. Probing lysine acetylation with a modification-specific marker ion using high-performance liquid chromatography/electrospray-mass spectrometry with collision-induced dissociation. *Anal Chem* 2002;**74**:5443–5449.

51 Evans CA, Ow SY, Smith DL, Corfe BM, Wright PC. Application of the CIRAD mass spectrometry approach for lysine acetylation site discovery. *Methods Mol Biol* 2013;**981**:13–23.

52 Mayne J, Ning Z, Zhang X, Starr AE, Chen R, Deeke S, Chiang CK, Xu B, Wen M, Cheng K, Seebun D. Bottom-up proteomics (2013–2015): keeping up in the era of systems biology. *Anal Chem* 2015;**88**:95–121.

53 Higgs RE, Butler JP, Han B, Knierman MD. Quantitative proteomics via high resolution MS quantification: capabilities and limitations. *Int J Proteomics* 2013;**2013**(674282).

54 Kirkpatrick DS, Gerber SA, Gygi SP. The absolute quantification strategy: a general procedure for the quantification of proteins and post-translational modifications. *Methods* 2005;**35**:265–273.

55 Picotti P, Aebersold R. Selected reaction monitoring-based proteomics: workflows, potential, pitfalls and future directions. *Nat Methods* 2012;**9**:555–566.

56 Carr SA, Abbatiello SE, Ackermann BL, Borchers C, Domon B, Deutsch EW, Grant RP, Hoofnagle AN, Hüttenhain R, Koomen JM, Liebler DC, Liu T, MacLean B, Mani DR, Mansfield E, Neubert H, Paulovich AG, Reiter L, Vitek O, Aebersold R, Anderson L, Bethem R, Blonder J, Boja E, Botelho J, Boyne M, Bradshaw RA, Burlingame AL, Chan D, Keshishian H, Kuhn E, Kinsinger C, Lee JS, Lee SW, Moritz R, Oses-Prieto J, Rifai N, Ritchie J, Rodriguez H, Srinivas PR, Townsend RR, Van Eyk J, Whiteley G, Wiita A, Weintraub S. Targeted peptide measurements in biology and medicine: best practices for mass spectrometry-based assay development using a fit-for-purpose approach. *Mol Cell Proteomics* 2014;**13**:907–917.

57 Colangelo CM, Chung L, Bruce C, Cheung KH. Review of software tools for design and analysis of large scale MRM proteomic datasets. *Methods* 2013;**61**:287–298.

58 Sherrod SD, Myers MV, Li M, Myers JS, Carpenter KL, Maclean B, Maccoss MJ, Liebler DC, Ham AJ. Label-free quantitation of protein modifications by pseudo selected reaction monitoring with internal reference peptides. *J Proteome Res* 2012;**11**:3467–3479.

59 Lin S, Wein S, Gonzales-Cope M, Otte GL, Yuan ZF, Afjehi-Sadat L, Maile T, Berger SL, Rush J, Lill JR, Arnott D, Garcia BA. Stable-isotope-labeled histone peptide library for histone post-translational modification and variant quantification by mass spectrometry. *Mol Cell Proteomics* 2014;**13**:2450–2466.

60 Lange V, Picotti P, Domon B, Aebersold R. Selected reaction monitoring for quantitative proteomics: a tutorial. *Mol Syst Biol* 2008;**4**:222–236.

61 Peterson AC, Russell JD, Bailey DJ, Westphall MS, Coon JJ. Parallel reaction monitoring for high resolution and high mass accuracy quantitative, targeted proteomics. *Mol Cell Proteomics* 2012;**11**:475–488.

62 Tang H, Fang H, Yin E, Brasier AR, Sowers LC, Zhang K. Multiplexed parallel reaction monitoring targeting histone modifications on the QExactive mass spectrometer. *Anal Chem* 2014;**86**:5526–5534.

63 Sowers JL, Mirfattah B, Xu P, Tang H, Park IY, Walker C, Wu P, Laezza F, Sowers LC, Zhang K. Quantification of histone modifications by parallel-reaction monitoring: a method validation. *Anal Chem* 2015;**87**:10006–10014.

64 Tsou CC, Avtonomov D, Larsen B, Tucholska M, Choi H, Gingras AC, Nesvizhskii AI. DIA-Umpire: comprehensive computational framework for data-independent acquisition proteomics. *Nat Methods* 2015;**12**:258–264.

65 Rardin MJ, Schilling B, Cheng LY, MacLean BX, Sorensen DJ, Sahu AK, MacCoss MJ, Vitek O, Gibson BW. MS1 peptide ion intensity chromatograms in MS2 (SWATH) data independent acquisitions. Improving post acquisition analysis of proteomic experiments. *Mol Cell Proteomics* 2015;**14**:2405–2419.

66 Geiger T, Cox J, Mann M. Proteomics on an Orbitrap benchtop mass spectrometer using all-ion fragmentation. *Mol Cell Proteomics* 2010;**9**:2252–2261.

67 Li G-Z, Vissers JPC, Silva JC, Golick D, Gorenstein MV, Geromanos SJ. Database searching and accounting of multiplexed precursor and product ion spectra from the data independent analysis of simple and complex peptide mixtures. *Proteomics* 2009;**9**:1696–1719.

68 Sidoli S, Lin S, Xiong L, Bhanu NV, Karch KR, Johansen E, Hunter C, Mollah S, Garcia BA. Sequential window acquisition of all theoretical mass spectra (SWATH) analysis for characterization and quantification of histone post-translational modifications. *Mol Cell Proteomics* 2015;**14**:2420–2428.

69 Krautkramer KA, Reiter L, Denu JM, Dowell JA. Quantification of SAHA-dependent changes in histone modifications using data-independent acquisition mass spectrometry. *J Proteome Res* 2015;**14**:3252–3262.

70 Srebalus B, Hilderbrand AE, Valentine SJ, Clemmer DE. Resolving isomeric peptide mixtures: a combined HPLC/ion mobility-TOFMS analysis of a 4000-component combinatorial library. *Anal Chem* 2002;**74**:26–36.

71 Gethings LA, Connolly JB. Simplifying the proteome: analytical strategies for improving peak capacity. *Adv Exp Med Biol* 2014;**806**:59–77.

72 Rodriguez-Suarez E, Hughes C, Gethings L, Giles K, Wildgoose J, Stapels M, Fadgen KE, Geromanos SJ, Vissers JP, Elortza F, Langridge JI. An ion mobility assisted data independent LC-MS strategy for the analysis of complex biological samples. *Curr Anal Chem* 2013;**9**:199–211.

73 Shliaha PV, Bond NJ, Gatto L, Lilley KS. Effects of traveling wave ion mobility separation on data independent acquisition in proteomics studies. *J Proteome Res* 2013;**12**:2323–2339.

74 Shliaha PV, Jukes-Jones R, Christoforou A, Fox J, Hughes C, Langridge J, Cain K, Lilley KS. Additional precursor purification in isobaric mass tagging experiments by traveling wave ion mobility separation (TWIMS). *J Proteome Res* 2014;**13**:3360–3369.

75 Helm D, Vissers JP, Hughes CJ, Hahne H, Ruprecht B, Pachl F, Grzyb A, Richardson K, Wildgoose J, Maier SK, Marx H, Wilhelm M, Becher I, Lemeer S, Bantscheff M, Langridge JI, Kuster B. Ion mobility tandem mass spectrometry enhances performance of bottom-up proteomics. *Mol Cell Proteomics* 2014;**13**:3709–3715.

76 Donohoe GC, Maleki H, Arndt JR, Khakinejad M, Yi J, McBride C, Nurkiewicz TR, Valentine SJ. A new ion mobility-linear ion trap instrument for complex mixture analysis. *Anal Chem* 2014;**86**:8121–8128.

77 Lermyte F, Williams JP, Brown JM, Martin EM, Sobott F. Extensive charge reduction and dissociation of intact protein complexes following electron transfer on a quadrupole-ion mobility-time-of-flight MS. *J Am Soc Mass Spectrom* 2015;**26**:1068–1076.

78 Shvartsburg AA, Zheng Y, Smith RD, Kelleher NL. Ion mobility separation of variant histone tails extending to the "middle-down" range. *Anal Chem* 2012;**84**:4271–4276.

79 Evertts AG, Zee BM, Dimaggio PA, Gonzales-Cope M, Coller HA, Garcia BA. Quantitative dynamics of the link between cellular metabolism and histone acetylation. *J Biol Chem* 2013;**288**:12142–12151.

80 Zheng Y, Sweet SM, Popovic R, Martinez-Garcia E, Tipton JD, Thomas PM, Licht JD, Kelleher NL. Total kinetic analysis reveals how combinatorial methylation patterns are established on lysines 27 and 36 of histone H3. *Proc Natl Acad Sci U S A* 2012;**109**:13549–13554.

81 Olsen JV, Vermeulen M, Santamaria A, Kumar C, Miller ML, Jensen LJ, Gnad F, Cox J, Jensen TS, Nigg EA, Brunak S, Mann M. Quantitative phosphoproteomics reveals widespread full phosphorylation site occupancy during mitosis. *Sci Signal* 2010;**3**:ra3.

82 Nakayasu ES, Wu S, Sydor MA, Shukla AK, Weitz KK, Moore RJ, Hixson KK, Kim JS, Petyuk VA, Monroe ME, Pasa-Tolic L, Qian WJ, Smith RD, Adkins JN, Ansong C. A method to determine lysine acetylation stoichiometries. *Int J Proteomics* 2014;**2014**:730725.

83 Hersman E, Nelson DM, Griffith WP, Jelinek C, Cotter RJ. Analysis of histone modifications from tryptic peptides of deuteroacetylated isoforms. *Int J Mass Spectrom* 2012;**312**:5–16.

84 Baeza J, Dowell JA, Smallegan MJ, Fan J, Amador-Noguez D, Khan Z, Denu JM. Stoichiometry of site-specific lysine acetylation in an entire proteome. *J Biol Chem* 2014;**289**:21326–21338.

85 Pesavento JJ, Mizzen CA, Kelleher NL. Quantitative analysis of modified proteins and their positional isomers by tandem mass spectrometry: human histone H4. *Anal Chem* 2006;**78**:4271–4280.

86 Tipton, J.D., Tran, J.C., Catherman, A.D., Ahlf, D.R., Durbin, K.R., Kelleher, N.L.: Analysis of intact protein isoforms by mass spectrometry. *J Biol Chem* 2011;**286**, 25451–25458.

87 Tran JC, Zamdborg L, Ahlf DR, Lee JE, Catherman AD, Durbin KR, Tipton JD, Vellaichamy A, Kellie JF, Li M, Wu C, Sweet SM, Early BP, Siuti N, LeDuc RD, Compton PD, Thomas PM, Kelleher NL. Mapping intact protein isoforms in discovery mode using top-down proteomics. *Nature* 2011;**480**:254–25.

88 Durbin KR, Tran JC, Zamdborg L, Sweet SM, Catherman AD, Lee JE, Li M, Kellie JF, Kelleher NL. Intact mass detection, interpretation, and visualization to automate top-down proteomics on a large scale. *Proteomics* 2010;**10**:3589–3597.

89 Xiu L, Valeja SG, Alpert AJ, Jin S, Ge Y. Effective protein separation by coupling hydrophobic interaction and reverse phase chromatography for top-down proteomics. *Anal Chem* 2014;**86**:7899–7906.

90 Cannon J, Lohnes K, Wynne C, Wang Y, Edwards N, Fenselau C. High-throughput middle-down analysis using an orbitrap. *J Proteome Res* 2010;**9**:3886–3890.

91 Wiesner J, Premsler T, Sickmann A. Application of electron transfer dissociation (ETD) for the analysis of posttranslational modifications. *Proteomics* 2008;**8**:4466–4483.

92 Sweredoski MJ, Moradian A, Raedle M, Franco C, Hess S. High resolution parallel reaction monitoring with electron transfer dissociation for middle-down proteomics. *Anal Chem* 2015;**87**:8360–8366.

93 Bryson BD, Del Rosario AM, Gootenberg JS, Yaffe MB, White FM. Engineered bromodomains to explore the acetylproteome. *Proteomics* 2015;**15**:1470–1475.

94 Andersen JL, Thompson JW, Lindblom KR, Johnson ES, Yang CS, Lilley LR, Freel CD, Moseley MA, Kornbluth S. A biotin switch-based proteomics approach identifies 14-3-3ζ as a target of Sirt1 in the metabolic regulation of caspase-2. *Mol Cell* 2011;**43**:834–842.

95 Choudhary C, Kumar C, Gnad F, Nielsen ML, Rehman M, Walther TC, Olsen JV, Mann M. Lysine acetylation targets protein complexes and co-regulates major cellular functions. *Science* 2009;**325**:834–840.

96 Mertins P, Qiao JW, Patel J, Udeshi ND, Clauser KR, Mani DR, Burgess MW, Gillette MA, Jaffe JD, Carr SA. Integrated proteomic analysis of post-translational modifications by serial enrichment. *Nat Methods* 2013;**10**:634–637.

97 Uhlmann T, Geoghegan VL, Thomas B, Ridlova G, Trudgian DC, Acuto O. A method for large-scale identification of protein arginine methylation. *Mol Cell Proteomics* 2012;**11**:1489–1499.

98 Plank M, Fischer R, Geoghegan V, Charles PD, Konietzny R, Acuto O, Pears C, Schofield CJ, Kessler BM. Expanding the yeast protein arginine methylome. *Proteomics* 2015;**15**:3232–3243.

99 Boisvert FM, Côté J, Boulanger MC, Richard S. A proteomic analysis of arginine-methylated protein complexes. *Mol Cell Proteomics* 2003;**2**:1319–1330.

100 Guo A, Gu H, Zhou J, Mulhern D, Wang Y, Lee KA, Yang V, Aguiar M, Kornhauser J, Jia X, Ren J, Beausoleil SA, Silva JC, Vemulapalli V, Bedford MT, Comb MJ. Immunoaffinity enrichment and mass spectrometry analysis of protein methylation. *Mol Cell Proteomics* 2014;**13**:372–387.

101 Cao XJ, Arnaudo AM, Garcia BA. Large-scale global identification of protein lysine methylation *in vivo*. *Epigenetics* 2013;**8**:477–485.

102 Carlson SM, Gozani O. Emerging technologies to map the protein methylome. *J Mol Biol* 2014;**426**:3350–3362.

103 Leech SH, Evans CA, Shaw L, Wong CH, Connolly J, Griffiths JR, Whetton AD, Corfe BM. Proteomic analyses of intermediate filaments reveals cytokeratin8 is highly acetylated – implications for colorectal epithelial homeostasis. *Proteomics* 2008;**8**:279–288.

104 Champleboux M, Govin J. Bromodomains shake the hegemony of pan-acetyl antibodies. *Proteomics* 2015;**15**:1457–1458.

105 Moore KE, Carlson SM, Camp ND, Cheung P, James RG, Chua KF, Wolf-Yadlin A, Gozani O. A general molecular affinity strategy for global detection and proteomic analysis of lysine methylation. *Mol Cell* 2013;**50**:444–456.

106 Carlson SM, Moore KE, Green EM, Martin GM, Gozani O. Proteome-wide enrichment of proteins modified by lysine methylation. *Nat Protoc* 2014;**9**:37–50.

107 Liu H, Galka M, Mori E, Liu X, Lin YF, Wei R, Pittock P, Voss C, Dhami G, Li X, Miyaji M, Lajoie G, Chen B, Li SS. A method for systematic mapping of protein lysine methylation identifies functions for HP1β in DNA damage response. *Mol Cell* 2013;**50**:723–735.

108 Jaffrey SR, Snyder SH. The biotin switch method for the detection of S-nitrosylated proteins. *Sci STKE* 2001;pll.

109 Thompson JW, Robeson A, Andersen JL. Identification of deacetylase substrates with the biotin switch approach. *Methods Mol Biol* 2013;**1077**:133–148.

110 Taouatas N, Mohammed S, Heck AJ. Exploring new proteome space: combining Lys-N proteolytic digestion and strong cation exchange (SCX) separation in peptide-centric MS-driven proteomics. *Methods Mol Biol* 2011;**753**:157–167.

111 Chen SH, Chen CR, Chen SH, Li DT, Hsu JL. Improved Nα-acetylated peptide enrichment following dimethyl labeling and SCX. *J Proteome Res* 2013;**12**:3277–3287.

112 Noble, W.S., MacCoss, M.J.: Computational and statistical analysis of protein mass spectrometry data. *PLoS Comput Biol* 2012;**8**:e1002296.

113 Kim MS, Zhong J, Pandey A. *Common errors in mass spectrometry-based analysis of post-translational modifications*. Proteomics; 2015. DOI: 10.1002/pmic.201500355.

114 Fu Y, Qian X. Transferred subgroup false discovery rate for rare post-translational modifications detected by mass spectrometry. *Mol Cell Proteomics* 2014;**13**:1359–1368.

115 Creasy DM, Cottrell JS. Error tolerant searching of uninterpreted tandem mass spectrometry data. *Proteomics* 2002;**2**:1426–1434.

116 Craig R, Beavis RC. TANDEM: matching proteins with tandem mass spectra. *Bioinformatics* 2004;**20**:1466–1467.

117 Kertesz-Farkas A, Keich U, Noble WS. Improved false discovery rate estimation procedure for shotgun proteomics. *J Proteome Res* 2015;**14**:3148–3161.

118 Yuan ZF, Lin S, Molden RC, Garcia BA. Evaluation of proteomic search engines for the analysis of histone modifications. *J Proteome Res* 2014;**13**:4470–4478.

119 Ting YS, Egertson JD, Payne SH, Kim S, MacLean B, Käll L, Aebersold R, Smith RD, Noble WS, MacCoss MJ. Peptide-centric proteome analysis: an alternative strategy for the analysis of tandem mass spectrometry data. *Mol Cell Proteomics* 2015;**14**:2301–2307.

120 Hornbeck PV, Zhang B, Murray B, Kornhauser JM, Latham V, Skrzypek E. PhosphoSitePlus, 2014: mutations, PTMs and recalibrations. *Nucleic Acids Res* 2015;**43**:D512–D520.

121 Li J, Jia J, Li H, Yu J, Sun H, He Y, Lv D, Yang X, Glocker MO, Ma L, Yang J. SysPTM 2.0: an updated systematic resource for post-translational modification. *Database* 2014:bau025.

122 Minguez P, Letunic I, Parca L, Garcia-Alonso L, Dopazo J, Huerta-Cepas J. Bork, P: PTMcode v2: a resource for functional associations of post-translational modifications within and between proteins. *Nucleic Acids Res* 2014;**43**(D1):D494–D502.

123 Matlock MK, Holehouse AS, Naegle KM. ProteomeScout: a repository and analysis resource for post-translational modifications and proteins. *Nucleic Acids Res* 2015;**43**(D1):D521–D530.

124 Shi SP, Xu HD, Wen PP, Qiu JD. 2015. Progress and challenges in predicting protein methylation sites. Mol. *BioSystems* 2015;**11**:2610–2619.

125 Hou T, Zheng G, Zhang P, Jia J, Li J, Xie L, Wei C, Li Y. LAceP: lysine acetylation site prediction using logistic regression classifiers. *PLoS One* 2014;**9**:e89575.

126 Cesaro L, Pinna LA, Salvi M. A comparative analysis and review of lysyl residues affected by post-translational modifications. *Curr Genomics* 2015;**16**:128–138.

127 Li T, Du Y, Wang L, Huang L, Li W, Lu M, Zhang X, Zhu WG. Characterization and prediction of lysine (K)-acetyl-transferase specific acetylation sites. *Mol Cell Proteomics* 2012;**11**:M111–011080.
128 Chuh KN, Pratt MR. Chemical methods for the proteome-wide identification of post-translationally modified proteins. *Curr Opin Chem Biol* 2015;**24**:27–37.
129 Blum G, Bothwell IR, Islam K, Luo M. Profiling protein methylation with cofactor analog containing terminal alkyne functionality. *Curr Protocols Chem Biol* 2013;**5**:67–88.
130 Yap MC, Kostiuk MA, Martin DD, Perinpanayagam MA, Hak PG, Siddam A, Majjigapu JR, Rajaiah G, Keller BO, Prescher JA, Wu P. Rapid and selective detection of fatty acylated proteins using ω-alkynyl-fatty acids and click chemistry. *J Lipid Res* 2010;**51**:1566–1580.
131 Hennrich ML, Gavin AC. Quantitative mass spectrometry of post-translational modifications: keys to confidence. *Sci Signal* 2015;**8**:re5.
132 Schmidt A, Forne I, Imhof A. Bioinformatic analysis of proteomics data. *BMC Syst Biol* 2014;**8**:S3.
133 Beltrao P, Bork P, Krogan NJ, van Noort V. Evolution and functional cross-talk of protein post-translational modifications. *Mol Syst Biol* 2013;**9**:714.
134 Lu, Z., Cheng, Z., Zhao, Y., Volchenboum, S.L.: Bioinformatic analysis and post-translational modification crosstalk prediction of lysine acetylation. *PLoS One* 2011;**6**:e28228.

5

Tyrosine Nitration

Xianquan Zhan[1], Ying Long[1] and Dominic M. Desiderio[2]

[1] Key Laboratory of Cancer Proteomics of Chinese Ministry of Health, Xiangya Hospital, Central South University, Changsha, Hunan, P. R. China
[2] The Charles B. Stout Neuroscience Mass Spectrometry Laboratory, Department of Neurology, College of Medicine, University of Tennessee Health Science Center, Memphis, Tennessee, USA

5.1 Overview of Tyrosine Nitration

Tyrosine nitration of a protein is an addition of a nitro group ($-NO_2$) to position-3 of the phenolic ring of a tyrosine residue in a protein. Tyrosine nitration is a relatively chemically stable oxidative/nitrative modification and is a marker of oxidative injuries. Tyrosine nitration alters the functions of proteins associated with multiple physiological and pathological processes such as cancer, inflammation disease, and neurodegenerative diseases [1–7]. Tyrosine nitration occurs in a normal physiological status and increases in a pathology [3, 8]. Tyrosine nitration changes physical and chemical properties relative to a tyrosine [2, 9, 10]. A nitro group is an electron-withdrawing group that decreases the electron density of the phenolic ring of a nitrotyrosine relative to a tyrosine. The decreased electron density decreases the affinity between enzyme–substrate, ligand–receptor, or antigen–antibody when tyrosine nitration occurs within these binding regions [2]. Tyrosine nitration also alters the pK_a of the phenolic hydroxyl group of a nitrotyrosine ($pK_a = \sim 7.1$) significantly lower than that of tyrosine ($pK_a = \sim 10$) [2, 9], and the spectrophotometric properties of a nitrotyrosine residue are different from tyrosine. A nitrotyrosine can be reduced to a stable aminotyrosine [10], which is useful for further study (discussed later). Tyrosine nitration is known to occur within a tyrosine kinase phosphorylation motif ([R/K]-XX-[D/E]-XXX-Y or [R/K]-XXX)-[D/E]-XX-Y; Y = the phosphorylation site) to impact on the tyrosine phosphorylation signaling system that is important in pathologies [11–16]. Moreover, an *in vivo* denitrase indicates that tyrosine nitration and denitration are a reversible

Analysis of Protein Post-Translational Modifications by Mass Spectrometry,
First Edition. Edited by John R. Griffiths and Richard D. Unwin.
© 2017 John Wiley & Sons, Inc. Published 2017 by John Wiley & Sons, Inc.

dynamic reaction similar to phosphorylation and dephosphorylation [17, 18]. Therefore, tyrosine nitration not only results from oxidative injuries but also alters protein structure and function. It involves multiple biological consequences, for example, sensitivity to proteolytic degradation, modification of enzymatic activities, impact on protein tyrosine phosphorylation, immunogenicity, and implication in disease [19–23].

Characterization of tyrosine nitration and accurate determination of each nitration site are essential to address biological functions and roles of tyrosine nitration [3]. However, it is very challenging to identify tyrosine nitration due to the varied mass spectrometry (MS) behaviors of a nitro group, its extreme low abundance *in vivo* [24, 25], and limited MS sensitivity [2, 26–29]. Selection of an appropriate MS, in combination with chemical derivation [27] and preferential enrichment [2, 3, 30], is needed to identify tyrosine nitration [3]. The varied MS behaviors of a nitro group involve a characteristic photodecomposition pattern of a nitro group in UV-laser-based matrix-assisted laser desorption/ionization (MALDI)-MS analysis of a nitroprotein [27–29] but not in electrospray ionization (ESI)-MS [27, 31–35]. This photodecomposition pattern of a nitro group decreases signal intensities of a nitropeptide and complicates the interpretation of a MALDI-MS spectrum. However, the characteristic photodecomposition pattern can confirm the existence of a nitro group with MALDI-MS [27]. A nitrotyrosine residue is easily reduced to a more stable aminotyrosine residue for MS analysis [10]. Thus, chemical derivation is helpful for MS analysis. The *in vivo* low abundance of endogenous nitrotyrosine sites (1 in 10^6 tyrosine residues) and the MS sensitivity requirement require a preferential enrichment of nitroproteins or nitropeptides from a biological extract before MS analysis [26, 36, 37].

Several chemical derivation and targeted enrichment approaches have been published [38]: (i) antinitrotyrosine antibody-based immunoaffinity enrichment of nitropeptides [39] or nitroproteins [2, 40]; (ii) use of selective chemoprecipitation and subsequent release of tagged species (conversion of nitro group to a small 4-formylbenzylamido tag) for liquid chromatography–tandem mass spectrometry (LC-MS/MS) analysis of nitropeptides [41]; (iii) conversion of a nitro group to an amino group coupled with targeted enrichment [42] by first acetylating free amines, followed by conversion of nitrotyrosine to aminotyrosine, and then biotinylation of aminotyrosine; (iv) conversion of a nitro group to an amino group coupled with derivation of the amino group [43]. Briefly, protection of α- and ε-amino groups in a protein or peptide with $^{13}C_0/^{13}C_4$- or D_0/D_6-acetic anhydride, reduction of nitrotyrosine to aminotyrosine with sodium dithionate (also known as sodium hydrosulfite), and derivation of aminotyrosine with 1-(6-methyl[D_0/D_3]nicotinoyloxy) succinimide; (v) reduction of the nitro group to an amino group and dansylate with dansyl chloride, followed by MS^n analysis [44, 45]; (vi) a new quantitative identification strategy used isobaric tags for relative and absolute quantification (iTRAQ) reagents to selectively label nitrotyrosine residues (not primary amines) coupled with MS analysis [46]; (vii) after the use of "light"- and "heavy"-labeled acetyl groups to block N-terminal and lysine residues of tryptic nitropeptides, reduction of nitrotyrosine to aminotyrosine with sodium dithionite and derivatization of light- and heavy-labeled aminotyrosine peptides with either tandem mass tags (TMT) or iTRAQ, respectively [47]; and (viii) combining fractional diagonal chromatography (COFRADIC) [48,49]-peptide sorting, which is based on a hydrophilic shift after the reduction of the nitro group to its amino counterpart, with ESI-MS [48] and MALDI-MS [49]. Except for the proteomics method based on antinitrotyrosine antibodies and gel-based separation, chemical derivation, precursor ion scanning, and multidimensional chromatography have been used to characterize and quantify tyrosine nitration in a protein and in its modification sites [37, 50].

In-depth analysis of tyrosine nitration in a protein is necessary to address fully the biological functions and roles of tyrosine nitration, and several aspects should be considered here [38]. It is important to be able to quantify tyrosine nitration in a protein in a pathological status and the degree of nitration with quantitative proteomics [47]. Quantification of body fluid biomarkers (nitroprotein, nitropeptide) is important for prediction, diagnosis, and prognosis of a disease with quantitative body fluid nitroproteomics and nitropeptidomics [26, 51]. It is also important to be able to locate nitrotyrosine sites within an important protein domain and motif with bioinformatics [2, 52], to clarify important protein system networks that involve nitroproteins with systems biology [26, 53], and to reveal the three-dimensional structure of a nitroprotein to address the influences of local primary structure on tyrosine nitration [54]. Finally, it is important to also discover the effects of tyrosine nitration on protein function toward development of a drug against tyrosine nitration [26, 55].

5.2 MS Behavior of Nitrated Peptides

MS is a key technique to identify tyrosine nitration in a protein and each modification site. However, the MS behaviors of a nitropeptide are obviously different between MALDI UV-laser MS and infrared-MALDI-Fourier transform ion cyclotron resonance mass spectrometry (IR-MALDI-FT-ICR-MS) [27–29, 56]; between MALDI UV-laser MS and ESI-MS [27–29]; among different fragmentation models, including collision-induced dissociation (CID), electron transfer dissociation (ETD), electron capture dissociation (ECD), and metastable atom-activated dissociation (MAD)-MS [57–59]; and among different types of CID-MS/MS instruments. The various MS behaviors complicate the interpretation of MS and MS/MS spectra of a nitropeptide. Recognition of these various MS behaviors of a nitropeptide can assist in accurate identification of tyrosine nitration in a peptide.

For MALDI UV-laser MS, a photochemical decomposition pattern ($[M+H]^+$, $[M+H-16]^+$, $[M+H-30]^+$, and $[M+H-32]^+$) of the nitro group ($-NO_2$) is induced by the high-energy laser at 337 nm to decrease the signal intensity of the precursor ion of a nitropeptide and complicate a MALDI-MS spectrum [27–29, 32]. Figure 5.1 summarizes the production of dityrosine and nitrotyrosine and likely products of photochemical decomposition of a nitrotyrosine [60].

Evidence from several experiments strongly supports this photochemical decomposition pattern of a nitro group with MALDI UV-laser and the complicated MS spectrum. Studies of a synthetic nitropeptide AAFGY($-NO_2$)AR ($[M+H]^+ = 800.4$) with MALDI UV-Laser MS [29] found a photochemical decomposition pattern ($[M+H]^+$, $[M+H-16]^+$, $[M+H-14]^+$, $[M+H-32]^+$, and $[M+H-30]^+$) in the MS spectrum that corresponded to m/z 800.4, 784.4, 786.4, 768.4, and 770.4, respectively (Figure 5.2).

The $[M+H]^+$ ion (m/z 800.4) represents the nitrotyrosine (Tyr-NO_2)-containing peptide; $[M+H-16]^+$ (m/z 784.4) the nitrosotyrosine (Tyr-NO)-containing peptide after loss of an oxygen atom from a nitro group; $[M+H-14]^+$ (m/z 786.4) the hydroxylaminotyrosine (Tyr-NHOH)-containing peptide after reduction of the nitroso (Tyr-NO) group; $[M+H-32]^+$ (m/z 768.4) the triplet nitrenetyrosine (Tyr-N)-containing peptide after loss of two oxygen atoms; and $[M+H-30]^+$ (m/z 770.4) the aminotyrosine (Tyr-NH_2)-containing peptide after reduction of triplet nitrene (Tyr-N) group. The $-30/32$ Da photodecomposition products (Tyr-NH_2 and Tyr-N) were obviously lower in abundance than the corresponding $[M+H]^+$ ion (Tyr-NO_2) and the $-14/16$ Da photodecomposition products

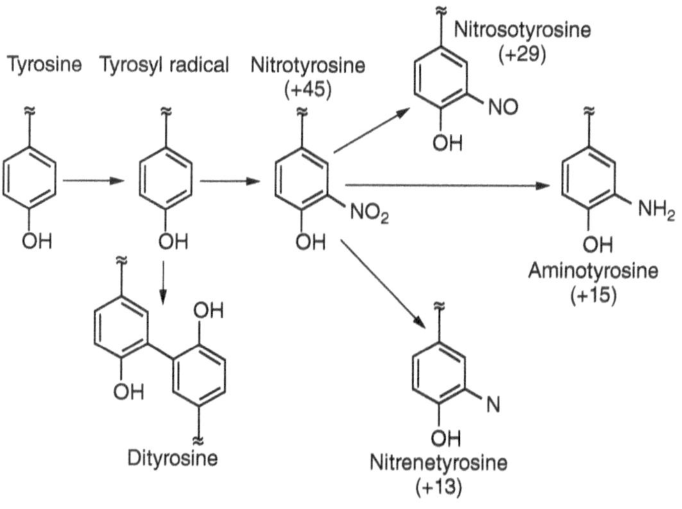

Figure 5.1 Generation of dityrosine and nitrotyrosine and likely products from nitrotyrosine photochemical decomposition. Source: Turko & Murad 2005 [60], Reproduced with permission of Elsevier, Desiderio [38]. Reproduced with permission of Wiley.

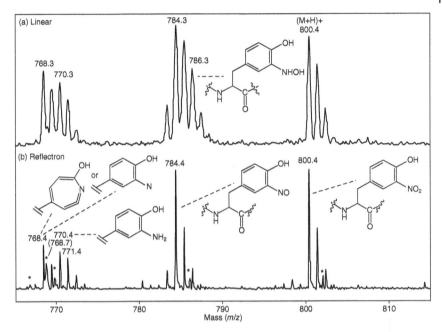

Figure 5.2 Photodecomposition pattern of the synthetic nitropeptide AAFGY(–NO₂)AR in the MALDI-TOF spectrum in the (a) linear mode and (b) reflectron mode. The structure of 3-nitrotyrosine and the proposed photodecomposition products are shown next to various ions. Several small ions (asterisk) might represent metastable peaks (see text for details). A slight increase in the abundance of the ion at m/z 771.4 over what would be expected for the ^{13}C isotope peak for the aminotyrosine products at m/z 770.4 in the linear and reflectron spectra suggests that a small amount of a catechol product might have formed as well. Source: Sarver 2001, [29]. Reproduced with permission of Elsevier, Desiderio, 2015. [38] Reproduced with permission of Wiley.

(Tyr-NHOH and Tyr-NO). These MALDI UV-induced photodecompositions were also confirmed with nitropeptides from tetranitromethane (TNM)-nitrated bovine serum albumin (BSA) [29], TNM-treated angiotensin II ([M+H]$^+$, m/z 1092.5) [28], and synthetic peptides, including leucine enkephalin (LE1: Y-G-G-F-L, molecular weight (MW) = 555.1818 Da), nitro-Tyr-leucine enkephalin [LE2: (3-NO₂)Y-G-G-F-L, MW = 600.0909 Da], and d₅-Phe-nitro-Tyr-leucine enkephalin [LE3: (3-NO₂)Y-G-G-(d₅)F-L, MW = 605.1818 Da] [27]. The base peak intensity of the [M+H]$^+$ ion of leucine enkephalin (LE1, NL = 1.01E5) was much higher than that of nitro-Tyr leucine enkephalin (LE2, NL = 3.25E4) and d(5)-Phe-nitro-Tyr leucine enkephalin (LE3, NL = 9.09E4) to demonstrate that photochemical decomposition decreased ion intensity and complicated the MS spectrum (Figure 5.3) [27].

For vMALDI-MS/MS analysis of LE1, LE2, and LE3, b- and a-ions were the most intense fragment ions compared with y-ions (Figure 5.4) [27]; these data

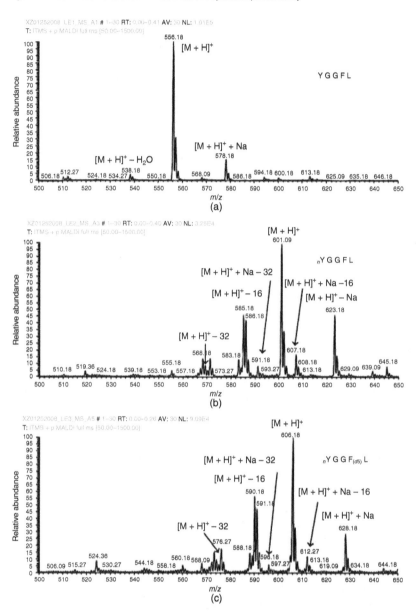

Figure 5.3 MALDI-MS spectra of LE1 (a), LE2 (b), and LE3 (c). nY = nitro-Tyr. F(d_5) = Phe residue with five ^{2}H (d) atoms. Source: Reproduced from Zhan and Desiderio [27], with permission from Elsevier Science, copyright 2009; Reproduced from Zhan et al. [38], with permission from Wiley-VCH, copyright 2015; and reproduced from Zhan et al. [26], with permission from Hindawi Publishing Corporation. Copyright 2013 remains with authors due to the open-access article under the Creative Commons Attribution License.

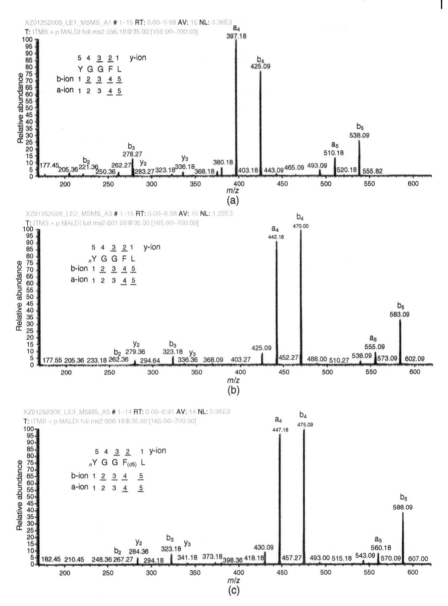

Figure 5.4 MS^2 spectra of LE1 (a), LE2 (b), and LE3 (c). nY = nitro-Tyr. F(d$_5$) = Phe residue with five ^{2}H (d) atoms. Source: Reproduced from Zhan and Desiderio [27], with permission from Elsevier Science, copyright 2009; Reproduced from Zhan et al. [38], with permission from Wiley-VCH, copyright 2015; and reproduced from Zhan et al. [26], with permission from Hindawi Publishing Corporation. Copyright 2013 remains with authors due to the open-access article under the Creative Commons Attribution License.

have been corroborated with MALDI-MS/MS analysis of nitrated angiotensin II [28].

Compared with the unmodified peptide (LE1), more collision energy was required for optimized fragmentation of the nitropeptide (Figure 5.5a) but increased the intensity of the a_4-ion and decreased the intensity of the b_4-ion (a-ion = loss of CO from a b-ion) (Figure 5.5b).

Furthermore, optimized laser fluence maximized fragmentation of the nitropeptide. Although MS3 analysis confirmed the MS^2-derived amino acid sequence, MS^3 analysis requires a higher amount of peptide relative to MS^2 [27]. Thus, MS^3 analysis might not be suitable for routine analysis of endogenous low-abundance nitroproteins. Only when a target is determined can MS^3 be used for confirmation.

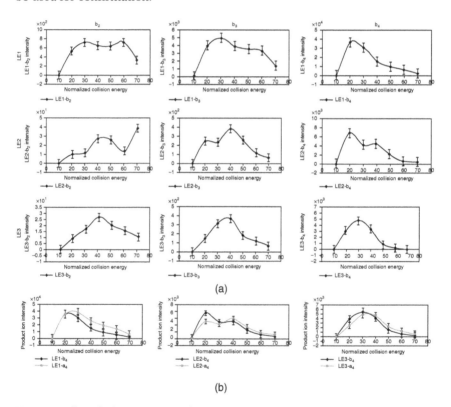

Figure 5.5 Effect of collision energy on fragmentation of nitropeptides. (a) Relationship between collision energy and product ion intensity ($n = 3$). (b) Relationship between collision energy and product ion b_4 and a_4 intensities ($n = 3$). Source: Reproduced from Zhan and Desiderio [27], with permission from Elsevier Science, copyright 2009; Reproduced from Zhan et al. [38], with permission from Wiley-VCH, copyright 2015; and reproduced from Zhan et al. [26], with permission from Hindawi Publishing Corporation. Copyright 2013 remains with authors due to the open-access article under the Creative Commons Attribution License.

To detect a nitropeptide, the amount of peptide must reach the sensitivity of a mass spectrometer; for synthetic nitropeptides, the sensitivity of vMALDI-LTQ was 1 fmol for MS detection and 10 fmol for MS^2 detection [27]. The precise reason why MALDI with laser light at 337 nm induces photodecompositions of a nitro group in a nitropeptide – but not with infrared light – and the structures of these decomposition products remain unknown [29]; however, it is probable that the higher energy content at 337 nm induces loss of one or two oxygen atoms of the nitro group in a nitropeptide [29]. These photochemical decomposition patterns with a MALDI laser can confirm the presence of nitrotyrosine residue in the analyzed sample. IR-MALDI-FT-ICR-MS is a highly efficient method to determine protein nitration but does not fragment [M+H]+ ions of nitrotyrosine peptides [56]. Moreover, for MALDI-MS analysis of a nitropeptide, the optimum matrix is sinapinic acid but not 2,5-dihydroxybenzoic acid [61].

For ESI-MS, no chemical decomposition of a nitro group is found in a spectrum [27–29, 31–35] as confirmed with ESI-MS analysis of TNM-nitrated angiotensin II [DRVY(–NO$_2$)IHPF; MW = 1090.76 Da] [28]. A mononitrated angiotensin II ion ([M+2H]$^{2+}$ m/z 546.38, [NO$_2$-Tyr]-angiotensin II) and dinitrated angiotensin II ion ([M+2H]$^{2+}$ m/z 568.85, [(NO$_2$)$_2$-Tyr]-angiotensin II) are also found in the ESI-MS spectrum without decomposition (Figure 5.6).

The fragmentation of [M+2H]$^{2+}$ precursor ions for mononitrated angiotensin II at m/z 546.38 and for dinitrated angiotensin II at m/z 568.85 (Figure 5.7) demonstrates characteristic immonium ions at m/z 181.06 for mononitrated tyrosine and at m/z 226.0 for dinitrated tyrosine in an ESI-MS/MS spectrum (Figure 5.7), which indicate the presence of a nitrotyrosine residue.

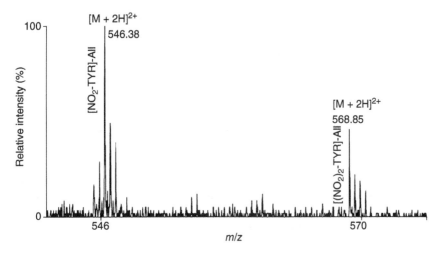

Figure 5.6 ESI-MS spectrum of nitrated angiotensin II to show mono- and dinitrated angiotensin II. Source: Desiderio, 2015. [38] Reproduced with permission of Wiley.

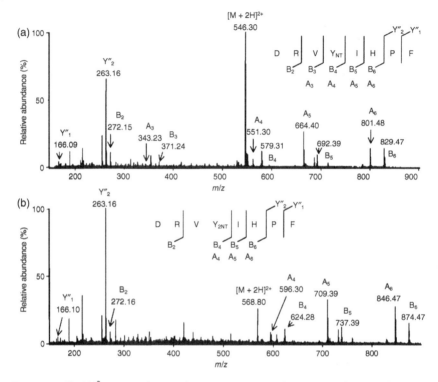

Figure 5.7 The MS^2 spectra of nitrated angiotensin II peptides. The doubly charged ions were selected as precursor ions for mononitrated angiotensin II at *m/z* 546.30 (a) and for the dinitrated angiotensin II at *m/z* 568.80 (b). Source: Desiderio, 2015. [38] Reproduced with permission of Wiley, Petersson 2001. [28]. Reproduced with permission of Wiley.

The characteristic immonium ion-based precursor ion scan spectrum accurately identifies nitropeptides in complex sample (Figure 5.8). The ESI-MS behavior of a nitropeptide and the precursor ion scans for an immonium ion at *m/z* 181.06 were further confirmed with ESI-MS analysis of TNM-nitrated BSA [28].

For MS/MS analysis of a nitropeptide, different fragmentation modes result in different fragmentation behaviors of precursor ions of a nitropeptide. Fragmentation methods include CID, ECD, ETD, and MAD [57–59]. Nitration does not appear to affect the CID behavior of peptides. However, for doubly charged peptides, production of ECD sequence fragments is severely inhibited with nitration, although ECD of triply charged nitrotyrosine-containing peptides produces some singly charged sequence fragments. ECD of nitropeptides was characterized by multiple losses of small neutral species, including hydroxyl radicals, water, and ammonia. The origin of neutral losses was investigated with activated ion (AI) ECD. Loss of ammonia appears to be the result

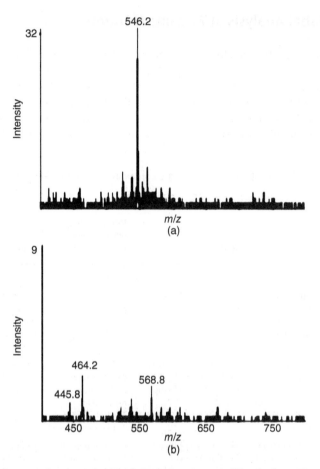

Figure 5.8 Precursor ion scan spectra of nitrated angiotensin II for the formation of immonium ion at *m/z* 181.06 for mononitrated tyrosine (a) and at *m/z* 226.0 for dinitrated tyrosine (b). Source: Desiderio, 2015. [38] Reproduced with permission of Wiley, Petersson 2001. [28]. Reproduced with permission of Wiley.

of noncovalent interactions between a nitro group and protonated lysine side chains [57, 58]. Further studies have found that high kinetic energy helium MAD produced extensive backbone fragmentation with significant retention of post-translation modifications (PTMs). Although the high electron affinity of a nitrotyrosine moiety quenched radical chemistry and fragmentation in ECD and ETD, MAD does produce numerous backbone cleavages in the vicinity of the nitration. Compared with CID, MAD produced more fragment ions and differentiated I/L residues in nitrated peptides. MAD induced radical ion chemistry, even in the presence of strong radical traps, therefore offers unique advantages to ECD, ETD, and CID to determine a nitropeptide [59].

5.3   Global Analysis of Tyrosine Nitration

Protein tyrosine nitration alters the bioactivity of a protein and plays important roles in multiple cellular, physiological, and pathological processes [62–64]. Clarification of endogenous tyrosine nitration at the proteome level under different pathophysiological conditions would elucidate the biological functions and roles of tyrosine nitration in pathology. The global proteomic methods for the identification of nitrated proteins (Table 5.1) mainly include

Table 5.1 Identification of *in vivo* nitrotyrosine-containing proteins in different pathological conditions.

References	Specimen	Methods	Nitroprotein and nitrotyrosine sites	Remark
(i) Inflammation-related disease				
Ghosh et al. [65]	Lung tissues from allergen-induced murine model of asthma	2D western blot and LC-ESI-MS/MS	Twenty-seven putative nitrated proteins were identified	Inflammation-related disease. No nitrotyrosine sites were identified
Lanone et al. [66]	Rectus abdominis muscle from the same control and septic patients	Western blot, MALDI-TOF-PMF, and molecular modeling	Inducible nitric oxide synthase (iNOS) was nitrated at Tyr299, Tyr336, Tyr446, and Tyr698	Analysis coupled with iNOS three-dimensional crystal model
Chatterjee et al. [67]	Spleens from LPS-induced systemic inflammation model of C57BL6/J mice	1D western blot and LC-ESI-MS/MS	Carboxypeptide B1 (CPB1) was nitrated at specific tyrosine sites	Inflammation-related disease
Dhiman et al. [68]	Plasma from patients with Chagas' disease	1D and 2D western blot, MALDI-PMF, and LC-ESI-MS/MS	Fifty differentially expressed/nitrated proteins were identified	Inflammation-related disease. No nitrotyrosine sites were identified
Ghesquiere et al. [48]	Serum proteome from a C57BL6/J mouse with septic shock	COFRADIC and ESI-MS	α2-Macroglobulin, apolipoprotein A-I, haptoglobin, and vitamin D–binding protein were nitrated at 6 specific tyrosine sites	Inflammation-related disease. Nitrotyrosine sites were identified

Table 5.1 (Continued)

References	Specimen	Methods	Nitroprotein and nitrotyrosine sites	Remark
		(ii) Aging and aging-related diseases		
Sharov et al. [69]	Skeletal muscle from 6-month-old and 34-month-old Fisher 344/Brown Norway F1 hybrid rats	1D western blot and HPLC-ESI-MS/MS	Phosphorylase b was found in the accumulation of 3-nitrotyrosine on Tyr113, Tyr161, and Tyr573. Nitration on Tyr 113 was detected in 6- and 34-month-old rat; nitration on Tyr161 and Tyr573 was detected only in 34-month-old rat	Endogenous. Nitration is accumulated with aging
Kanski, Hong, and Schoneich [70]	Skeletal muscle from 34-month-old Fisher 344/Brown Norway F1 rats	IEF, 1D western blot, and ESI-MS/MS	Eleven nitroproteins and 12 nitrotyrosine sites were identified	Endogenous
Marshall et al. [71]	Liver from young (19–22 weeks) and old (24 months) C57/BL6 male mice	1D western blot and LC-ESI-MS/MS	Six putative nitrated proteins were identified	Nitration is associated with aging. No nitrotyrosine site was identified
Kanski et al. [72]	Heart from 5- and 26-month-old Fisher 344/BN F1 hybrid rats	1D and 2D western blot, ESI-MS/MS	Forty-eight putative nitrated proteins. Nitration at Tyr105 of the electron transfer flavoprotein was identified	Endogenous. Heart homogenate and heart mitochondria. Nitration is the effect of biological aging. Not every protein was identified in its nitrotyrosine site
		(iii) Tumor		
Nakagawa et al.[73]	C6 rat glioma cell line	HPLC, MALDI-PMF	Cytochrome c was nitrated at Tyr 48, Tyr67, and Tyr74	*In vitro* nitrated with peroxynitrite

(*Continued*)

Table 5.1 (Continued)

References	Specimen	Methods	Nitroprotein and nitrotyrosine sites	Remark
Fiore et al. [74]	Human glioma tissues	Immunohistochemistry, 1DE-MALDI-PMF	Tubulin was nitrated at Tyr224 in glioma grade IV but not in grade I and noncancerous brain tissue	Endogenous. Laser-induced decomposition
Zhan and Desiderio [1]	Human pituitary postmortem tissue	2D western blot and vMALDI-MS/MS	Four nitroproteins and 4 nitrotyrosine sites were identified	Endogenous. Laser-induced decomposition
Zhan and Desiderio [75]	Human pituitary postmortem tissue	2D western blot and vMALDI-MS/MS	Four nitroproteins and 4 nitrotyrosine sites were identified	Endogenous. Laser-induced decomposition
Zhan and Desiderio [2]	Human nonfunctional pituitary adenoma tissue	NTAC-vMALDI-MS/MS	Nine nitroproteins, 10 nitrotyrosine sites, and 3 nitroprotein-interacting proteins were identified	Endogenous. Laser-induced decomposition
(iv) Neurodegenerative diseases				
Sacksteder et al. [76]	Brain from C57BL/6J mice	SCX-LC-ESI-MS/MS	Twenty-nine nitroproteins and 31 nitrotyrosine sites	Endogenous. Links to neurodegenerative disease
Zhang et al. [78]	Brain from C57BL/6J mice	LC-ESI-MS/MS		Endogenous
Danielson et al. [79]	Dox-inducible MAO-B PC12 cells	LC-ESI-MS/MS	α-synuclein was nitrated at Tyr39	Model of Parkinson's disease
Casoni et al. [80]	Spinal cord from Tg SOD1 G93A mice and Tg SOD1 WT mice	2D western blot and MALDI-PMF	Thirty-two nitroproteins and 16 nitrotyrosine sites	Endogenous. Laser-induced decomposition. Familial amyotrophic lateral sclerosis
Yoon et al. [81]	Mouse hippocampal cell line HT22	2D western blot and MALDI-PMF	Thirteen nitroproteins were detected	Glutamate-treated HT22 cells

Table 5.1 (Continued)

References	Specimen	Methods	Nitroprotein and nitrotyrosine sites	Remark
(v) Cardiovascular system and related diseases				
Liu et al. [82]	Male C57BL/6 mice myocardial ischemia–reperfusion injury (I/R) model	1D or 2D western blot, LC-ESI-MS/MS	Twenty-three nitroproteins were identified. Ten of them were from mitochondria	Endogenous. No nitrotyrosine sites were identified
Chen et al. [84]	Sprague Dawley rat *in vivo* myocardial regional ischemia–reperfusion model	1DE-LC-ESI-MS/MS	Flavin subunit is nitrated at Tyr56 and Tyr142	Endogenous. Mitochondrial complex II in the postischemic myocardium
Ai et al. [85]	Endothelial cell of human coronary arteries	LC-ESI-MS/MS	LDL was nitrated	
(vi) The neurovisual system				
Justilien et al. [86]	Mouse posterior eyecups	NTAC, SDS-PAGE, LC-ESI-MS/MS	Eight nitroproteins and nine nitrotyrosine sites	Endogenous. SOD2 knockdown mouse model of early AMD
Murdaugh et al. [87]	Human Bruch's membrane	HPLC, ESI-MS/MS	A2E was nitrated	Endogenous. Nitro-A2E is a specific biomarker of nitrosative stress in Bruch's membrane and its concentration is directly related to tissue age
Palamalai et al. [88]	Photoreceptor rod outer segments of cyclic light-reared rats treated or not with the antioxidant	2D western blot and LC-ESI-MS/MS	Ten putative nitroproteins were identified	Endogenous. No nitrotyrosine sites were identified

(Continued)

Table 5.1 (Continued)

References	Specimen	Methods	Nitroprotein and nitrotyrosine sites	Remark
		(vii) Diabetes		
Kato et al.[90]	Healthy and diabetic human urine	LC-ESI-MS/MS	Urine nitrotyrosine	Endogenous
		(viii) Kidney disease		
Piroddi et al. [92]	Plasma from kidney disease patients	2DE and LC-ESI-MS/MS	Fourteen tentative nitroproteins and 7 nitrotyrosine sites were identified	Endogenous
		(ix) Plant diseases		
Chaki et al. [93]	Sunflower hypocotyls	2D western blot and LC-ESI-MS/MS	Twenty-one putative nitroproteins were identified	No nitrotyrosine sites were identified
		(x) Others		
Aslan et al. [95]	Liver and kidney from sickle cell disease mouse	Western blot and precipitation, MALDI-PMF, LC-ESI-MS/MS	Actin was nitrated at Tyr91, Tyr198, and Tyr240	Endogenous
Lee et al. [35]	Hippocampus from smoke inhalation rat model	2D western blot and MALDI-PMF or MALDI-MS/MS	Five nitroproteins of mitochondrial proteins were identified	Endogenous. No nitrotyrosine sites were identified
Reed et al. [96]	Traumatic brain-injured rats	2D western blot and MALDI-PMF	Several nitroprotein such as GSH were identified	
Ulrich et al. [98]	Human lung tissues and blood samples, animal granule protein preparation	Western blot and MALDI-PMF	Six nitroproteins and nitrotyrosine sites at Tyr349 in eosinophil peroxidase (EPO) and Tyr33 in bother eosinophil cationic protein (ECP) and eosinophil-derived neurotoxin (EDN)	Endogenous
Webster, Brockman, and Myatt [97]	Human placenta	1D western blot and MALDI-PMF	p38 MAPK was nitrated	No nitrotyrosine sites were identified

Table 5.1 (Continued)

References	Specimen	Methods	Nitroprotein and nitrotyrosine sites	Remark
Casanovas et al. [99]	Lipoprotein lipase of bovine and rat	2D western blot and LC-ESI-MS/MS	Lipoprotein lipase was nitrated at Tyr95, Tyr164, Tyr 316	Endogenous
Hamilton et al. [101]	Human plasma	LC-ESI-MS/MS	Low-density lipoprotein (LDL) was nitrated at Tyr276, Tyr666, and Tyr720 of LDL-α1, Tyr2524 of LDL-α2, Tyr4141 of LDL-α3, Tyr3139, Tyr3205, and Tyr3489 of LDL-β2	
Sharov et al. [102]	Rabbit muscle	LC-ESI-MS/MS	Glycogen phosphorylase b was nitrated at 28 nitrotyrosine sites	
Zhu et al. [103]	Liver from SOD1$^{-/-}$ and WT C57BL/6 mice	1DE, LC-ESI-MS/MS	Ten candidate nitrated proteins were identified	No nitrotyrosine sites were identified
Chen and Chen [104]	Human blood samples from smokers and nonsmokers	LC-ESI-MS/MS under the selected reaction monitoring (SRM) mode	Hemoglobin was nitrated at Tyr24 and Tyr 42 (α-globin), Tyr130 (β-globin)	Nitration of human hemoglobin is associated with cigarette smoking
Sekar et al. [105]	Mast cells	2D western blot and LC-ESI-MS/MS	Aldolase was nitrated	No nitrotyrosine sites were identified
Ohma and Brautigan [106]	Human peripheral blood mononuclear cells	LC-ESI-MS/MS	Protein phosphatase 2A was nitrated at Tyr284	
Redondo-Horcajo et al. [107]	Endothelial cells from bovine aortas and mouse lung	LC-ESI-MS/MS	Manganese superoxide dismutase (MnSOD) was nitrated at Tyr34	*In vitro* nitrated with cyclosporine A

Note: Reproduced from Zhan et al. [38], with permission from Wiley-VCH, copyright 2015.

one-/two-dimensional (1D/2D) western blot coupled with MALDI peptide mass fingerprint (PMF) or LC-ESI-MS/MS, nitrotyrosine affinity column (NTAC)-MS/MS, iTRAQ-strong ion exchange (SCX)-LC-ESI-MS/MS, selected reaction monitoring (SRM), and COFRADIC coupled with ESI-MS.

Endogenous nitroproteins and nitration sites have been identified in multiple different pathophysiological conditions with proteomics (Table 5.1). Tyrosine nitration is involved in inflammation-related diseases. Nitroproteins were identified in bronchial epithelial cells and bronchoalveolar lavage with asthma [65], a septic patient's rectus abdominis muscle [66], experimental sepsis [67], Chagas' disease [68], and a serum sample of a C57BL6/J mouse model with septic shock [48]. Tyrosine nitration is involved in aging and aging-related diseases. Nitroproteins were discovered in aging rat skeletal muscle [69, 70], mouse liver [71], and rat heart [72]. Tyrosine nitration is clearly involved in tumorigenesis. Nitroproteins were identified in rat glioma cell lines [73] and human gliomas [74]. Eight nitroproteins were identified from human pituitary control tissue [1, 75], and nine nitroproteins, three nitroprotein-interacted proteins, and ten nitrotyrosine sites were identified from a pituitary adenoma [2] (Table 5.2). Tyrosine nitration is involved in neurodegenerative diseases. Nitroproteins were identified in mouse brain [76–78], Parkinson's disease [79], spinal cords of a mouse model of familial amyotrophic lateral sclerosis [80], and HT22 hippocampal cells [81]. Tyrosine nitration is involved in the cardiovascular system and related diseases. Nitroproteins have been identified in ischemia–reperfusion injury [82–84], vascular [85], and mouse heart [77, 78]. Tyrosine nitration is involved in the neurovisual system. Nitroproteins were identified in SOD2 knockdown mouse eyecup [86], human Bruch's membrane [87], and photoreceptor rod outer segments [88]. Tyrosine nitration is involved in diabetes. Nitroproteins have been discovered in diabetic rats [89], a diabetic patient's urine [90], and diabetic mellitus patients [91]. Tyrosine nitration is also involved in kidney disease, where nitroproteins have been identified in a kidney disease patient's plasma [92], diseases in plants [93, 94], and in sickle cell disease [95]; rat hippocampus after acute inhalation of combustion smoke [35]; traumatic brain-injured rats [96]; placenta/preeclampsia [97]; eosinophil granule toxins [98]; hypertriglyceridemia [99]; human plasma [100, 101]; rabbit muscle [102]; murine liver [103]; human hemoglobin associated with cigarette smoking [104]; mast cells [105]; human peripheral blood mononuclear cells [106]; and endothelial cells [107].

5.4 Enrichment Strategies

MS is the key technique to identify nitropeptides and nitroproteins and to accurately locate each nitration site within the amino acid sequence of a nitroprotein [1, 2, 75]. However, because of unique MS behaviors of nitropeptides,

Table 5.2 Nitroprotein and unnitrated protein identified from pituitary adenoma [2] and control tissue [1,75].

Pituitary adenoma		Pituitary control	
Protein name	nY site	Protein name	nY site
Nitrated protein		*Nitrated protein*	
Rho-GTPase-activating 5 [Q13017] (ARHGAP5)	Y^{550}	Synaptosomal-associated protein (SNAP91)	Y^{237}
Leukocyte immunoglobulin-like receptor A4 [P59901]	Y^{404}	Igα Fc receptor [P24071] (FCAR)	Y^{223}
Zinc finger protein 432 [O94892]	Y^{41}	Actin [P03996] (ACTA2, ACTG2, ACTC1)	Y^{296}
PKA β regulatory subunit [P31321] (PRKAR1B)	Y^{20}	PKG 2 [Q13237] (PRKG2)	Y^{354}
Sphingosine-1-phosphate lyase 1 [O95470]	Y^{356}, Y^{366}	Mitochondrial cochaperone protein HscB [Q8IWL3]	Y^{128}
Centaurin β1 [Q15027]	Y^{485}	Stanniocalcin 1[P52823] (STC1)	Y^{159}
Proteasome subunit α type 2 [P25787] (PSMA2)	Y^{228}	Proteasome subunit α type 2 (PSMA2)	Y^{228}
Interleukin 1 family member 6 [Q9UHA7] (IL1F6)	Y^{96}	Progestin and adipoQ receptor family member III [Q6TCH7] (PAQR3)	Y^{33}
Rhophilin 2 [Q8IUC4] (RHPN2)	Y^{258}		
Nitroprotein-interacted protein			
Interleukin-1 receptor-associated kinase-like 2 (IRAK-2) [O43187] (IRAK2)			
Glutamate receptor interacting protein 2 [Q9C0E4] (GRIP2)			
Ubiquitin [P62988] (UBB or UBC)			

Source: Zhan & Desiderio, 2006 [2]. Reproduced with permission of Elsevier; Zhan & Desiderio [1]. Reproduced with permission of Elsevier; Zhan & Desiderio [75]. Reproduced with permission of Elsevier.

Note: nY = nitrotyrosine. Modified from Zhan and Desiderio [1,2,75], with permission from Elsevier Science, copyright 2004, 2006, and 2007; Reproduced from Zhan, Wang, Desiderio [38], with permission from Wiley-VCH, copyright 2015.

extremely low abundance (1 in $\sim10^6$ tyrosines) of nitrotyrosine sites in an *in vivo* proteome [24, 25], and the limited MS sensitivity (high femtomole to low picomole level) to detect a nitropeptide [3, 9], it is necessary to isolate and preferentially enrich endogenous nitroproteins or nitropeptides before MS analysis [31, 33–35, 75].

Enrichment strategies have been developed for *in vitro* and *in vivo* nitrated samples. However, it is very important to realize that, even though most enrichment methods work well for *in vitro* nitrated samples, only a few enrichment methods work well for *in vivo* endogenous nitroproteins.

Because of easy availability of *in vitro* synthetic or nitrated nitropeptides or nitroproteins, some isolation and enrichment methods have been developed to analyze *in vitro* nitroproteins or nitropeptides in order to transit them to analyze *in vivo* nitroproteins and nitropeptides. Angiotensin II, ovalbumin (OVA), and BSA are the commonly used standard peptides and proteins nitrated *in vitro* with liquid TNM [10, 28, 29, 50, 108], peroxynitrite [109], or gaseous nitrogen dioxide and ozone ($NO_2 + O_3$) [50]. Some proteomes, for example, human plasma, were nitrated *in vitro* with liquid TNM and followed by nitroproteomics analysis to simulate *in vivo* proteome conditions [41]. In addition, because of more stable amino ($-NH_2$) group relative to the nitro ($-NO_2$) group during MS analysis, these nitroproteins or nitropeptides were reduced to aminoprotein or aminopeptide with reducing agent ($Na_2S_2O_4$) before MS analysis [10, 29, 110]. Most methods for nitroprotein enrichment have been developed based on immunoaffinity isolation and chemical derivation plus targeted enrichment with these prepared standard nitropeptides, nitroproteins, and nitroproteome samples *in vitro*, including nitroproteomic methods based on anti-3-nitrotyrosine antibody immunopurification and gel-based separations [28, 37] and methods that involve multidimensional chromatography, diagonal chromatography [48, 49], precursor ion scanning [28], and/or chemical derivation, which might characterize and quantify protein tyrosine nitration sites [37]. Among them, chemical derivation methods are most extensively studied in the analysis of *in vitro* nitropeptides and nitroproteins as modes of sample preparation before MS [36]. However, all these chemical derivation methods basically include the following procedure: reduction of the nitro to an amino group, derivation of the amino group with specific reagents, and followed by preferential enrichment. Several various strategies for chemical derivatization plus targeted enrichments have been described, including the conversion of a nitro to a more stable amino group in nitrotyrosine residue with reduction to readily distinguish aminopeptides from nonnitrated peptides with an easy-to-interpret peptide mass spectrum [10]. Due to photodecomposition of the nitro group with a MALDI UV-laser, a strategy has also been developed to acetylate N-terminal amines and ε-amines of lysine residues with acetic anhydride, reduce nitro to amino groups with sodium hydrosulfite, derivatize amino groups with 1-(6-methyl[D_0/D_3]nicotinoyloxy) succinimide, and followed by MALDI-TOF MS analysis [43]. Improved chemical-labeling methods have been designed to enrich nitropeptides independent of sequence context. In this procedure (Figure 5.9), all amines are blocked with acetylation, followed by conversion of a nitro to an amino group and biotinylation of the resultant aminotyrosine [42].

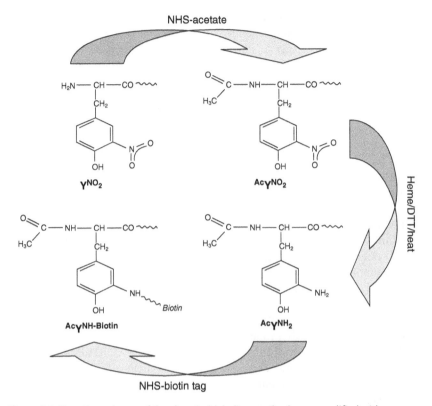

Figure 5.9 Reaction scheme of the chemical-labeling method as exemplified with an N-terminal nitrotyrosine residue. All amines were blocked with acetylation with acetic acid *N*-hydroxysuccinimide ester (NHS-acetate). Nitrotyrosine was reduced to aminotyrosine with heme and DL-dithiothreitol in a boiling water bath. The reaction sequence was completed with biotinylation of aminotyrosine with NHS-biotin. Source: From Desiderio, 2015. [38] Reproduced with permission of Wiley, Abello. (2010) [42], Reproduced with permission of Elsevier.

This entire reaction can be carried out in a single buffer without any sample cleanup or pH changes to reduce sample loss. Free biotin was removed with a strong cation exchanger, labeled peptides were subsequently enriched using an immobilized avidin column, and enriched peptides were analyzed with LC-MS/MS [42]. This method has been approved for *in vitro* nitrated samples [19, 42]. A method that specifically enriches nitropeptides to identify unambiguously nitropeptides and nitration sites with LC-MS/MS includes the conversion of nitrotyrosine to *N*-thioacetyl-aminotyrosine, followed by high-efficiency enrichment of sulfhydryl-containing peptides with thiopropyl sepharose beads [30]. Acetylation with acetic anhydride to block all primary amines, reduction of the nitro group to an amino group, derivatization of this amino group with

N-succinimidyl *S*-acetylthioacetate, and deprotection of the *S*-acetyl on *S*-acetylthioacetate will subsequently form free sulfhydryl groups [30]. This method was used to study *in vitro* nitrated BSA, human histone H1.2, and mouse brain tissue samples [30]. An alternative and quantitative strategy, which combines precursor isotopic labeling and isobaric tagging (cPILOT), increased the multiplexing capability to quantify a nitroprotein among 12 or 16 samples with TMT or iTRAQ. For this method, light- and heavy-labeled acetyl groups were used to block N-termini and lysine residues of tryptic peptides. Reduction of a nitro to an amino group with sodium dithionite is followed by derivatization of light- and heavy-labeled aminopeptides with either multiplex TMT or iTRAQ reagents [47]. This method demonstrated proof of principle to analyze *in vitro* nitrated BSA and mouse splenic proteins [47]. A new strategy, also based on the reactive and quantitative properties of iTRAQ reagents coupled with MS analysis, involves selective labeling of nitrotyrosine residues [46] to simultaneously localize and quantify nitration sites in model proteins and biological systems [46]. This method overcomes the drawback of iTRAQ quantitative proteomics that is limited to primary amines. COFRADIC [48, 49], a form of diagonal chromatography, has also been employed following the reduction of the nitro to an amino group with sodium dithionite. This method sorts peptides with reversed-phase chromatography based on a hydrophilic shift from nitropeptide (more hydrophilic) to aminopeptide (more hydrophobic) followed by ESI-MS [48] and MALDI-MS [49] identification. COFRADIC identified tyrosine nitration in a TNM-nitrated BSA and peroxynitrite-nitrated proteome of human Jurkat cells [48, 49]. Furthermore, one study has also used dansyl chloride to label nitration sites, followed by MS/MS plus a precursor ion scan [44, 45] to identify tyrosine nitration sites.

The interpretation of MS and MS/MS data of endogenous nitropeptides is very challenging. To avoid any risk of linking MS/MS spectra to an incorrect amino acid sequence, one group has combined the reduction of the nitro to an amino group and the use of the Peptizer algorithm to inspect MS/MS quality-related assumptions [111]. However, the optimal approach to determine the amino acid sequence of a nitropeptide remains a manual approach [2].

For the isolation and preferential enrichment of the much more difficult endogenous nitroproteins and nitropeptides from a biological proteome, several protocols have been developed, which are as follows: (i) two-dimensional electrophoresis (2DE)-based western blotting with antinitrotyrosine antibodies (Figure 5.10) [1, 26, 75, 112–114]. (ii) The use of nitrotyrosine affinity column (NTAC) (Figure 5.11) to enrich nitroproteins [2, 26, 86] and nitropeptides [115]. (iii) The use of COFRADIC to sort peptides according to the hydrophilic shift after the reduction of a nitro group to an amino group, followed by ESI or MALDI-MS [48]. (iv) The use of dansyl chloride to label nitration sites followed by a precursor ion scan and MS3 analysis [44, 45]. (v) After acetylation of all primary amines and reduction of nitro to amino group, derivation of amino

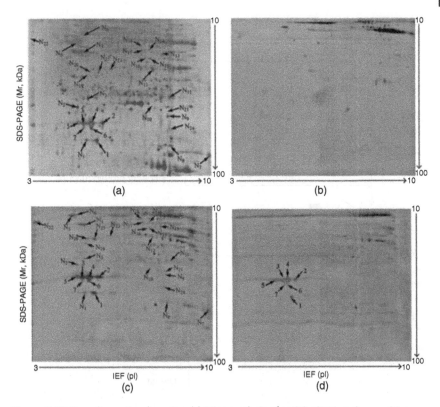

Figure 5.10 Two-dimensional western blotting analysis of anti-3-nitrotyrosine-positive proteins in a human pituitary (70 μg protein per 2D gel). (a) Silver-stained image on a 2D gel before transfer of proteins onto a PVDF membrane. (b) Silver-stained image on a 2D gel after transfer of proteins onto a PVDF membrane. (c) western blot image of anti-3-nitrotyrosine-positive proteins (anti-3-nitrotyrosine antibodies + secondary antibody). (d) Negative control of a western blot to show the cross-reaction of the secondary antibody (only the secondary antibody; no anti-3-nitrotyrosine antibody). Source: Desiderio (2007) [75], Reproduced with permission of Elsevier Science, Zhan, Wang, Desiderio [38], Reproduced with permission of Wiley-VCH, Zhan, Wang, & Desiderio (2013) [26], reproduced with permission of Hindawi Publishing Corporation.

group into a free sulfhydryl group followed by enrichment of sulfhydryl peptides with thiopropyl sepharose beads [30]. (vi) After acetylation of all primary amines in a nitropeptide, conversion of nitro to amino group, and followed by enrichment with biotinylation of an aminotyrosine (Figure 5.9) [10, 36, 42]. (vii) The use of a new tagging reagent, (3R,4S)-1-(4-(aminomethyl)phenylsulfonyl) pyrrolidine-3,4-diol (APPD) for selectively fluorogenic derivation of a nitro group in nitropeptides (after reduction to aminotyrosine) followed by boronate affinity enrichment [116]. (viii) Quantitative identification of nitroproteins and

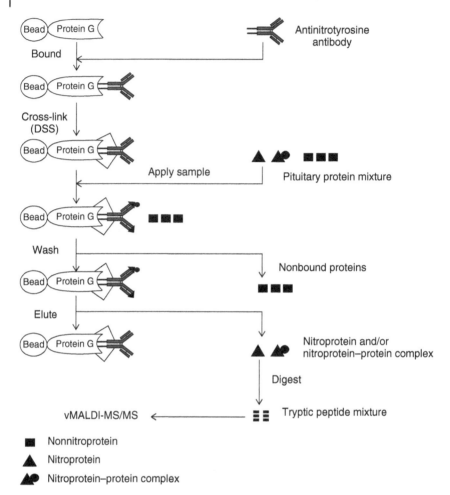

Bound

Cross-link (DSS)

Antinitrotyrosine antibody

Apply sample

Pituitary protein mixture

Wash

Nonbound proteins

Elute

Nitroprotein and/or nitroprotein–protein complex

Digest

vMALDI-MS/MS

Tryptic peptide mixture

■ Nonnitroprotein
▲ Nitroprotein
◢◤ Nitroprotein–protein complex

Figure 5.11 Experimental flowchart to identify nitroprotein and nitroprotein–protein complexes with NTAC-based MALDI-LTQ MS/MS. The control experiment (without any anti-3-nitrotyrosine antibody) was performed in parallel with the NTAC-based experiments. Source: Reproduced from Zhan and Desiderio [2], with permission from Elsevier Science, copyright 2006; Reproduced from Zhan et al. [38], with permission from Wiley-VCH, copyright 2015; and reproduced from Zhan et al. [26], with permission from Hindawi Publishing Corporation. Copyright 2013 remains with authors due to the open-access article under the Creative Commons Attribution License.

nitropeptides with TMT or iTRAQ [46, 47]; for this protocol, after the use of the "light"- and "heavy"-labeled acetyl groups to block N-termini and lysine residues of tryptic peptides, the nitro group was reduced to amino group and followed by derivation of light- and heavy-labeled aminopeptides with either multiplex TMT or iTRAQ reagents [46, 47]. This method can relatively enrich and quantitatively identify nitroproteins and nitropeptides.

Protocols (i)–(iii) have been used to identify endogenous nitration sites [1, 2, 48, 75, 86]. Protocols (iii)–(vii) have succeeded mainly in an *in vitro* nitrated peptide, protein, and proteome [10, 30, 44, 48, 49] and provide promise for studies of endogenous nitroproteins. All protocols focus on identification of nitropeptides, nitroproteins, and nitration sites. However, for discovery of disease-related nitroproteins, except for characterization of nitration sites and nitroproteins, quantitative identification of nitroproteins is needed. Protocol (viii) holds promise for that goal because it can relatively enrich nitropeptides, quantify nitroproteins, and identify nitration sites, and it has enhanced sample multiplexing capabilities [46, 47].

5.5 Concluding Remarks

Oxidative/nitrative-induced tyrosine nitration in a protein is an important modification and is associated with multiple pathophysiological processes [1–3, 26, 27, 75]. In addition, studies found a denitrase in mammalian cells and tissues. This enzyme implies that the reciprocal processes of nitration and denitration might modulate biological events and regulate cell signaling events [17, 18]. Nitration dynamically alters protein function [117], including activation or inactivation [118–120]. MS is the key technique used to identify nitroproteins and nitration sites and to understand the biological roles of tyrosine nitration [121–123]. However, characterization of endogenous nitroproteins and nitration sites is a very challenging issue because of extreme low abundance of nitration in biological samples and various MS behaviors among MALDI UV-laser-, ESI-, CID-, ECD-, ETD-, and MAD-MS. It is necessary to preferentially enrich endogenous nitroproteins and nitropeptides before MS analysis. The immunoaffinity enrichment, biotin affinity enrichment, and COFRADIC are currently developed enrichment strategies for the analysis of endogenous nitroproteins, and nitration sites have been found in different pathophysiological status. TMT- or iTRAQ-based quantitative nitroproteomics are promising methods to quantify the key nitroproteins and nitropeptides in a disease. Furthermore, protein domain/motif analysis, systems pathway analysis, and structural biological analysis of nitroproteins [38, 124] are needed to completely clarify the biological functions of tyrosine nitration.

However, it is very important to realize clearly that no high-sensitivity, high-reproducibility, and highly reliable methods currently exist for the analysis of the extreme low-abundance endogenous tyrosine nitration in a proteome [38, 125]. Many different approaches remain under development. Currently, antinitrotyrosine antibody-based immunoaffinity methods, for example, 2D western blotting and NTAC, have all been used to identify endogenous tyrosine nitration sites; however, an overwhelming amount of nonnitrated tryptic peptides negatively impacts the identification of tryptic nitropeptides in these studies. Therefore, it would be more effective to develop immunoaffinity enrichment of tryptic nitropeptides, but not nitroproteins, before MS analysis. Until now, most chemical derivation-based target enrichment methods have succeeded in *in vitro* experiments but not in *in vivo* endogenous tyrosine nitration site analysis. The COFRADIC-based methods succeeded in the identification of endogenous nitropeptides in a serum proteome; however, its throughput and sensitivity were very low, and it has not been used extensively in endogenous tissue nitroproteomes. Therefore, a better method is needed to analyze endogenous tyrosine nitration sites.

The following aspects are worth considering alone or in combination including (i) derivation of a nitro to amino group to stabilize MS behaviors, (ii) developing specific amino group tags to enrich nitrotyrosine peptides, (iii) enriching nitrotyrosine or aminotyrosine peptides but not proteins for sensitivity, (iv) choosing the appropriate ion source and collision model to fragment nitropeptide or aminopeptides, (v) developing super high-sensitivity mass spectrometers, (vi) improving liquid chromatography isolation, and (vii) developing reliable software for data analysis. We recommend the combined multiple strategies among items i–vii to maximize the coverage of endogenous tyrosine nitration sites in a proteome.

Acknowledgements

The authors acknowledge the financial support from the National Natural Science Foundation of China (Grant No. 81272798 and 81572278 to X.Z.), the Hunan Provincial Natural Science Foundation of China (Grant No. 14JJ7008 to X. Z.), and the Xiangya Hospital Funds for Talent Introduction (to X.Z.) and China "863" Plan Project (Grant No. 2014AA020610-1 to X. Z.).

Abbreviations

APPD	(3*R*,4*S*)-1-(4-(aminomethyl)phenylsulfonyl)pyrrolidine-3,4-diol
BSA	bovine serum albumin
CID	collision-induced dissociation
COFRADIC	combined fractional diagonal chromatography
cPILOT	combined precursor isotopic labeling and isobaric tagging
ECD	electron capture dissociation
ESI	electrospray ionization
ETD	electron transfer dissociation
FT-ICR	Fourier transform ion cyclotron resonance
IR	infrared

iTRAQ	isobaric tags for relative and absolute quantification
LC	liquid chromatography
LE1	leucine enkephalin
LE2	nitro-Tyr leucine enkephalin
LE3	d(5)-Phe-nitro-Tyr leucine enkephalin
MAD	metastable atom-activated dissociation
MALDI	matrix-assisted laser desorption/ionization
MS	mass spectrometry
MS/MS	tandem mass spectrometry
m/z	mass-to-charge ratio
NTAC	nitrotyrosine affinity column
3NT	3-nitrotyrosine residue
1D/2D	one/two dimensional
OVA	ovalbumin
PMF	peptide mass fingerprint
SCX	strong cation exchange
SRM	selected reaction monitoring
TMT	tandem mass tags
TNM	tetranitromethane
TOF	time of flight
2DE	two-dimensional electrophoresis

References

1 Zhan X, Desiderio DM. The human pituitary nitroproteome: detection of nitrotyrosyl-proteins with two-dimensional western blotting, and amino acid sequence determination with mass spectrometry. *Biochem Biophys Res Commun* 2004;**325**:1180–1186.

2 Zhan X, Desiderio DM. Nitroproteins from human pituitary adenoma tissue discovered with a nitrotyrosine affinity column and tandem mass spectrometry. *Anal Biochem* 2006;**354**:279–289.

3 Zhan X, Desiderio DM. Mass spectrometric identification of *in vivo* nitrotyrosine sites in the human pituitary tumor proteome. *Methods Mol Biol* 2009a;**566**:137–163.

4 Peinado MA, Hernandez R, Peragon J, Ovelleiro D, Pedrosa JA, Blanco S. Proteomic characterization of nitrated cell targets after hypobaric hypoxia and reoxygenation in rat brain. *J Proteomics* 2014;**109C**:309–321.

5 Franco MC, Estevez AG. Tyrosine nitration as mediator of cell death. *Cell Mol Life Sci* 2014;**71**:3939–3950.

6 Yeo WS, Kim YJ, Kabir MH, Kang JW, Ahsan-Ul-Bari MD, Kim KP. Mass spectrometric analysis of protein tyrosine nitration in aging and neurodegenerative diseases. *Mass Spectrom Rev* 2015;**34**:166–183.

7 Abdelmegeed MA, Song BJ. Functional roles of protein nitration in acute and chronic liver diseases. *Oxid Med Cell Longev* 2014;**2014**:149627.

8 Scaloni A. Mass spectrometry approaches for the molecular characterization of oxidatively/nitrosatively modified proteins. In: Dalle-Donne I, Scaloni A, Butterfield DA, editors. *Redox Proteomics: From Protein Modification to Cellular Dysfunction and Diseases*. Hoboken, NJ: Wiley; 2006. p 59–100.

9 Yee CS, Seyedsayamdost MR, Chang MC, Nocera DG, Stubbe J. Generation of the R2 subunit of ribonucleotide reductase by intein chemistry: insertion of 3-nitrotyrosine at residue 356 as a probe of the radical initiation process. *Biochemistry* 2003;**42**:14541–14552.

10 Ghesquiere B, Goethals M, Van Damme J, Staes A, Timmerman E, Vandekerchhove J, Gevaert K. Improved tandem mass spectrometric characterization of 3-nitrotyrosine sites in peptides. *Rapid Commun Mass Spectrom* 2006;**20**:2885–2893.

11 Guo T, Wang X, Li M, Yang H, Li L, Peng F, Zhan X. Identification of glioblastoma phosphotyrosine-containing proteins with two-dimensional western blotting and tandem mass spectrometry. *BioMed Res Int* 2015;**2015**:134050.

12 Hunter T. Synthetic peptide substrates for a tyrosine protein kinase. *J Biol Chem* 1982;**257**:4843–4848.

13 Cooper JA, Esch FS, Taylor SS, Hunter T. Phosphorylation sites in enolase and lactate dehydrogenase utilized by tyrosine protein kinase *in vivo* and *in vitro*. *J Biol Chem* 1984;**259**:7835–7841.

14 Patschinsky T, Hunter T, Esch FS, Cooper JA, Sefton BM. Analysis of the sequence of amino acids surrounding sites of tyrosine phosphorylation. *Proc Natl Acad Sci U S A* 1982;**79**:973–977.

15 Low IC, Loh T, Huang Y, Virshup DM, Pervaiz S. Ser70 phosphorylation of Bcl-2 by selective tyrosine nitration of PP2A-B56δ stabilizes its antiapoptotic activity. *Blood* 2014;**124**:2223–2234.

16 Joshi MS, Mihm MJ, Cook AC, Schanbacher BL, Bauer JA. *Alterations in connexin 43 during diabetic cardiomyopathy: competition of tyrosine nitration versus phosphorylation*. J Diabetes; 2014. DOI: 10.1111/1753-0407.12164.

17 Smallwood HS, Lourette NM, Boschek CB, Bigelow DJ, Smith RD, Pasa-Tolic L, Squier TC. Identification of a denitrase activity against calmodulin in activated macrophages using high-field liquid chromatography—FTICR mass spectrometry. *Biochemistry* 2007;**46**:10498–10505.

18 Deeb RS, Nuriel T, Cheung C, Summers B, Lamon BD, Gross SS, Hajja DP. Characterization of a cellular denitrase activity that reverses nitration of cyclooxygenase. *Am J Physiol Heart Circ Physiol* 2013;**305**:H687–H698.

19 Abello N, Kerstjens HA, Postma DS, Bischoff R. Protein tyrosine nitration: selectivity, physicochemical and biological consequences, denitration, and proteomics methods for the identification of tyrosine-nitrated proteins. *J Proteome Res* 2009;**8**:3222–3238.

20 Molina-Jijon E, Rodriguez-Munoz R, Namorado Mdel C, Pedraza-Chaverri J, Reyes JL. Oxidative stress induces claudin-2 nitration in experiment type 1 diabetic nephropathy. *Free Radic Biol Med* 2014;**72**:162–175.

21 Cabassi A, Binno SM, Tedeschi S, Ruzicka V, Dancelli S, Rocco R, Vicini V, Coghi P, Regolisti G, Montanari A, Fiaccadori E, Govoni P, Piepoli M, de Champlain J. Low serum ferroxidase I activity is associated with mortality in heart failure and related to both peroxynitrite-induced cysteine oxidation and tyrosine nitration of ceruloplasmin. *Circ Res* 2014;**114**:1723–1732.

22 DiDonato JA, Aulak K, Huang Y, Wagner M, Gerstenecker G, Topbas C, Gogonea V, DiDonato AJ, Tang WH, Mehl RA, Fox PL, Plow EF, Smith JD, Fisher EA, Hazen SL. Site-specific nitration of apolipoprotein A-I at tyrosine 166 is both abundant within human atherosclerotic plaque and dysfunctional. *J Biol Chem* 2014;**289**:10276–10292.

23 Uzasci L, Bianchet MA, Cotter RJ, Nath A. Identification of nitrated immunoglobulin variable regions in the HIV-infected human brain: implications in HIV infection and immune response. *J Proteome Res* 2014;**13**:1614–1623.

24 Haddad IY, Pataki G, Hu P, Galliani C, Beckman JS, Matalon S. Quantitation of nitrotyrosine levels in lung sections of patients and animals with acute lung injury. *J Clin Invest* 1994;**94**:2407–2413.

25 Shigenaga MK, Lee HH, Blunt BC, Christen S, Shigeno ET, Yip H, Ames BN. Inflammation and NO(X)-induced nitration: assay for 3-nitrotyrosine by HPLC with electrochemical detection. *Proc Natl Acad Sci U S A* 1997;**94**:3211–3216.

26 Zhan X, Wang X, Desiderio DM. Pituitary adenoma nitroproteomics: current status and perspectives. *Oxid Med Cell Longev* 2013;**2013**:580710.

27 Zhan X, Desiderio DM. MALDI-induced fragmentation of leucine enkephalin, nitro-Tyr leucine enkephalin, and d(5)-Phe-nitro-Tyr leucine enkephalin. *Int J Mass Spectrom* 2009;**287**:77–86.

28 Petersson AS, Steen H, Kalume DE, Caidahl K, Roepstorff P. Investigation of tyrosine nitration in proteins by mass spectrometry. *J Mass Spectrom* 2001;**36**:616–625.

29 Sarver A, Scheffler K, Shetlar MD, Gibson BW. Analysis of peptides and proteins containing nitrotyrosine by matrix-assisted laser desorption/ ionization mass spectrometry. *J Am Soc Mass Spectrom* 2001;**12**:439–448.

30 Zhang Q, Qian WJ, Knyushko TV, Clauss TR, Purvine SO, Moore RJ, Sacksteder CA, Chin MH, Smith DJ, Camp DG 2nd, Bigelow DJ, Smith RD. A method for selective enrichment and analysis of nitrotyrosine-containing peptides in complex proteome samples. *J Proteome Res* 2007;**6**:2257–2268.

31 Yeo WS, Lee SJ, Lee JR, Kim KP. Nitrosative protein tyrosine modifications: biochemistry and functional significance. *BMB Rep* 2008;**41**:194–203.

32 Lee SJ, Lee JR, Kim YH, et al. Investigation of tyrosine nitration and nitrosylation of angiotensin II and bovine serum albumin with electrospray ionization mass spectrometry. *Rapid Commun Mass Spectrom* 2007;**21**:2797–2804.

33 Kim JK, Lee JR, Kang JW, Lee SJ, Shin GC, Yeo WS, Kim KH, Park HS, Kim KP. Selective enrichment and mass spectrometric identification of nitrated peptides using fluorinated carbon tags. *Anal Chem* 2011;**83**:157–163.

34 Lee JR, Lee SJ, Kim TW, Kim JK, Park HS, Kim DE, Kim KP, Yeo WS. Chemical approach for specific enrichment and mass analysis of nitrated peptides. *Anal Chem* 2009;**81**:6620–6629.

35 Lee HM, Reed J, Greeley GH Jr, Englander EW. Impaired mitochondrial respiration and protein nitration in the rat hippocampus after acute inhalation of combustion smoke. *Toxicol Appl Pharmacol* 2009;**235**:208–215.

36 Dekker F, Abello N, Wisastra R, Bischoff R. Enrichment and detection of tyrosine-nitrated proteins. *Curr Protoc Protein Sci* 2012;Chapter 14: Unit 14.13.

37 Freeney MB, Schoneich C. Proteomic approaches to analyze protein tyrosine nitration. *Antioxid Redox Signal* 2013;**19**:1247–1256.

38 Zhan X, Wang X, Desiderio DM. Mass spectrometry analysis of nitrotyrosine-containing proteins. *Mass Spectrom Rev* 2015;**34**:423–448.

39 Gusanu M, Petre BA, Przybylski M. Epitope motif of an anti-nitrotyrosine antibody specific for tyrosine-nitrated peptides revealed by a combination of affinity approaches and mass spectrometry. *J Pept Sci* 2011;**17**:184–191.

40 Sultana R, Reed T, Butterfield DA. Detection of 4-hydroxy-2-nonenal- and 3-nitrotyrosine-modified proteins using a proteomics approach. *Methods Mol Biol* 2009;**519**:351–361.

41 Prokai-Tatrai K, Guo J, Prokai L. Selective chemoprecipitation and subsequent release of tagged species for the analysis of nitropeptides by liquid chromatography-tandem mass spectrometry. *Mol Cell Proteomics* 2011;**10**(8):M110.002923.

42 Abello N, Barroso B, Kerstjens HAM, Postma DS, Bischoff R. Chemical labeling and enrichment of nitrotyrosine-containing peptides. *Talanta* 2010;**80**:1503–1512.

43 Tsumoto H, Taguchi R, Kohda K. Efficient identification and quantification of peptides containing nitrotyrosine by matrix-assisted laser desorption/ionization time-of-flight mass spectrometry after derivation. *Chem Pharm Bull* 2010;**58**:488–494.

44 Amoresano A, Chiappetta G, Pucci P, D'Ischia M, Marino G. Bidimensional tandem mass spectrometry for selective identification of nitration sites in proteins. *Anal Chem* 2007;**79**:2109–2117.

45 Amoresano A, Chiappetta G, Pucci P, Marino G. A rapid and selective mass spectrometric method for the identification of nitrated proteins. *Methods Mol Biol* 2008;**477**:15–29.

46 Chiappetta G, Corbo C, Palmese A, Galli F, Piroddi M, Marino G, Amoresano A. Quantitative identification of protein nitration sites. *Proteomics* 2009;**9**:1524–1537.

47 Robinson RA, Evans AR. Enhanced sample multiplexing for nitrotyrosine-modified proteins using combined precursor isotopic labeling and isobaric tagging. *Anal Chem* 2012;**84**:4677–4686.

48 Ghesquiere B, Colaert N, Helsens K, Dejager L, Vanhaute C, Verleysen K, Kas K, Timmerman E, Goethals M, Libert C, Vandekerckhove J, Gevaert K. *In vitro* and *in vivo* protein-bound tyrosine nitration characterized by diagonal chromatography. *Mol Cell Proteomics* 2009;**8**:2642–2652.

49 Larsen TR, Bache N, Gramsbergen JB, Roepstorff P. Identification of nitrotyrosine containing peptides using combined fractional diagonal chromatography (COFRADIC) and off-line nano-LC-MALDI. *J Am Soc Mass Spectrom* 2011;**22**:989–996.

50 Zhang Y, Yang H, Posch IU. Analysis of nitrated proteins and tryptic peptides by HPLC-chip-MS/MS: site-specific quantification, nitration degree, and reactivity of tyrosine residues. *Anal Bioanal Chem* 2011;**399**:459–471.

51 Zhan X, Desiderio DM. The use of variations in proteomes to predict, prevent, and personalize treatment for clinically nonfunctional pituitary adenomas. *EPMA J* 2010b;**1**:39–459.

52 Zhan X, Desiderio DM. Nitroproteins identified in human ex-smoker bronchoalveolar lavage fluid. *Aging Dis* 2011;**2**:00–115.

53 Zhan X, Desiderio DM. Signaling pathway networks mined from human pituitary adenoma proteomics data. *BMC Med Genomics* 2010a;**3**:13.

54 Seeley KW, Stevens SM Jr. Investigation of local primary structure effects on peroxynitrite-mediated tyrosine nitration using targeted mass spectrometry. *J Proteomics* 2012;**75**:1691–1700.

55 Palamalai V, Miyagi M. Mechanism of glyceraldehyde-3-phosphate dehydrogenase inactivation by tyrosine nitration. *Protein Sci* 2010;**19**:255–262.

56 Petre BA, Youhnovski N, Lukkari J, Weber R, Przybylski M. Structural characterization of tyrosine-nitrated peptides by ultraviolet and infrared matrix-assisted laser desorption/ionization Fourier transform ion cyclotron resonance mass spectrometry. *Eur J Mass Spectrom (Chichester, Eng)* 2005;**11**:513–518.

57 Jones AW, Cooper HJ. Probing the mechanisms of electron capture dissociation mass spectrometry with nitrated peptides. *Phys Chem Chem Phys* 2010;**12**:13394–13399.

58 Jones AW, Mikhailov VA, Iniesta J, Cooper HJ. Electron capture dissociation mass spectrometry of tyrosine nitrated peptides. *J Am Soc Mass Spectrom* 2010;**21**:268–277.

59 Cook SL, Jackson GP. Characterization of tyrosine nitration and cysteine nitrosylation modifications by metastable atom-activation dissociation mass spectrometry. *J Am Soc Mass Spectrom* 2011;**22**:221–232.

60 Turko IV, Murad F. Mapping sites of tyrosine nitration by matrix-assisted laser desorption/ionization mass spectrometry. *Methods Enzymol* 2005;**396**:266–275.

61 Sheeley SA, Rubakhin SS, Sweedler JV. The detection of nitrated tyrosine in neuropeptides: a MALDI matrix-dependent response. *Anal Bioanal Chem* 2005;**382**:22–27.

62 Li B, Held JM, Schilling B, Danielson SR, Gibson BW. Confident identification of 3-nitrotyrosine modifications in mass spectral data across multiple mass spectrometry platforms. *J Proteomics* 2011;**74**:2510–2521.

63 Dalle-Donne I, Scaloni A, Giustarini D, Cavarra E, Tell G, Lungarella G, Colombo R, Rossi R, Milzani A. Proteins as biomarkers of oxidative/nitrosative stress in diseases: the contribution of redox proteomics. *Mass Spectrom Rev* 2005;**24**:55–99.

64 Dalle-Donne I, Scaloni A, Butterfield DA. *Redox Proteomics: From Protein Modifications to Cellular Dysfunction and Diseases.* Hoboken, New Jersey: John Wiley & Sons, Inc.; 2006.

65 Ghosh S, Janocha AJ, Aronica MA, Swaidani S, Comhair SAA, Xu W, Zheng L, Kaveti S, Kinter M, Hazen SL, Erzurum SC. Nitrotyrosine proteome survey in asthma identifies oxidative mechanism of catalase inactivation. *J Immunol* 2006;**176**:5587–5597.

66 Lanone S, Manivet P, Callebert J, Launay JM, Payen D, Aubier M, Boczkowski J, Mebazaa A. Inducible nitric oxide synthase (NOS2) expressed in septic patients is nitrated on selected tyrosine residues: implications for enzymic activity. *Biochem J* 2002;**366**:399–404.

67 Chatterjee S, Lardinois O, Bonini MG, Bhattacharjee S, Stadler K, Corbett J, Deterding LJ, Tomer KB, Kadiiska M, Mason RP. Site-specific carboxypeptidase B1 tyrosine nitration and pathophysiological implications following its physical association with nitric oxide synthase-3 in experimental sepsis. *J Immunol* 2009;**183**:4055–4066.

68 Dhiman M, Nakayasu ES, Madaiah YH, Reynolds BK, Wen JJ, Almeida IC, Garg NJ. Enhanced nitrosative stress during *Trypanosoma cruzi* infection causes nitrotyrosine modification of host proteins: implications in Chagas' disease. *Am J Pathol* 2008;**173**:728–740.

69 Sharov VS, Galeva NA, Kanski J, Williams TD, Schoneich C. Age-associated tyrosine nitration of rat skeletal muscle glycogen phosphorylase b: characterization by HPLC-nanoelectrospray-tandem mass spectrometry. *Exp Gerontol* 2006;**41**:407–416.

70 Kanski J, Hong SJ, Schoneich C. Proteomic analysis of protein nitration in aging skeletal muscle and identification of nitrotyrosine-containing sequences *in vivo* by nanoelectrospray ionization tandem mass spectrometry. *J Biol Chem* 2005;**280**:24261–24266.

71 Marshall A, Lutfeali R, Raval A, Chakravarti DN, Chakravarti B. Differential hepatic protein tyrosine nitration of mouse due to aging-effect on mitochondrial

energy metabolism, quality control machinery of the endoplasmic reticulum and metabolism of drugs. *Biochem Biophys Res Commun* 2013;**430**:231–235.

72 Kanski J, Behring A, Pelling J, Schoneich C. Proteomic identification of 3-nitrotyrosine-containing rat cardiac proteins: effects of biological aging. *Am J Physiol Heart Circ Physiol* 2005;**288**:H371–H381.

73 Nakagawa H, Komai N, Takusagawa M, Miura Y, Toda T, Miyata N, Ozawa T, Ikota N. Nitration of specific tyrosine residues of cytochrome *C* is associated with caspase-cascade inactivation. *Biol Pharm Bull* 2007;**30**:15–20.

74 Fiorce G, Di Cristo C, Monti G, Amoresano A, Columbano L, Pucci P, Cioffi FA, Cosmo AD, Palumbo A, d'Ischia M. Tubulin nitration in human gliomas. *Neurosci Lett* 2006;**394**:57–62.

75 Zhan X, Desiderio DM. Linear ion-trap mass spectrometric characterization of human pituitary nitrotyrosine containing proteins. *Int J Mass Spectrom* 2007;**259**:96–104.

76 Sacksteder CA, Qian WJ, Knyushko TV, Wang HW, Chin MH, Lacan G, Melega WP, Camp DG II, Smith RD, Smith DJ, Squier TC, Bigelow DJ. Endogenously nitrated proteins in mouse brain: links to neurodegenerative disease. *Biochemistry* 2006;**45**:8009–8022.

77 Zhang X, Monroe ME, Chen B, Chin MH, Heibeck TH, Schepmoes AA, Yang F, Petritis BO, Camp DG 2nd, Pounds JG, Jacobs JM, Smith DJ, Bigelow DJ, Smith RD, Qian WJ. Endogenous 3,4-dihydroxyphenylalanine and dopaquinone modifications on protein tyrosine: links to mitochondrially derived oxidative stress via hydroxyl radical. *Mol Cell Proteomics* 2010;**9**:1199–1208.

78 Danielson SR, Held JM, Schilling B, Oo M, Gibson BW, Andersen JK. Preferentially increased nitration of alpha-synuclein at tyrosine-39 in a cellular oxidative model of Parkinson's disease. *Anal Chem* 2009;**81**:7823–7828.

79 Casoni F, Basso M, Massignan T, Gianazza E, Cheroni C, Salmona M, Bendotti C, Bonetto V. Protein nitration in a mouse model of familial amyotrophic lateral sclerosis: possible multifunctional role in the pathogenesis. *J Biol Chem* 2005;**280**:16295–16304.

80 Yoon SW, Kang S, Ryu SE, Poo H. Identification of tyrosine-nitrated proteins in HT22 hippocampal cells during glutamate-induced oxidative stress. *Cell Prolif* 2010;**43**:584–593.

81 Liu B, Tewari AK, Zhang L, Green-Church KB, Zweier JL, Chen YR, He G. Proteomic analysis of protein tyrosine nitration after ischemia reperfusion injury: mitochondria as the major target. *Biochem Biophys Acta* 2009;**1974**:476–485.

82 Chen CL, Chen J, Rawale S, Varadharaj S, Kaumaya PPT, Zweier JL, Chen YR. Protein tyrosine nitration of the Flavin subunit is associated with oxidative modification of mitochondrial complex II in the post-ischemic myocardium. *J Biol Chem* 2008;**283**:27991–28003.

83 Ai L, Rouhanizadeh M, Wu JC, Takabe W, Yu H, Alavi M, Chu Y, Miller J, Heistad DD, Hsiai TK. Shear stress influences spatial variations in vascular Mn-SOD expression. *Am J Physiol Cell Physiol* 2008;**294**:C1576–C1585.

84 Justilien V, Pang JJ, Renganathan K, Zhan X, Crabb JW, Kim SR, Sparrow JR, Hauswirth WW, Lewin AS. SOD2 knockdown mouse model of early AMD. *Invest Ophthalmol Vis Sci* 2007;**48**:4407–4420.

85 Murdaugh LS, Wang Z, Del Priore LV, Dillon J, Gaillard ER. Age-related accumulation of 3-nitrotyrosine and nitro-A2E in human Bruch's membrane. *Exp Eye Res* 2010;**90**:564–571.

86 Palamalai V, Darrow RM, Organisciak DT, Miyagi M. Light-induced changes in protein nitration in photoreceptor rod outer segments. *Mol Vis* 2006;**12**:1543–1551.

87 Kato Y, Dozaki N, Nakamura T, Kitamoto N, Yoshida A, Naito M, Kitamura M, Osawa T. Quantification of modified tyrosines in healthy and diabetic human urine using liquid chromatography/tandem mass spectrometry. *J Clin Biochem Nutr* 2009;**44**:67–78.

88 Piroddi M, Palmese A, Pilolli F, Amoresano A, Pucci P, Ronco C, Galli F. Plasma nitroproteome of kidney disease patients. *Amino Acids* 2011;**40**:653–667.

89 Chaki M, Valderrama R, Fernández-Ocaña AM, Carreras A, López-Jaramillo J, Luque F, Palma JM, Pedrajas JR, Begara-Morales JC, Sánchez-Calvo B, Gómez-Rodríguez MV, Corpas FJ, Barroso JB. Protein targets of tyrosine nitration in sunflower (Helianthus annuus L.) hypocotyls. *J Exp Bot* 2009;**60**:4221–4234.

90 Aslan M, Ryan TM, Townes TM, Coward L, Kirk MC, Barnes S, Alexander CB, Rosenfeld SS, Freeman BA. Nitric oxide-dependent generation of reactive species in sickle cell disease. Actin tyrosine induces defective cytoskeletal polymerization. *J Biol Chem* 2003;**278**:4194–4204.

91 Reed TT, Owen J, Pierce WM, Sebastian A, Sullivan PG, Butterfield DA. Proteomic identification of nitrated brain proteins in traumatic brain-injured rats treated postinjury with gamma-glutamylcysteine ethylester: insights into the role of elevation of glutathione as a potential therapeutic strategy for traumatic brain injury. *J Neurosci Res* 2009;**87**:408–417.

92 Ulrich M, Petre A, Youhnovski N, Prömm F, Schirle M, Schumm M, Pero RS, Doyle A, Checkel J, Kita H, Thiyagarajan N, Acharya KR, Schmid-Grendelmeier P, Simon HU, Schwarz H, Tsutsui M, Shimokawa H, Bellon G, Lee JJ, Przybylski M, Döring G. Post-translational tyrosine nitration of eosinophil granule toxins mediated by eosinophil peroxidase. *J Biol Chem* 2008;**283**:28629–28640.

93 Webster RP, Brockman D, Myatt L. Nitration of p38 MAPK in the placenta: association of nitration with reduced catalytic activity of p38 MAPK in pre-eclampsia. *Mol Hum Reprod* 2006;**12**:677–685.

94 Casanovas A, Carrascal M, Abián J, López-Tejero MD, Llobera M. Lipoprotein lipase is nitrated *in vivo* after lipopolysaccharide challenge. *Free Radic Biol Med* 2009;**47**:1553–1560.

95 Hamilton RT, Asatryan L, Nilsen JT, Isas JM, Gallaher TK, Sawamura T, Hsiai TK. LDL protein nitration: implication for LDL protein unfolding. *Arch Biochem Biophys* 2008;**479**:1–14.

96 Sharov VS, Galeva NA, Dremina ES, Williams TD, Schöneich C. Inactivation of rabbit muscle glycogen phosphorylase b by peroxynitrite revisited: does the nitration of Tyr613 in the allosteric inhibition site control enzymatic function? *Arch Biochem Biophys* 2009;**484**:155–166.

97 Zhu JH, Zhang X, Roneker CA, McClung JP, Zhang S, Thannhauser TW, Ripoll DR, Sun Q, Lei XG. Role of copper, zinc-superoxide dismutase in catalyzing nitrotyrosine formation in murine liver. *Free Radic Biol Med* 2008;**45**:611–618.

98 Chen HJ, Chen YC. Reactive nitrogen oxide species-induced post-translational modifications in human hemoglobin and the association with cigarette smoking. *Anal Chem* 2012;**84**:7881–7890.

99 Sekar Y, Moon TC, Slupsky CM, Befus AD. Protein tyrosine nitration of aldolase in mast cells: a plausible pathway in nitric oxide-mediated regulation of mast cell function. *J Immunol* 2010;**185**:578–587.

100 Ohama T, Brautigan DL. Endotoxin conditioning induces VCP/p97-mediated and inducible nitric-oxide synthase-dependent Tyr284 nitration in protein phosphatase 2A. *J Biol Chem* 2010;**285**:8711–8718.

101 Redondo-Horcajo M, Romero N, Martínez-Acedo P, Martínez-Ruiz A, Quijano C, Lourenço CF, Movilla N, Enríquez JA, Rodríguez-Pascual F, Rial E, Radi R, Vázquez J, Lamas S. Cyclosporine A-induced nitration of tyrosine 34 MnSOD in endothelial cells: role of mitochondrial superoxide. *Cardiovasc Res* 2010;**87**:356–365.

102 Bigelow DJ, Qian WJ. Quantitative proteome mapping of nitrotyrosines. *Methods Enzymol* 2008;**440**:191–205.

103 Tao RR, Huang JY, Shao XJ, Ye WF, Tian Y, Liao MH, Fukunaga K, Lou YJ, Han F, Lu YM. Ischemic injury promotes Keap1 nitration and disturbance of antioxidative responses in endothelial cells: a potential vasoprotective effect of melatonin. *J. Pineal. Res.* 2013;**54**:271–281.

104 Lu N, Zhang Y, Li H, Gao Z. Oxidative and nitrative modifications of alpha-enolase in cardiac proteins from diabetic rats. *Free Radic Biol Med* 2010;**48**:873–881.

105 Safinowski M, Wilhelm B, Reimer T, Weise A, Thomé N, Hänel H, Forst T, Pfützner A. Determination of nitrotyrosine concentrations in plasma samples of diabetes mellitus patients by four different immunoassays leads to contradictive results and disqualifies the majority of the tests. *Clin Chem Lab Med* 2009;**47**:483–488.

106 Cecconi D, Orzetti S, Vandelle E, Rinalducci S, Zolla L, Delledonne M. Protein nitration during defense response in Arabidopsis thaliana. *Electrophoresis* 2009;**30**:2460–2468.

107 Hui Y, Wong M, Zhao SS, Love JA, Ansley DM, Chen DD. A simple and robust LC-MS/MS method for quantification of free 3-nitrotyrosine in human plasma from patients receiving on-pump CABG surgery. *Electrophoresis* 2012;**33**:697–704.

108 Sokolovsky M, Riordan JF, Vallee BL. Tetranitromethane. A reagent for the nitration of tyrosyl residues in proteins. *Biochemistry* 1966;**5**:3582–3589.

109 Fujigaki H, Saito K, Lin F, Fujigaki S, Takahashi K, Martin BM, Chen CY, Masuda J, Kowalak J, Takikawa O, Seishima M, Markey SP. Nitration and inactivation of IDO by peroxynitrite. *J Immunol* 2006;**176**:372–379.

110 Sokolovsky M, Riordan JF, Vallee BL. Conversion of 3-nitrotyrosine to 3-aminotyrosine in peptides and proteins. *Biochem Biophys Res Commun* 1967;**27**:20–25.

111 Ghesquiere B, Helsens K, Vandekerckhove J, Gevaert K. A stringent approach to improve the quality of nitrotyrosine peptide identifications. *Proteomics* 2011;**11**:1094–1098.

112 Aulak KS, Miyagi M, Yan L, West KA, Massillon D, Crabb JW, Stuehr DJ. Proteomic method identifies proteins nitrated *in vivo* during inflammatory challenge. *Proc Natl Acad Sci USA* 2001;**98**:12056–12061.

113 Miyagi M, Sakaguchi H, Darrow RM, Yan L, West KA, Aulak KS, Stuehr DJ, Hollyfield JG, Organisciak DT, Crabb JW. Evidence that light modulates protein nitration in rat retina. *Mol Cell Proteomics* 2002;**1**:293–303.

114 Butt YK, Lo SC. Detecting nitrated proteins by proteomic technologies. *Methods Enzymol* 2008;**440**:17–31.

115 Petre BA, Ulrich M, Stumbaum M, Bernevic B, Moise A, Doring G, Przybylski M. When is mass spectrometry combined with affinity approaches essential? A case study of tyrosine nitration in proteins. *J Am Soc Mass Spectrom* 2012;**23**:1831–1840.

116 Dremina ES, Li X, Galeva NA, Sharov VS, Stobaugh JF, Schoneich CA. Methodology for simultaneous fluorogenic derivatizaiton and boronate affinity enrichment of 3-nirotyrosine containing peptides. *Anal Biochem* 2011;**418**:184–196.

117 Mani AR, Moore KP. Dynamic assessment of nitration reactions *in vivo*. *Methods Enzymol* 2005;**396**:151–159.

118 Lin HL, Myshkin E, Waskell L, Hollenberg PF. Peroxynitrite inactivation of human cytochrome P450 2B6 and 2E1: heme modification and site-specific nitrotyrosine formation. *Chem Res Toxicol* 2007;**20**:1612–1622.

119 Lin HL, Kenaan C, Zhang H, Hollenberg PF. Reaction of human cytochrome P450 3A4 with peroxynitrite: nitrotyrosine formation on the proximal side impairs its interaction with NADPH-cytochrome P450 reductase. *Chem Res Toxicol* 2012;**25**:2642–2653.

120 Yamakura F, Kawasaki H. Post-translational modifications of superoxide dismutase. *Biochim Biophys Acta-Proteins Proteom* 2010;**1804**:318–325.

121 Kanski J, Schoneich C. Protein nitration in biological aging: proteomic and tandem mass spectrometric characterization of nitrated sites. *Methods Enzymol* 2005;**396**:160–171.

122 Spickett CM, Pitt AR. Protein oxidation: role in signaling and detection by mass spectrometry. *Amino Acids* 2012;**42**:5–21.

123 Tsikas D. Analytical methods for 3-nitrotyrosine quantification in biological samples: the unique role of tandem mass spectrometry. *Amino Acids* 2012;**42**:45–63.

124 Cheng S, Lian B, Liang J, Shi T, Xie L, Zhao YL. Site selectivity for protein tyrosine nitration: insights from features of structure and topological network. *Mol Biosyst* 2013;**9**:2860–2868.

125 Tsikas D, Duncan MW. Mass spectrometry and 3-nitrotyrosine: strategies, controversies, and our current perspective. *Mass Spectrom Rev* 2014;**33**:237–276.

6

Mass Spectrometry Methods for the Analysis of Isopeptides Generated from Mammalian Protein Ubiquitination and SUMOylation

Navin Chicooree and Duncan L. Smith

Biological Mass Spectrometry, Cancer Research UK Manchester Institute, The University of Manchester, Manchester, UK

6.1 Overview of Ub and SUMO

6.1.1 Biological Overview of Ubiquitin-Like Proteins

In addition to the ubiquitin (Ub) protein, ubiquitin-like proteins (Ubls) are a related super family of proteins that are involved in the post-translational modification (PTM) of target proteins. There are at least 14 Ubls including; small ubiquitin-like modifier (SUMO) proteins that have been identified in all systems spanning from prokaryotes to mammals. With the exception of the Pups, they share a similar three-dimensional core structure but are significantly different in their primary structure [1]. The main focus of this chapter is to review recent developments in applicable LC-MS/MS methods that improve bottom-up proteomic analyses of isopeptides derived from ubiquitin- and SUMO-modified proteins.

Activated forms of these two types of PTMs predominantly modify proteins through a mechanism of attachment by forming a covalent isopeptide bond between ε-amino groups of lysine residues on proteins targeted for PTM. Although it occurs significantly less frequently, ubiquitination can also occur through a peptide covalent bond between the C-termini of activated Ub/Ubls and N-terminal α-amino groups of proteins [2]. In addition, ubiquitination has been reported to occur by covalent attachment at the hydroxyl group of target acceptor threonine and serine residues or through the thiol group of target acceptor cysteine residues [3, 4]. Similarly, SUMOylation has been reported to predominantly occur via the isopeptide bond of ε-amino group of the target acceptor lysine. This chapter focuses on methods enabling the analysis of lysine ubiquitination and SUMOylation, which has occurred via the isopeptide bond of ε-amino group of the target acceptor lysine.

Analysis of Protein Post-Translational Modifications by Mass Spectrometry,
First Edition. Edited by John R. Griffiths and Richard D. Unwin.
© 2017 John Wiley & Sons, Inc. Published 2017 by John Wiley & Sons, Inc.

6.1.2 Biological Overview of Ub and SUMO

The mechanism for the formation of the isopeptide bond between these two PTMs and their respective target protein is an ATP-dependent biochemically assisted process predominantly occurring via a triple enzyme cascade. There are three types of enzymes involved: an E1-activating enzyme, an E2-conjugating enzyme, and an E3-ligating enzyme. The process begins with the activation of one of these two Ub/Ubl proteins, which occurs by the formation of a thioester bond between the thiol group of a cysteine residue on the active site of the E1 enzyme and the glycine residue at C-terminus of the Ub/Ubl protein. This forms the so-called E1-Ub/Ubl complex. The E2 enzyme is involved in a transesterification reaction with the E1-Ub/Ubl complex and covalently bonds itself to the C-terminal of the Ubl via the formation of a thioester bond, forming the E2-Ub/Ubl complex. The E3 ligase enzyme catalyzes the formation of the isopeptide bond between the ε-amine group of the acceptor lysine on the target protein and the glycine residue at the C-terminus of the Ub/Ubl protein. This process results in the formation of a Ub/Ubl-target protein complex. It is also possible that the E2 enzyme can act independently of the E3 enzyme and specifically facilitate protein SUMOylation. This occurs through the E2 enzyme being able to recognize and establish noncovalent stabilizing interactions with the presence of specific types of amino acids surrounding the acceptor lysine of the target protein within a region known as SUMOylation consensus sites. A typical example of a consensus site would be ΨKXD/E, where Ψ represents a hydrophobic amino acid, X is any amino acid, and D/E are acid residues [5, 6]. It is important to state that the process of SUMOylation in general also occurs on target proteins where no known consensus motif is present [7, 8]. There are a number of different types and classes of E3 ligases that have been identified for both ubiquitination and SUMOylation [9, 10]. E3 ligases govern the specificity of the type of proteins that will be targeted for Ub/Ubl modification. The formation of covalent isopeptide bonds is reversible and hydrolysis of these isopeptide bonds occurs biochemically through deubiquitinating enzymes and deSUMOylating enzymes. [11, 12]. The functions of these types of enzymes are essential and they are central to the function and regulation of these two PTMs [13, 14].

6.1.3 Biological Functions of Ub and SUMO

These two PTMs are highly regulated and can target a protein and assemble a monomodification or assemble more complex polymeric modifications consisting of branched and mixed chains comprising different lengths and linkages. For example, the process of ubiquitination can involve the targeting of a protein for monomodification with subsequent extension by the covalent addition of other ubiquitin proteins to the ε-amino group of one of ubiquitin's seven internal lysines (Lys 6, 11, 27, 29, 33, 48, and 63) or the α-amino group of its N-terminal methionine [15, 16]. The type of ubiquitin chain assembly on a modified protein directly impacts the protein-specific biological function within the cell. For example, modification of a protein by polyubiquitination chain assembly through the internal lysine 48-amino-acid residues of ubiquitin results in the protein being targeted for degradation in the 26S proteosome complex [17]. Similarly, these different types of chain assemblies impact the specific biological function of the SUMOylated proteins [18, 19].

These two PTMs can be involved in a multitude of important and diverse biological functions: (i) independently, such as in protein endocytosis in the case of ubiquitination [20, 21] and transcriptional activity in the case of SUMOylation [22, 23]; (ii) in combination with each other, known as cross talk, such as in DNA damage, replication, and repair [24–26]; and (iii) cross talk with other PTMs, for example, cross talk between ubiquitination and phosphorylation such as in regulation of E3 ligase activity [27] and cross talk between acetylation and SUMOylation in regulating p53-dependent gene transcription by the promotion and inhibition of p53 binding to DNA and chromatin [28]. Dysregulation of these ubiquitination and SUMOylation pathways has a substantial impact on the progression of a number of disease states including a number of cancers [29, 30] and neurodegenerative diseases [31, 32].

6.2 Mass Spectrometry Behavior of Isopeptides

6.2.1 Terminology of a Ub/Ubl isopeptide

Before we advance further on the discussion of isopeptide products generated from target proteins that have been ubiquitinated or SUMOylated, it is important to simplify their terminology:

1) In the literature, isopeptides are commonly also referred to as ubiquitin peptides, ubiquitinated peptides, ubiquitinated isopeptides, SUMO peptides, SUMOylated peptides, and SUMOylated isopeptides. However, in order to simplify things, the isopeptide products that are generated from target proteins that have been ubiquitinated are referred to as *Ub-isopeptides*. Isopeptide products that have been generated from target proteins that have been SUMOylated are referred to as follows: (i) *SUMO(x)-isopeptides*, with "*x*" referring to the specific type of SUMOylation PTM that the isopeptide has been derived from SUMO(1), SUMO(2), or SUMO(3), or (ii) *SUMO-isopeptides* when generalizing.
2) The truncated version of the Ub/Ubl modification that is covalently attached to the peptide backbone of the isopeptide via the isopeptide bond

postdigestion is commonly referred to in literature as a remnant, a branch, an iso-tag, or an iso-chain. However, in order to simplify things, this is referred to in this chapter as an *iso-chain*.

6.2.2 Mass Spectrometry Analysis of SUMO-Isopeptides Derived from Proteolytic Digestion

A key aspect in the proteomic workflow for the LC-nESI-MS/MS, direct ESI-MS/MS, or MALDI-MS/MS analysis of these two PTMs is the implementation of an effective sample preparation strategy. In the succeeding text, we will introduce the importance of digestion strategies and discuss the impact that different digestion strategies have on the type of Ub-isopeptide and SUMO-isopeptide that can be generated from the proteins that have been ubiquitinated and SUMOylated and on their amenability to mass spectrometric analysis.

An important aspect that needs to be taken into consideration in amenability of SUMO-isopeptides and Ub-isopeptides to mass spectrometric analysis is the length and amino acid composition of their iso-chains. The length of the iso-chain is dependent upon the biological proteolytic or chemical digestion technique used in relation to which amino acids are preferentially (i) typically or (ii) atypically cleaved. (i) A typical amino acid cleavage site on a protein is one which the proteolytic enzyme or chemical being used preferentially cleaves at. (ii) An atypical amino acid cleavage site on a protein is one which the proteolytic enzyme or chemical being used preferentially cleaves at less frequently. It is important to remember that they both occur and can be of importance during the stages of mass spectrometric analysis and data interpretation. There are three types of digestion strategy that have been employed for the sample preparation of Ub-isopeptides and SUMO-isopeptides: (i) proteolytic enzyme digestion, (ii) dual proteolytic enzyme digestion, and (iii) proteolytic enzyme and/or chemical digestion.

6.2.3 Analysis of SUMO-Isopeptides with Typical Full-Length Tryptic Iso-chains

Using trypsin for proteolytic enzyme digestion of a protein that has been SUMO(1)ylated, to generate a SUMO(1)-isopeptide results in typical C-terminal cleavage of the lysine residue at position 78, resulting in the generation of a full-length tryptic iso-chain with a length of 19 amino acids. Typical tryptic cleavage C-terminal to the arginine at position 61 of the SUMO(2) iso-chain and position 60 of the SUMO(3), resulting in the generation of SUMO(2)/(3)-isopeptides containing a full-length tryptic iso-chain of 32 amino acids. The iso-chains generated on SUMO-isopeptides from typical tryptic cleavage are much longer than the iso-chains generated in Ub-isopeptides (discussed later). The SUMO(1)-isopeptides and SUMO-(2/3)-isopeptides derived from tryptic digestion contain full-length iso-chains consisting of 19 (ELGMEEEDVIEVYQEQTGG) and 32 (FDGQPINETDTPAQLEMEDEDTIDVFQQQTGG) amino acids, respectively. These large iso-chains are highly charged, therefore rendering the SUMO-isopeptide a highly charged species for electrospray-based MS analysis. The MS-based approaches to analyze these highly charged SUMO-isopeptides with these full-length tryptic iso-chains provide strategies that have been developed and applied to improve the analytical performance of these analytes in targeted approaches.

Hybrid linear ion-trap Fourier transform-based mass spectrometry with low-energy ion-trap CID has been coupled to liquid chromatography via nano-electrospray sources to analyze highly charged SUMO-isopeptides containing a full-length tryptic iso-chain in a targeted approach involving two parts [33]. The first part of this work flow utilized the high-resolution mass spectrometer and high mass accuracy measurement capabilities of Orbitrap and FT-ICR mass spectrometers to identify their naturally higher precursor ion charge states of a set of full-length SUMO(1)-isopeptides and SUMO(2)-isopeptides generated from independent post-tryptic and Lys-C digestion of polymeric-SUMO protein chains produced in *in vitro* SUMOylation assays. This set of SUMO-isopeptides was then analyzed by low-energy CID in the ion trap and the *m/z* of the predominantly multiply charged product ions were measured in only the ion trap when the LTQ-FT-ICR was used or in both the ion trap and the Orbitrap when using the LTQ Orbitrap. Manual analysis of the highly complex CID MS/MS spectra was difficult due to the domination of multiply charged b′- and/or y′-type product ions characteristically generated under low-energy CID conditions from the SUMO-isopeptides with highly charged full-length tryptic SUMO iso-chain. (Note: b and y product ions from the iso-chain and backbone of the SUMO-isopeptides are distinguished as follows: b′-/y′-type and c′- and z′-type product ions refer to ions from the iso-chain of SUMO- and Ub-isopeptides and b-/y-type and c- and z-type product ions refer to ions from the backbone of SUMO- and Ub-isopeptides). Consequently, there were a limited number of product ions identified from the peptide backbone.

In order to analyze the resulting CID MS/MS spectra of these highly charged SUMO-isopeptides, the authors used a specialist software, which transformed them into a "virtual" SUMO-isopeptide, and subsequent *in silico* fragmentation enabled the product ions of the iso-chain to be calculated and identified on the CID MS/MS spectra. The second part of the approach involved using the analytical information such as the retention times, their naturally higher precursor ion charge states, and their fragmentation behavior to successfully target and identify the same set of SUMO-isopeptides in a more complex sample from cultured mammalian cells. The specialist software and other available software [34, 35] used to assist with the interpretation of the complex SUMO-isopeptide CID MS/MS spectra suffer from:

1) An input of poor spectral quality populated by dominant multiply charged product ions from the SUMO-isopeptide's highly charged iso-chain, further suffering from limited sequence coverage of product ions generated from the isopeptide backbone, thereby limiting the backbone's comprehensive structural elucidation and identification

2) Limited consideration for the impact of multiple variable modifications that may occur along the iso-chain, such as multiple events of deamidation from the presence of two to five glutamine residues.

Another targeted approach used to analyze full-length tryptic SUMO-isopeptides utilized the capability of specialized Fourier transform ion cyclotron resonance (FT-ICR)-based mass spectrometry to provide high-resolution mass and high mass accuracy measurements. In addition, FT-ICR is capable of utilizing a range of specialized low-energy-based fragmentation techniques: electron capture dissociation (ECD), associated activated-ion electron capture dissociation (AI-ECD), and infrared multiphoton dissociation (IRMPD) in MS/MS mode. ESI-FT-ICR with ECD and IRMPD has been used to directly analyze a tryptic digest containing SUMO(1)-isopeptides bearing full-length tryptic SUMO-1 iso-chains, SUMO(1)ylated RanGap1protein fragment, and SUMO(1) ylated RanB2 protein fragments [36]. The IRMPD MS/MS spectrum generated from the analysis of the SUMO(1)-isopeptide derived from the SUMO(1)ylated RanGap1 generated a series of only singly charged b/b'- and y/y'-type product ions (b'/y' iso-chain ions labeled with b/y (S) in Figure 6.1a) from the backbone and iso-chain of the isopeptide (Figure 6.1a). The generation of only singly charged products greatly reduces the complexity of the IRMPD MS/MS spectrum. This is in contrast to the predominance of multiply charged product ions observed in CID MS/MS spectra, which are typically generated from the iso-chain of SUMO(1)/(2)-isopeptides. The shift from the generation of multiply charged to only singly charged product ions is indicative of IRMPD MS/MS spectra [37]. Although IRMPD fragmentation of the precursor ion is less efficient than CID fragmentation, the reduction in complexity observed in the IRMPD MS/MS spectrum of the SUMO(1)-isopeptide is advantageous and enables improved structural elucidation of both the backbone and the iso-chain. The additional series of predominantly b'- and y'-type-related neutral loss ions generated from the neutral loss of H_2O from SUMO(1) iso-chain of the isopeptide can be attributed to secondary dissociation events of either initial b'- and y'-type product ions or the dehydrated precursor ion, typically indicative of the IRMPD fragmentation technique and resulting IRMPD MS/MS spectra of peptides [37]. The generation of these product ions arising from

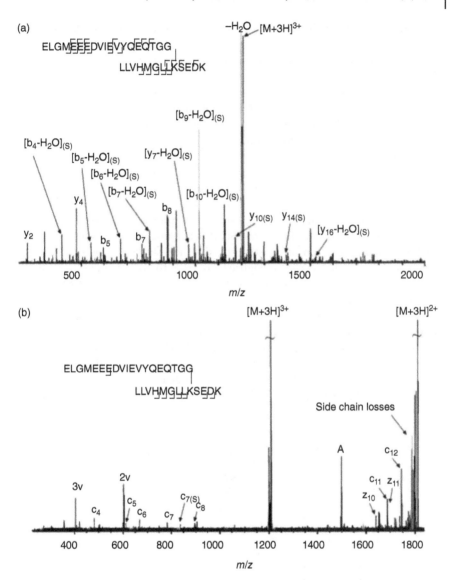

Figure 6.1 (a) IRMPD MS/MS spectrum of a SUMO(1)-isopeptide tryptically derived from a SUMO(1)ylated RanGAP1418-587 protein fragment. (b) An ECD MS/MS of a SUMO(1)-isopeptide tryptically derived from a SUMO(1)ylated RanGAP1418-587 protein fragment. (S) refers to product ions generated from the SUMO iso-chain. v = harmonic and (A) = artifact not removed in quadrupole/SWIFT isolation. Source: Cooper 2005, [36]. Reproduced with permission from American Chemical Society.

the neutral loss could also be favored due to the presence of multiple glutamic acids within the SUMO(1) iso-chain. This could account for their under-fragmentation at the N-terminal position of these types of related y'-ions and the C-terminal position of b'-ions generated from the SUMO(1) iso-chain of the isopeptide, which is a fragmentation behavior observed under low-energy CID conditions of peptides [38]. The complementary ECD MS/MS spectrum of the same SUMO(1)-isopeptide generated a series of both singly charged c- and z-type product ions from the backbone of the isopeptide (Figure 6.1).

The characteristic lower fragmentation efficiency of ECD (also observed with its analogous fragmentation technique, ETD) compromises the abundance of product ions; however, the characteristic sequence-independent fragmentation pattern that occurs under ECD fragmentation of peptides can reduce the complexity of the spectra, enabling the c- and z-type product ions to be identified on the ECD MS/MS spectrum of the SUMO(1)-isopeptide backbone. By contrast, only the generation of 1 c'-type product ion (note: c'-type product ion from iso-chain labeled as c (S) in Figure 6.2b) from the iso-chain of the isopeptide was observed. It was suggested that this was most likely due to the predominantly acidic physicochemical nature of the iso-chain and the lack of a basic residue to enable efficient ECD fragmentation to occur. Limited fragmentation under AI-ECD conditions also occurred along the iso-chain of the SUMO(1)-isopeptide generated from a SUMO(1)ylated RanB2 protein fragment. The limited fragmentation of the SUMO(1) iso-chain under ECD conditions could indicate that the physicochemical nature of the full-length SUMO(1) iso-chains may prevent SUMO(1)-isopeptides from being amenable

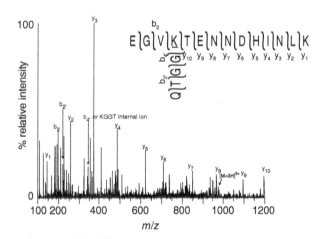

Figure 6.2 An example of a collision-cell CID MS/MS spectrum of a SUMO(2)-isopeptide atypically tryptically derived from a SUMO(2)ylated SUMO(2) protein. Source: Chicooree, 2013 [39]. Reproduced with permission of Wiley.

to full sequence structural elucidation by ECD. However, the resulting few iso-chain ions that are generated from limited ECD fragmentation of the iso-chain could be used as iso-chain diagnostic ions. This provides scope to observe iso-chain-specific diagnostic fragmentation patterns and subsequently characterize product ions, provided the MS/MS spectra are reproducible. Although these specific fragmentation techniques and the associated MS instrumentation are specialized, the limited complexity and additional improvement in product-ion coverage of the SUMO-isopeptide backbone that is observed in the resulting MS/MS spectra using these ECD and IRMPD to complement each other is a contrast to the high complexity of those observed from CID collision-cell and trap-type analysis of SUMO(1/2)-isopeptide also bearing full-length tryptic iso-chains [33]. The reduced complexity of the product-ion spectra benefits *de novo* sequencing and enables easier interpretation, structural elucidation, and assignment of the product ions by comparison with *in silico* digestion of the SUMO(1)-isopeptides without the need to employ specialist software.

Another approach has been developed using a combination of QTOF-based nanoLC-nESI-MS/MS and MALDI-TOF/TOF MS/MS analysis to analyze SUMO-isopeptides bearing full-length tryptic iso-chains. nanoLC-nESI-MS/MS was used to utilize its high resolution and mass accuracy in order to screen for potential SUMO-isopeptide ions based on the presence of quadruply charged monoisotopic signal clusters from a tryptic digest sample generated from SUMOylated proteins produced *in vitro* [7]. These proteins included a Ubc 9 protein that had been SUMOylated by a SUMO(2) protein and human centromere protein C10 and C28 fragments that had been SUMOylated by SUMO(1) and SUMO(2) proteins. The potential SUMO-isopeptide ions were targeted for analysis using MALDI-TOF/TOF MS/MS in order to utilize its ability to induce characteristic fragmentation patterns relating to preferential C-terminal cleavage of aspartic acid residues and generate singly charged product ions from the singly charged SUMO-isopeptide ions. Full-length tryptic SUMO(2) (and SUMO(3)) iso-chains of the singly charged tryptic SUMO(2)-isopeptide ion contain multiple aspartic acid residues and full-length tryptic SUMO(1) iso-chains of the singly charged tryptic SUMO(1)-isopeptides contain a single aspartic acid residue. These aspartic acid residues can often be exploited under MALDI-TOF/TOF high-energy CID by facilitating the fragmentation of a characteristic series of highly abundant b'- and/or y'-type product ions from the iso-chain due to their preferential C-terminal fragmentation.

This type of fragmentation is described as the "aspartic acid effect" and occurs via a charge-remote fragmentation pathway. An in-depth explanation on this fragmentation mechanism and its origins from high-energy CID experiments can be found in Paizs and Suhai (2005) [38]. Although the occurrence of this type of preferential fragmentation and subsequent release of abundant ions prevents complete sequence coverage of each amino acid within the iso-chain or backbone, the aspartic acid residues are distributed along the full-length of the SUMO-(2/3) iso-chain to the extent that their b'- and/or y'-type iso-chain-specific product ions are representative of the large portions of the iso-chain to a good degree whereby structural elucidation can be facilitated. In addition to the generation of these iso-chain-specific product ions, the SUMO(2)-isopeptides analyzed by Chung et al. produced high-energy CID MS/ MS spectra containing both b- and y-type product ions from the isopeptide backbone. Although the SUMO(1) iso-chain contains one aspartic acid residue, the advantageous nature of TOF/TOF high-energy CID MS/MS spectra being highly reproducible [40] resulted in additional y-type product ions being generated from both the backbone and iso-chain of the SUMO(1)-isopeptides. The generation of characteristic product ions from the iso-chain of all the isopeptides analyzed enabled structural elucidation of the isopeptides from *in vitro* samples without the need for specialized bioinformatic algorithms. Although the authors indicated that comprehensive bioinformatic algorithms could be developed for global analysis based on the characteristic fragmentation patterns and additional observations of SUMO(2)/(3) iso-chains, they have presented in their targeted approach. The generation of only singly charged product ions under high-energy CID conditions used in MALDI-TOF/TOF is also an advantage in that clean SUMO-isopeptide tandem MS/MS spectra are observed with less complexity in comparison to the SUMO-isopeptide MS/MS spectra produced using low-energy CID, which contain predominantly multiply charged product ions generated from neutral losses and internal ions. However, there is an initial reliance on two instrument platforms to apply this targeted approach.

In summary, the targeted MS approaches discussed have provided strategies to overcome the inherently challenging physicochemical nature of the SUMO iso-chains and improve the overall analytical performance of these challenging analytes. However, all these approaches are reliant on multistage workflows, specialist software programs, specialized fragmentation techniques, and multi-instrument platforms.

6.2.4 Analysis of SUMO-Isopeptides with Atypical Tryptic Iso-chains and Shorter Iso-chains Derived from Alternative Digestion Strategies

6.2.4.1 SUMO-Isopeptides with Atypical Iso-chains Generated from Tryptic Digestion

The inherently challenging physicochemical nature of the full-length tryptic SUMO iso-chains renders these highly charged SUMO-isopeptides refractory to general proteomic analysis. Therefore, a number of research groups have used proteomic strategies that are directed toward further improving the MS-based proteomic amenability of these SUMO-isopeptides with a view to enabling the high-throughput analytical analysis of putative SUMO-isopeptides from more complex biological samples or samples reflecting biological complexity. A set of such approaches (which are not discussed at great length in this chapter) involves mutagenesis to deplete the SUMO protein of lysine residues and substitute arginine into the iso-chain in order to generate SUMO-isopeptides with shorter iso-chains [41, 42]. While these mutagenic approaches mark the improvement in the analytical analysis of SUMO-isopeptides and the number of SUMO-isopeptides identified, the adverse biological impact that lysine depletion and multiple amino acid substitutions have on the SUMOylation of target proteins cannot be overlooked [8]. In order to avoid the adverse biological impact of mutagenic approaches, a strategy that involves the MS analysis of SUMO(2/3)-isopeptides that have been derived from atypical tryptic cleavage of wild-type SUMO(2/3)ylated proteins and SUMO(1)-isopeptides that have been derived from the independent use of alternative proteolytic digestion of SUMO(1)ylated proteins has been devised.

Mass spectrometry sequence-grade trypsin and Lys-C can also independently generate less frequent atypical cleavages along the SUMO iso-chain, which are much shorter than those generated from typical full-length tryptic and Lys-C cleavages. For example, the iso-chains that are generated from atypical tryptic cleavage C-terminal to glutamine residues at positions 88/89 SUMO(2) iso-chains and 89/90 SUMO(3) iso-chains contain 4 and 3 amino acids of identical amino acid composition, QTGG and TGG, which demonstrated more amenability to analytical analysis than the full-length SUMO(2/3)-isopeptides. In addition, alternative enzymes such as elastase used for independent digestion of a SUMO(1)ylated protein also generated SUMO(1)-isopeptides with a short GG iso-chain, again demonstrating amenability to overall MS-based proteomic analyses [39].

QTOF and LTQ Orbitrap-based LC-MS/MS utilizing low-energy collision-cell CID and low-energy linear ion-trap CID, combined with an unbiased database searching strategy, has been used in an approach to analyze these types of analytically amenable isopeptides generated from independent post-tryptic digestion of di-SUMO(2/3)ylated proteins in simple and more complex systems [39]. First, QTOF-based LC-MS/MS was used to develop an approach to analyze the di-SUMO(2)-isopeptides and di-SUMO(3)-isopeptides from simple digestion samples. The conventional Mascot bioinformatic search algorithm commonly used for MS-based proteomic analyses [43] was used to analyze the data by considering the attachment of consecutive additions of the first 10-amino-acid residues from the C-terminus of the SUMO(2/3) iso-chains as consecutive variable modifications to all available lysine residues. This bioinformatic process was termed "consecutive residue additions to lysine" (CRA(K)). Mascot generated a list of putative SUMO(2)/(3)-isopeptides, and all isopeptide ion hits were considered. The MS/MS spectra generated from low-energy collision-cell CID analysis of the putative isopeptides were manually interpreted using theoretical backbone fragment ions generated from Mascot, along with theoretically calculated iso-chain ions to confirm the identity of the putative SUMO(2)/(3)-isopeptides. The CID MS/MS spectra displayed were of advantageously limited complexity, which is a function of the short TGG and QTGG iso-chains. A comprehensive series of predominantly singly charged y-type product ions from the isopeptide backbone was observed in the CID MS/MS spectra along with the evidence of additional singly charged b′-type product ions generated from the TGG and QTGG iso-chains, enabling comprehensive structural elucidation of the SUMO(2)/(3)-isopeptides (Figure 6.2).

The CRA(K) method developed to analyze these SUMO(2)/(3)-isopeptides from simple digestion samples was validated on a more complex sample: an anti-Ub pulldown of an HEK293-T cell lysate. The much faster LTQ Orbitrap-based LC-MS/MS was used to analyze a tryptic digest of the more complex Ub/Ubl enriched protein sample and submitted for bioinformatic analyses using the Mascot-enabled CRA(K) approach. Mascot generated a list of putative SUMO(2)/(3)-isopeptides, and all isopeptide hits were considered. The MS/MS spectra generated from low-energy linear ion-trap CID analysis of the putative isopeptides were manually interpreted to confirm the identity of the putative SUMO(2)/(3)-isopeptides. The QTOF-based LC-MS/MS analysis of a SUMO(1)-isopeptide derived from the elastase digestion of a SUMO(1) ylated RanGap protein fragment was also performed. The Mascot-enabled CRA(K) approach was applied in the same way, although relaxed enzyme specificity was selected due to the use of elastase. Mascot generated the putative SUMO(1)-isopeptide containing a GG iso-chain. Consequently, GG is also the iso-chain generated from tryptic digestion of Ub-isopeptides derived from ubiquitinated proteins and not fully diagnostic to SUMO; however, the SUMO(1)ylated RanGap protein fragment was subjected to tryptic digestion and did not result in the generation of any subsequent or similar SUMO(1)-isopeptide. The MS/MS spectra generated from low-energy collision-cell CID analysis of the putative isopeptides were manually interpreted to confirm the identity of this putative SUMO(1)-isopeptide. The CID MS/MS spectra displayed a comprehensive series of predominantly singly charged b- and y-type product ions from the backbone of the SUMO(1)-isopeptide (Figure 6.3), enabling its comprehensive structural elucidation and confirmation of the presence of the GG iso-chain. Interestingly, the series of higher-order b-type product ions observed are uncharacteristic of collision-cell CID data [44] and is likely due to the isopeptide backbone containing a very basic gas-phase histidine amino acid residue with a basic side chain at its N-terminus, allowing for b-type product-ion stabilization in addition to y-type product-ion stabilization in the collision cell in a similar way to the product-ion spectra observed when having a lysine at the N-terminus when using Lys-N [45].

The CRA(K) approach does not require specialist software algorithms for analysis and instead uses widely available Mascot software thereby maximizing the successful LC-MS/MS analysis and subsequent data analysis of analytically amenable SUMO(1)/(2)/(3)-isopeptides generated from independent proteolytic digestion of SUMO(1)/(2)/(3)ylated protein.

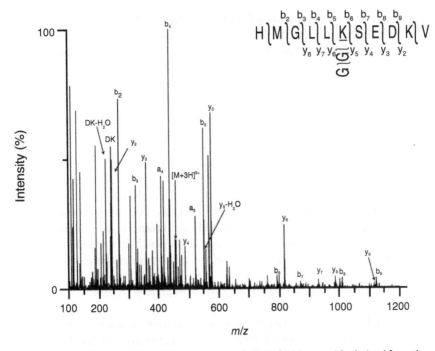

Figure 6.3 A collision-cell CID MS/MS spectrum of a SUMO(1)-isopeptide derived from the elastase digestion of SUMO(1)ylated RanGap1 418–587 protein fragment. Source: Chicooree, 2013 [39]. Reproduced with permission of Wiley.

6.2.4.2 Dual Proteolytic Enzyme Digestion with Trypsin and Chymotrypsin

Dual proteolytic digestion is an approach that involves the sequential use of two enzymes. Chymotrypsin and trypsin can be used to generate SUMO-isopeptides with shorter SUMO iso-chains in comparison to the full-length iso-chains generated from typical tryptic digestion [46]. First, trypsin is used to digest a poly-SUMO(2)ylated protein resulting in SUMO(2)-isopeptides with a SUMO(2) iso-chain containing 32 amino acids. Second, chymotrypsin is used to cleave C-terminal to the phenylalanine residue at position 87 resulting in the SUMO(2) iso-chains containing 6 amino acids, QQQTGG. Chymotrypsin has a C-terminal cleavage preference for a number of amino acid residues including phenylalanine, tyrosine, tryptophan, and leucine. QTOF ion mobility-based mass spectrometry with low-energy CID analysis was used to analyze SUMO(2)-isopeptides from dual digestion of poly-SUMO(2)ylated protein (data not shown). The use of ion mobility facilitated the successful separation of the predominantly higher charge state SUMO(2)-isopeptide ions under a narrow drift-time distribution from the predominantly lower charge state linear peptide ions generated under nESI conditions. Subsequent CID analysis of the SUMO(2)-isopeptide ions typically facilitated CID MS/MS spectra, which contained predominantly y-type ions from the backbone of the SUMO(2)-isopeptides as expected with additional b′-type-related product ions generated from the neutral loss of NH_3 from the SUMO(2) iso-chain, which could be manually interpreted by matching theoretical ions generated *in silico*. This approach to the analysis of SUMO(2)-isopeptides was applied to a simple digestion mixture; it would benefit from demonstrating applicability in a more complex digestion mixture such as one enriched for more SUMO-isopeptides, where its robustness could be assessed. The QQQTGG iso-chains generated from this dual-enzyme digestion strategy may benefit from utility in LC-MS/MS CRA(K)-based approaches [39]. However, the length of this iso-chain is susceptible to multiple events of glutamine-related deamidation, which would thereby add additional complexity to bioinformatic analyses, in comparison to the QTGG and TGG iso-chains, which offer fewer permutations of deamidation and, therefore, a decrease in complexity.

6.2.4.3 Proteolytic Enzyme and Chemical Digestion with Trypsin and Acid

An elegant dual-digestion strategy involving trypsin and then microwave-assisted acid chemical digestion with 12.5% acetic acid to selectively target cleavages C-terminal to aspartic acid residues by acid hydrolysis has been developed [47]. The selectivity involved in this approach facilitates a reduction in the length of these SUMO-(1)/(2)/(3) iso-chains of their respective SUMO-isopeptides in order to generate SUMO-isopeptides amenable to analytical analysis. Cleavage at the aspartic acid 86 position on the SUMO(1) iso-chain results in the generation of an iso-chain containing 11 amino acids, VIEVYQEQTGG. This is also the case when C-terminal cleavage of the aspartic acid residues at positions 85 and 82 on the SUMO(2) iso-chain and at positions 84 and 81 on the SUMO(3) iso-chains takes place, resulting in the generation of iso-chains containing either 8 (VFQQQTGG) or 11 amino acids (TIDVFQQQTGG). nLC-nESI-LTQ Orbitrap with low-energy linear ion-trap CID or ETD-based LC-MS/MS and MALDI-TOF/TOF with high-energy CID-based MS and MS/MS analysis have been employed to analyze these three isopeptides: di-SUMO(2)-isopeptides, di-SUMO(3)-isopeptides, and the SUMO(1)-isopeptide derived from the dual digestion of a SUMO(1)ylated E2-25K protein prepared *in vitro*. LC-MS/MS with ETD analysis of these three types of isopeptides facilitated ETD MS/MS spectra amenable to manual interpretation by matching theoretical ions generated from protein prospector. Since ETD fragmentation patterns occur in a predominantly sequence-independent manner, the ETD MS/MS displayed the generation of a characteristically comprehensive series of abundant c/c′- and z/z′-type product ions from both the backbone and iso-chains. This enabled the comprehensive structural elucidation of SUMO(1)/(2)/(3)-isopeptides (Figure 6.4).

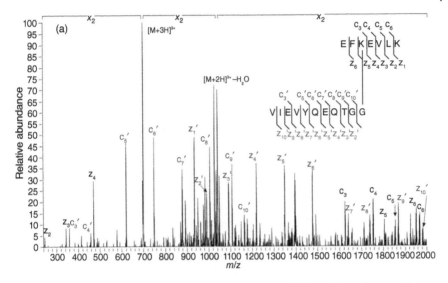

Figure 6.4 An ETD MS/MS spectrum of a SUMO(1)-isopeptide derived from the aspartic acid cleavage/trypsin digestion of the SUMO(1)ylated E2-25K protein. Source: Omoruyi, 2012 [47]. Reproduced with permission from Wiley.

By comparison, LC-MS/MS with low-energy CID and MALDI-TOF/TOF high-energy CID analysis of all three types of isopeptides facilitated CID MS/MS with less complex spectra than observed when analyzing similar SUMO-isopeptides with full-length iso-chains with CID, enabling the comprehensive structural elucidation of these isopeptides. Both instrument platforms and subsequent fragmentation techniques were very effective and generated MS/MS spectra with low complexity and predominantly facilitated a good balance of fragmentation of product ions between the backbone of the isopeptide and their respective iso-chains, critical for comprehensive structural elucidation. The results render this dual-digestion process using trypsin and 12.5% acetic acid in combination with LC-MS/MS with ETD and CID and MALDI-TOF/TOF with high-energy CID as an effective approach to analyzing these SUMO(1)/(2)/(3)-isopeptides in simple digestion mixtures. Assessment of the diagnostic value of product-ion fragmentation from the iso-chains for postacquisition extraction could enable the demonstration of this strategy in more complex digestion mixtures where unknown SUMO-isopeptides are present. The selectivity of the acid hydrolysis can result in the generation of the multiple forms of a particular SUMO-isopeptide. This may have an adverse impact on the abundance of these isopeptides by diluting the ion current between multiple signals in data-dependent acquisitions (DDA) of more complex samples. In addition, the isopeptides generated using this approach contain multiple glutamine residues. The number of deamidation events available for input into a bioinformatic algorithm represents an inherent compromise. Therefore, if this approach was applied to the analysis of putative SUMO-isopeptides in more complex samples, it may benefit from data-independent acquisition (DIA) approaches where data acquisition is unbiased toward precursor ion signal abundance.

6.2.5 MS Analysis of Modified Ub- and SUMO-Isopeptides under CID Conditions

Typical tryptic cleavage C-terminal to arginine residues at position 74 and 72 on the ubiquitin iso-chain occurs and results in the generation of a Ub-isopeptide containing an iso-chain of 2 or 4 amino acids: GG or LRGG. It has been reported that under efficient and complete digestion conditions the GG iso-chain is predominantly formed over the miscleaved LRGG form [48]. However, under time-controlled digestion conditions where partial digestion is used, the LRGG form is prevalent [49]. As we have discussed, complex fragmentation patterns and generation of a number of large product ions that can arise from the full-length SUMO iso-chains under CID analysis of isopeptides render MS/MS spectra difficult to decipher and interpret with high confidence. Analysis of Ub-isopeptides bearing the GG iso-chain and their subsequent MS/MS spectra has historically suffered from issues with false-positive identifications by bioinformatic software algorithms due to (i) misassignment of a flanking asparagine amino acid residue and alkylation of two cysteine residues, which are isobaric to the m/z of the GG iso-chain and (ii) general difficulty in interpreting MS/MS spectra manually due to the GG iso-chain of the Ub modification providing limited informative iso-chain product ions. Approaches using ETD with ion-trap-based instruments [50] and ECD on FT-ICR-based mass spectrometry [51] demonstrated more confident assignment of Ub-isopeptides due to the preservation of the GG iso-chain on the backbone of product ions related to the specific site of modification, enabled by the non-ergodic nature of the fragmentation process. However, there was still a requirement for the generation of GG or LRGG iso-chain Ub-modification-specific product ions to improve CID analysis of Ub-isopeptides. An approach developed to assess this requirement for the LRGG iso-chain modification-specific product ions involved the use of QTOF-based ESI-MS/MS and LC-nESI-MS/MS with low-energy CID analysis. It was used to analyze a synthetic model Ub-isopeptide comprising a calmodulin-related amino acid backbone. This Ub-isopeptide had been subjected to a time-controlled partial-digestion strategy using trypsin, which facilitated the generation of a Ub iso-chain containing 4 amino acids: LRGG [49]. The CID MS/MS spectrum of the Ub-isopeptide containing the LRGG iso-chain displayed predominantly y-type product ions from its backbone and two b′-type product ions from its iso-chain; b_2' (LR) and b_4' (GG), in addition to an internal ion, LRGGK-28 Da. The combination of these three ions indicated diagnostic confirmation of the LRGG iso-chain although their robust diagnostic utility was not assessed beyond the model peptide stage. More effective strategies that have been developed focus on using N-terminal-specific chemical derivatization, which selectively modify the α-amino groups of N-terminal and iso-N-terminal groups, along with the ε-amino groups of lysine side chains. The modification of the N-terminus and iso-N-terminus imparts favorable fragmentation behavior on Ub-isopeptides (and atypically derived SUMO(2)/(3)-isopeptides with TGG and QTGG iso-chains) under CID conditions enabling the generation of (i) characteristic diagnostic product ions from the Ub-isopeptide backbone, which indicate the site of the specific GG iso-chain Ub modification or (ii) diagnostic ions specific to the GG iso-chain Ub-modification, thus enabling complete identification of the GG iso-chain and the specific site of modification on the Ub-isopeptide backbone.

6.2.6 SPITC Modification

4-Sulfophenyl isothiocyanate (SPTIC) reagent has been used to sulfonate the α-amino group of the N-termini and the iso-N-termini of singly charged Ub-isopeptides to enhance their analysis using MALDI-TOF/TOF-based mass spectrometry high-energy CID [52]. The protons from sulfonic acid groups of the sulfonation modification of these two N-termini results in Edman-type cleavage of the modification and first amino acid present at the N-terminus, which consequently facilies the fragmentation of a unique cluster of characteristic signature ions at higher m/z values. This enables the identification of a Ub-isopeptide with (i) two N-termini amino acids and (ii) the GG iso-chain. Characteristic signature fragmentation ions resulting from the sulfonation tags of the Ub-isopeptides include abundant ion signals representing the m/z of (i) the loss of a tag and the first N-terminal amino acid of the Ub-isopeptide backbone from the m/z of the singly charged Ub-isopeptide molecular ion and (ii) the loss of the second tag and the first iso-N-terminal amino acid from the iso-chain from the m/z of the singly charged Ub-isopeptide molecular ion. This indicates the presence of a glycine amino acid residue at the iso-N-terminus and thus confirms the presence of the GG iso-chain. In addition to the cluster of characteristic signature fragmentation ions, sequence-specific y-type product ion are generated, which enable further confirmation of the GG iso-chain attached to the acceptor lysine residue. The acidic nature, which the sulfonation modification imparts on the Ub-isopeptide is significant. The fragmentation mechanism of the sulfonated Ub-isopeptides under high-energy CID conditions results in the formation of b-type product and related ions that are neutralized or negative under positive-ion mode. This produces an exclusive series of y-type product ions on the CID MS/MS spectrum. This method also showed utility on three tryptic Ub-isopeptides generated from a tryptic digest of a ubiquitinated C-terminal Hsc70 interacting protein generated *in vitro* [53].

6.2.7 Dimethyl Modification

Deuterated formaldehyde and sodium cyanoborohydride have been used in a process called reductive methylation of ubiquitinated isopeptides (RUbI) [54]. This produces one deuterated dimethyl group at the α-amino group of N-terminus and the iso-N-terminus of multiply charged Ub-isopeptides enhancing their analysis using QTOF-based LC-MS/MS with low-energy collision-cell CID [54]. The presence of two methyl groups replacing two hydrogen atoms on the amine at the N-terminus of a peptide imparts favorable fragmentation behavior of an N-terminal dimethylated peptide and N-terminal product-ion generation under low-energy collision-cell CID by (i) preventing the proton transfer that would usually occur from a non-dimethylated amine, from the a_1 ion to the y_x ions in the a_1-y_x peptide fragmentation pathway, resulting in the generation of a highly abundant a_1'-type ion [55, 56]; (ii) improved coverage of predominantly b/b' and also y/y' product ions due to the dimethyl modification reducing the α-amino group's basic character in the gas phase, enabling more random protonation of the peptide backbone, facilitating fragmentation by CID [57]; (iii) the electron-donating nature of the dimethyl groups stabilizes the a_1 ion resulting in its enhancement abundance postfragmentation of the N-terminal $C_{alpha}-C_{amide}$ bond of the peptide in a similar way to the a_1-y_x peptide fragmentation pathway [56]; and (iv) Harrison demonstrated that N-terminal acetylated peptides prevent sequence scrambling of peptides under low-energy collision-cell CID conditions and advantageously improved the presence of higher-order b-ions [44]. This could suggest that the presence of the dimethyl group similarly provides steric hindrance at the N-terminus of a Ub-isopeptide, advantageously resulting in the presence of higher-order b-type product ions. The three advantages that the dimethyl groups impart on the fragmentation of peptides (and N-terminal product-ion formation) were also observed from both N-termini of Ub-isopeptides [54]. The dimethyl groups at each N-termini of a Ub-isopeptide result in the addition of 32 Da at each N-termini, thereby increasing the mass of the Ub-isopeptide by 64 Da. The CID MS/MS spectra of a dimethylated di-Ub-K48-isopeptide ion resulted in the generation of comprehensive b- and y-type product ions and the generation of (i) a characteristic abundant a_1 ion from its N-terminus, which identifies the first N-terminal amino acid to be leucine/isoleucine at m/z 118.16; (ii) a characteristic abundant a_1' ion from its iso-N-terminus, which identifies the first iso-N-terminal amino acid of the iso-chain to be G at m/z 62.09; and (iii) a b_2' ion at m/z 147.11 ion from the iso-N-terminus, thus confirming the full identity of the iso-chain attached to the acceptor lysine of the Ub-isopeptide to be GG. The comprehensive a_1 product ion, b- and y-type product ions generated from the Ub-isopeptide enabled manual structural elucidation of its backbone and the a_1' and b_2' ions enabled the manual spectrometric structural elucidation of its GG iso-chain (Figure 6.5a). By comparison, the CID MS/MS spectrum of the equivalent dimethylated K48-peptide, that is, the peptide without the GG iso-chain covalently attached to the acceptor lysine (Figure 6.5b), only shows the presence of the a_1 ion at m/z 118.16 from the N-terminus of the peptide but does not show the additional a_1' and b_2', which represent the GG iso-chain as expected.

The combined diagnostic utility of the a_1' and b_2' ions was assessed by extraction postacquisition from LC-MS/MS data representing the tryptic digest of an anti-Ub pulldown of HEK293-T cell lysate samples spiked with a tryptic digest of six-protein mix. Extracting the a_1' and b_2' ions results in a significant reduction in chromatographic data complexity toward enhancing the analysis of potential Ub-isopeptide candidates. The retention times of peak profiles from the two extracted ion chromatogram (XIC) traces that perfectly co-eluted were identified as potential Ub-isopeptides. The unequivocal identity of all these Ub-isopeptides was confirmed by manual interrogation of their CID MS/MS spectra. The observation of the a_1' and b_2' ions thus enabled

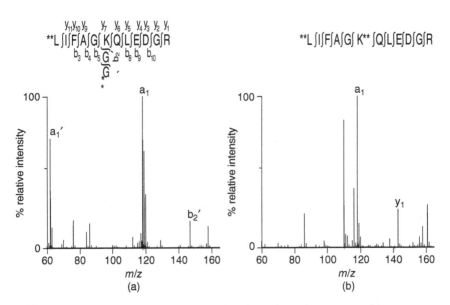

Figure 6.5 (a) The mass range between m/z 60 and 280 shows the presence of the diagnostic a_1'-type product ion (m/z 62.09) and the b_2'-type product ion (m/z 147.11) generated from the GG iso-chain. (b) These two diagnostic ions are shown to be absent in the linear version of the peptide without the GG iso-chain modification. Source: Chicooree, 2013 [54]. Reproduced with permission from Springer.

comprehensive spectrometric structural elucidation of the GG iso-chain, in addition to the presence of a comprehensive series of b- and y-type product ions, enabling the structural elucidation of the backbone sequence of the Ub-isopeptide. This demonstrated the robust diagnostic utility and enhanced selectivity of the a_1' and b_2' ions toward the analysis of Ub-isopeptides in complex samples. From the Ub-isopeptides identified, the selectivity of these diagnostic a_1' and b_2' ions in the RUbI approach enabled the identification of a "false-negative" large Ub-isopeptide which carried 6+ charges and contained two miscleavages within its peptide sequence and its site-specific GG iso-chain, which was proved to be beyond the searching and identification capabilities of Mascot when these dimethylated groups were added as variable modifications. Furthermore, Mascot suggested the presence of a Ub-isopeptide that was not detected during the RUbI approach. However, when analyzing the CID MS/MS spectra of the Ub-isopeptide there were no diagnostic a_1' and b_2' ions to indicate the presence of a GG iso-chain attached to the acceptor lysine, thereby confirming the Ub-isopeptide identified by Mascot to be a false-positive result. A minor caveat to the RUbI approach is that the a_1' and b_2' diagnostic ions would also be generated from tryptic linear peptides with a GG at their N-terminus; however, experimental and theoretical agreement is that this type of false-positive result would occur at a rate of approximately 0.5%. This demonstrates the important role the information-rich MS/MS spectrum has to play as it facilitates structural confirmation or rebuttal of the conclusions made based on extraction of the a_1' and b_2' ions [54]. Furthermore, a proportion of this 0.5% would also contain linear peptides that may be derived from target proteins that have undergone N-terminal protein ubiquitination, which occurs significantly less frequently than lysine protein ubiquitination.

The RUbI approach has also found utility in the enhanced analysis of SUMO(2/3)-isopeptides atypically generated from tryptic digestion of simple di-SUMO(2/3)-ylated proteins that have been spiked into a mixture of six protein tryptic peptides [58]. It was found that the presence of these dimethyl groups at the N-termini and iso-N-termini of these SUMO(2/3)-isopeptides also imparted favorable N-termini fragmentation behavior of the isopeptide and N-termini product-ion generation under low-energy collision-cell CID conditions. This resulted in the generation of (i) a set of modification-specific diagnostic a_1', b_2', and b_3' ions from the iso-N-termini of a SUMO(2/3)-isopeptide with an atypically cleaved TGG iso-chain and (ii) a set of modification-specific diagnostic a_1', b_2', and b_4' ions from the iso-N-termini of a SUMO(2/3)-isopeptide with an atypically cleaved QTGG iso-chain. For example, the CID MS/MS spectra of a dimethylated di-SUMO(2)-isopeptide ion (Figure 6.6) resulted in the generation of comprehensive b- and y-type product ions and the generation of (i) a characteristic abundant a_1 ion from its N-terminus, which identifies the first N-terminal amino acid to be glutamic acid at m/z 134.10; (ii) a characteristic abundant a_1' ion from its iso-N-terminus,

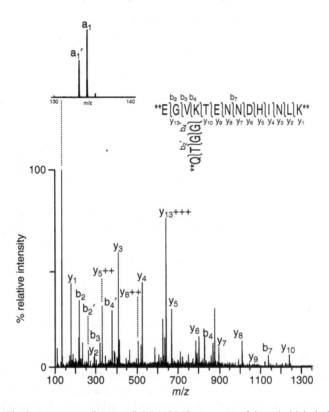

Figure 6.6 The low-energy collision-cell CID MS/MS spectrum of dimethyl-labeled SUMO(2)-isopeptide bearing a QTGG iso-chain derived from atypical tryptic digestion of a SUMO(2)ylated SUMO(2) protein, showing the presence of abundant diagnostic ions generated from the QTGG iso-chain; a_1' (*m/z* 133.13) (shown in the region zoomed in between *m/z* 130 and 140), b_2' (*m/z* 262.17) and b_4' (*m/z* 376.22) ions annotated. Source: Chicooree, 2013 [58]. Reproduced with permission from Springer.

which identifies the first iso-N-terminal amino acid of the iso-chain to be Q at *m/z* 133.13; (iii) a b_2' ion at *m/z* 262.17 from the iso-N-terminus, which identifies the first two amino acids of the iso-chain to be QT; and (iv) a b_4' ion at *m/z* 376.22 from the iso-N-terminus, thus confirming the full identity of the iso-chain attached to the acceptor lysine of the SUMO(2)-isopeptide to be QTGG. The comprehensive a_1 product ion and b- and y-type product ions generated from the SUMO(2)-isopeptide enabled manual structural elucidation of its backbone and the a_1' and b_2' ions enabled the manual spectrometric structural elucidation of its QTGG or TGG iso-chain.

XICs of the modification-specific diagnostic iso-N-termini ions result in a significant reduction in chromatographic data complexity. Perfect co-elution of the XIC traces enabled the identification of potential SUMO(2)-isopeptides from simple and semi-complex digestion mixtures. This demonstrated the robust diagnostic utility and enhanced selectivity of the a_1' and b' ions toward the analysis of SUMO(2/3)-isopeptides in semi-complex digestion mixtures.

6.2.8 m-TRAQ Modification

Since the proof of principle of enhanced Ub- and SUMO-isopeptide analysis was established using the RUbI approach [54, 58], mTRAQ reagent [59] has been used to chemically derivatize the α-amino groups of the N-termini and iso-N-termini on a suite of both synthetic tryptic Ub-isopeptides and atypically tryptic SUMO(2/3)-isopeptides to further facilitate an enhancement in their analysis. This was shown to enhance qualitative analysis and introduce relative quantitation across three channels in both SWATH precursor and product-ion acquisition modes. Performance was demonstrated on a QTOF-based LC-MS/MS mass spectrometer with low-energy collision-cell CID, in a DIA SWATH MS approach termed mass spectral enhanced detection of Ubls using SWATH acquisition (MEDUSA) [60]. Derivatization with mTRAQ imparts similar characteristic fragmentation behavior, product ion, and unique iso-N-terminal modification-specific product-ion generation on Ub- and SUMO(2/3)-isopeptides as dimethylation. However, the chemical composition and size of the mTRAQ modification combined with the presence of a carbonyl group within its structure assist in providing additional advantages over dimethylation by facilitating the generation of (i) additional N-terminal b_1-type and diagnostic iso-N-terminal b_1'-type product ions, (ii) more abundant diagnostic b'-type ions from the iso-N-terminus, and (iii) b-ions from the iso-N-terminus in a higher m/z range, which may benefit mass spectrometers where the transmission and observation of low mass ions is problematic. The advantages of improved product b'-type ion generation from the iso-N-terminus of the Ub/SUMO modification specifically combined with the ability of SWATH MS DIA to data capture all product ions from all detectable precursor ions in a single scan cycle, enables improved (i) comprehensive identification of the specific modification (ii) and enhancement toward the analysis of Ub- and SUMO-isopeptides, and (iii) provides greater quantitative selectivity of both Ub- and SUMO-isopeptides in SWATH product-ion acquisition mode. Figure 6.7 illustrates an overlay of the perfect coelution of the XICs of the diagnostic iso-N-terminal modification-specific a_1' and b' ions required to indicate and subsequently enable comprehensive diagnostic identification. The most abundant b'-ion from each isotag confirms the presence of the full GG isotag representing a Ub modification or the full TGG and QTGG isotags representing the presence of a SUMO(2/3) modification on three different synthetic isopeptides.

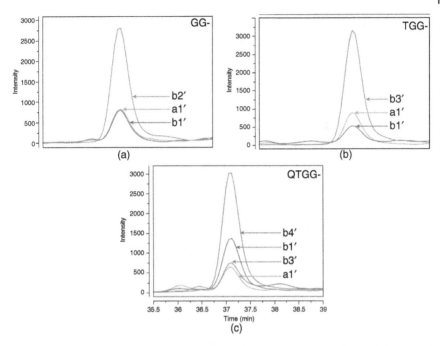

Figure 6.7 Postacquisition extracted diagnostic a′-type and b′-type product ions generated during SWATH acquisition of the synthetic isopeptides with the backbone sequence, NSSYVLL(<u>K</u>)TGK bearing the following iso-chains on its internal lysine residue; (<u>K</u>), after labeling with mTRAQ 8. (a) GG iso-chin with $a_1' = m/z$ 178.1430, $b_1' = m/z$ 206.1379, and $b_2' = m/z$ 263.1594; (b) TGG iso-chain with $a_1' = m/z$ 222.1692, $b_1' = m/z$ 277.1750, and $b_3' = m/z$ 364.2070; and (c) QTGG iso-chain with $a_1' = m/z$ 249.1801, $b_1' = m/z$ 277.1750, $b_3' = m/z$ 435.2442, and $b_4' = m/z$ 492.2656. Source: Griffiths, 2014 [60]. Reproduced with permission from Springer.

The use of triplex mTRAQ reagents – Δ0, Δ4, and Δ8 – in the MEDUSA approach enabled the quantification of Ub- and SUMO(2/3)-isopeptides across three channels in the SWATH MS mode. The high-quality and the most abundant diagnostic b′ product ion generated from the full iso-N-terminal Ub/SUMO(2/3) modification enables quantification in SWATH MS/MS mode. Quantification of synthetic Ub- and SUMO(2)/(3)-isopeptides using the MEDUSA approach was achieved by spiking two categories of predetermined theoretical ratios of these synthetic isopeptides – (i) 1(Δ0):2(Δ4):10(Δ8) and (ii) 1(Δ0):5(Δ4):10(Δ8) – into a background of *Escherichia coli* peptides, which had also been labeled at a predetermined theoretical ratio of 1(Δ0):1(Δ4):1(Δ8). Figure 6.8 shows an example of the quantification of both Ub-isopeptides. Figure 6.8a is an overlay of the three integrated and extracted monoisotopic precursor ion chromatograms from the TOF MS, which represent the three channels the synthetic Ub-isopeptide had been spiked in to the *E. coli* background at a known ratio of 1(Δ0):5(Δ4):10(Δ8). The calculated peak areas for each of the three peaks enabled the relative quantification of a ratio that was equivalent to 1(Δ 0):5.3(Δ 4):10.7(Δ8), with a high degree of accuracy in comparison to the predetermined theoretical ratio. The same Ub-isopeptide at the same known ratio was quantified in the SWATH-MS/MS product-ion acquisition mode, Figure 6.8b is an overlay of the three integrated and extracted dominant diagnostic b_2' ion representing the GG isotag from each of the three channels; m/z 255.1452 (Δ0), m/z 259.1523 (Δ4), and m/z 263.1594 (Δ8). The calculated peak areas for each of the three peaks enabled the relative quantification with a ratio equivalent to 1(Δ0):6.6(Δ4):18.8(Δ8) to a lower degree of accuracy in comparison to the predetermined theoretical ratio. The MEDUSA approach successfully enabled a suite of synthetic Ub- and SUMO(2)/(3)-isopeptides to be relatively quantified across three channels using (i) the extracted monoisotopic precursor ion peak areas from the SWATH-MS spectra or (ii) the most abundant diagnostic product-ion peak areas from the SWATH-MS/MS spectra. As expected, the MEDUSA approach has the same minor caveat to the RUbI approach in that the iso-N-terminal diagnostic ions would also be generated from tryptic linear peptides with a GG at their N-terminus; however, experimental and theoretical agreement is that this type of false-positive result would occur at a rate of 0.5% [54, 60]. Depending on the prevalence of N-terminal protein ubiquitination, which occurs significantly less frequently than lysine ubiquitination, the percentage of 0.5% can be further reduced due to the superior b- and b'-type product-ion coverage generated from m-TRAQ-labeled isopeptides in comparison to RUbI, enabling us to distinguish between two isomeric species, a false-positive

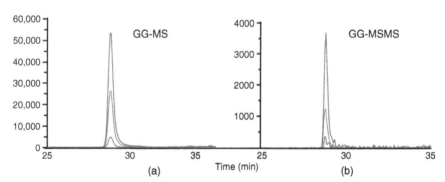

(a)   Time (min)   (b)

Figure 6.8 MS and MS/MS-based quantitation of isopeptides with the backbone sequence, EGV(K̲)TENNDHINLK bearing the GG iso-chain on its internal lysine residue; (K̲), based upon extracted ion peak areas. Of the three co-eluting extracted monoisotopic precursor ion chromatogram traces on (a) and (b); the trace of lowest intensity corresponds to Δ0-tagged (100 fmol), the trace of mid-range intensity relates to Δ4-tagged (500 fmol), and the trace of highest intensity represent Δ8-tagged (1 pmol) isopeptides. Source: Griffiths, 2014 [60]. Reproduced with permission from Springer.

linear peptide with a GG at its N-terminus and its equivalent Ub-isopeptide. This was possible due to the presence of a set of a/a'- and b/b'-type product-ion pairs in the MS/MS spectrum. This set of product-ion pairs was absent from the MS/MS spectrum of the linear peptide containing an N-terminal GG, which only contained the a- and b-type product ion with further consecutive b-type product ions from the backbone of the linear peptide confirming the remainder of its N-terminal sequence [60].

6.3 Enrichment and Global Analysis of Isopeptides

6.3.1 Overview of Enrichment Approaches

The dynamically reversible PTM of mammalian substrate proteins by wild-type mammalian ubiquitination and SUMOylation occurs at sub-stoichiometric levels. These sub-stoichiometric levels greatly limit the analytical coverage of putative Ub- and SUMO-isopeptides in a background of thousands of more abundant peptides generated postproteolytic digestion of complex biological samples. This subsequently limits the number of putative modification sites that can be analyzed and uniquely identified on a large or global scale by MS analyses. In order to circumvent this, enrichment strategies have been developed and incorporated into Ub- and SUMO-specific proteomic work flows with the ultimate aim of enriching for Ub- and SUMO-isopeptides. These enrichment strategies are implemented at the following stages: the pre-analytical stage:

i) protein level; this level focuses on the isolation and purification of Ub and/or SUMO proteins for subsequent analytical analyses, which are of interest to this book or alternative types of analyses but beyond the scope of this chapter;

ii) isopeptide level; this level focuses on the enrichment of isopeptides containing an internal lysine (K) bearing a GG iso-chain ((K)-GG iso-chain), the analytical stage;

iii) chromatographic level; this level focuses on the enrichment of generated Ub-isopeptides bearing a single G amino acid, through the use of automated chromatography using the elegant application of combined *fra*ctional *dia*gonal chromatography (COFRADIC).

Of course, these isopeptides are in contrast to the classical Ub-isopeptides that bear the classical GG iso-chain from direct tryptic of digestion of ubiquitinated proteins. For simplicity, we refer to these enrichment-approach-generated isopeptides as Ub-G-isopeptides or N-terminal Ub-G-peptides (if an N-terminal G-Boc modification is present). This enrichment strategy has been demonstrated for the effective analysis of Ub-G-isopeptides and enhanced analysis of N-terminal Ub-G-isopeptides, with a view to being applied to the analysis enrichment-approach-generated SUMO-isopeptides.

Inclusion of enrichment strategies for wild-type ubiquitinated proteins or Ub-isopeptides at the

i) protein level involves the use of (a) affinity- or epitope-tagged ubiquitin proteins [61], ubiquitin-binding domain (UBD) proteins [62], and tandem ubiquitin-binding entities (TUBES) proteins [63], resulting in subsequent purification of ubiquitinated and polyubiquitinated proteins; (b) anti-ubiquitin antibodies that have the capability to recognize ubiquitinated and polyubiquitinated proteins [48, 61] although additional PTMs involved in possible cross-talk with ubiquitination such as SUMO(2/3)ylation can also be recognized [39].

ii) isopeptide-level enrichment involves the use of monoclonal antibodies that recognize and are highly specific to the (K)-GG iso-chain [64, 65], which is internally present on isopeptides that have been generated post-tryptically from substrate proteins that have been post-translationally modified by ubiquitin, Nedd, or ISG15.

iii) analytical analysis–level enrichment involves the use of automated reversed-phase high-performance liquid chromatography (RP-HPLC) using COFRADIC. The application of COFRADIC enables the enrichment of Ub-G-isopeptides and N-terminal Ub-G-peptides by utilizing changes in their hydrophilicity in response to a selective change in their chemical structure [66].

A combination of enrichment strategies, known as dual-enrichment strategies, using types of TUBES known as Trypsin Resistant-TUBES (TR-TUBES) at the protein level and the monoclonal K-ε-GG antibody at the isopeptide level have demonstrated an enhancement in the analysis of Ub-isopeptides related to ligase substrates and activity [67]. Ultimately each level of enrichment that is incorporated into the proteomic workflows is conducted with a view to maximize the analytical and bioinformatic analyses of protein ubiquitination or SUMOylation. Each of the enrichment strategies has its advantages and disadvantages, although the progression of the enrichment strategies over time has resulted in vastly improving the depth of our biological knowledge of these PTMs. A few selected examples from these enrichment strategies that have been developed at the isopeptide level and analytical level for Ub-isopeptides and the protein level for protein SUMOylation and isopeptide level for SUMO-isopeptides are discussed further.

6.3.2 K-GG Antibody

Antibody-based enrichment strategies using monoclonal antibodies have been developed and effectively utilized to directly enrich for isopeptides generated from post-tryptic digestion, which contain an internal (K)-GG iso-chain with high selectivity [64, 65]. Although it is accepted in large- and global-scale

Ub-isopeptide analysis, and site mapping this (K)-GG iso-chain is predominantly referred to as being from a Ub-isopeptide; however, it is not unique to the Ub-isopeptide. The (K)-GG iso-chain is also present on tryptic isopeptides that have been derived from substrate proteins that have undergone PTM by two other ubiquitin-like PTMs: NEDDylation or ISG15ylation. However, with biological expertise and methodologies, it has been possible to determine the regulatory state of these two PTMs under certain conditions [68, 69] and subsequently select an appropriate biological technique to distinguish them [65]. Although the use of these biological techniques is governed by their complexity of implementation and availability, it is generally assumed that the majority of isopeptides containing a (K)-GG iso-chain identified and/or quantified are predominantly from the Ub-isopeptides with an understanding and appreciation that a percentage of these isopeptides are derived from NEDDylation or ISG15ylation. Aside from this caveat of (K)-GG isopeptide iso-chain selectivity, an additional caveat to the K-ε-GG antibodies is that they demonstrate a preference for certain amino acids adjacent to the lysine residue bearing the GG iso-chain. For example, the clone GX41 monoclonal antibody [64] demonstrates a preference toward amino acid residues including leucine, isoleucine, and tyrosine, whereas the rabbit monoclonal antibody [65] demonstrates a degree of preference for aspartic and glutamic acid residues [70]. Enrichment of Ub-isopeptides post-tryptic digestion in combination with subsequent LC-MS/MS and bioinformatic analyses enables a substantial advantage in that it has greatly increased the discovery number toward thousands to tens of thousands of ubiquitination sites on hundreds and thousands of proteins in identification and quantification studies on mammalian cell and tissue samples on a global scale [70–76]. By comparison, in terms of numbers, these are substantial when considering a study conducted on mammalian cell samples using a double-affinity-tagged ubiquitin protein-level enrichment in combination with LC-MS/MS analysis and bioinformatic analyses resulted in the identification of 753 ubiquitination sites on 471 proteins [77]. In addition, by further comparison, the first large-scale ubiquitination site–mapping study conducted on a yeast sample using an affinity-tagged ubiquitin protein-level enrichment strategy in combination with LC-MS/MS resulted in the identification of 110 ubiquitination sites on 72 ubiquitinated protein conjugates from a total of 1075 proteins identified [78]. The vast increase in the numbers of ubiquitination sites that have been identified due to this type of antibody enrichment strategy render it a powerful tool in the proteomic workflow of Ub-isopeptide analysis. To further enhance Ub-analysis, modification of isopeptides using approaches such as the RUbI [54] or MEDUSA [60] approach could be implemented post-antibody enrichment, thereby enabling comprehensive structural elucidation of the GG iso-chains. However, this would be dependent on the development of appropriate bioinformatic software to accommodate global analysis LC-MS/MS data.

6.3.3 COFRADIC

Applying COFRADIC as an enrichment strategy to analyze Ub-isopeptides is the most recent approach to have emerged in effective Ub-G-isopeptide enrichment [66]. This enrichment strategy requires two stages: (i) sample preparation and (ii) application of COFRADIC.

i) The sample preparation stage involves (i) first, blocking all free primary protein amino groups via acetylation with NHS-acetate. (ii) Second, the use of the catalytic core of a USP2 DUB enzyme – USP2cc. USP2cc is used to deubiquitinate ubiquitinated proteins by cleaving the isopeptide bond formed between the C-terminus of the ubiquitin protein and the ε-amino group of the side chain of a target lysine residue from a protein or the α-amino group of an N-terminal target lysine residue from a protein (formed from N-terminal ubiquitination of protein targets), resulting in the reintroduction of the primary protein amino groups of these target lysines. (iii) Third, selective chemical modification of these reintroduced primary amino groups via acylation with Gly-Boc-OSu to introduce a Gly-Boc group (G-Boc). (d) Tryptic digestion of the sample results in isopeptides with an internal lysine bearing a G-Boc iso-chain or peptides with N-terminal lysine bearing a G-Boc at its N-terminus.

ii) Applying COFRADIC involves (i) the tryptic sample being run on a primary RP-HPLC separation where fractions are collected and pooled; (ii) typically, a pooled fraction is then treated with 10% TFA to cleave off the Boc group from the G-Boc modification, resulting in a chemical change to the structure of the Ub-G-isopeptide or N-terminal Ub-G-peptide, specifically evoking a hydrophilic retention time shift of these Ub-G-isopeptides or N-terminal Ub-G-peptides, which is observed in (iii). (iii) The samples are then subjected to a secondary RP-HPLC separation under identical conditions to the primary RP-HPLC separation. The hydrophilic retention time shift that had been evoked from the cleavage of the Boc group in (ii) is observed during the secondary RP-HPLC run with the Ub-G-isopeptides and N-terminal Ub-G-peptides chromatographically separating from peptides, which did not undergo a hydrophilic retention time shift. This enables fractions of enriched Ub-G-isopeptides and N-terminal Ub-G-peptides to be collected and prepared for subsequent LC-MS/MS analysis. A detailed scheme of this COFRADIC-based enrichment strategy is depicted in Figure 6.9. (Note: Ub-G-isopeptides and N-terminal Ub-G-peptides are referred to in the reproduced Figure 6.9 as Gly-BOC-peptides.)

This COFRADIC-based enrichment strategy was effective in enabling the identification of 7504 ubiquitinated lysines on 3338 proteins along with 9 ubiquitinated protein N-termini from human Jurkat cells [66]. It was reported that the number of ubiquitinated lysines identified was 43% higher than those

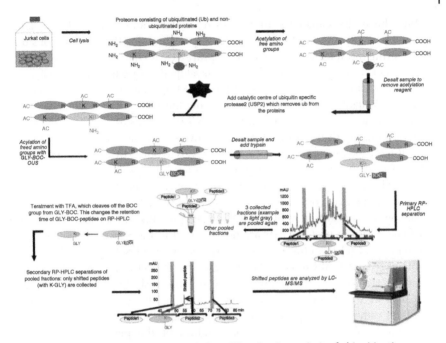

Figure 6.9 An illustration of the COFRADIC workflow for the analysis of ubiquitination. Source: Stes 2014 [66]. Reproduced with permission of American Chemical Society.

reported in studies using epitope-tagged ubiquitin [77] and the (K)-GG iso-chain-specific antibodies [74]. Furthermore, COFRADIC enabled the identification of sites of N-terminal ubiquitination, which would not have been possible with the (K)-GG iso-chain-specific antibodies due to their specificity for the internal (K)-GG iso-chain. A caveat to this enrichment strategy is that the acetylation reaction was determined to reach a completion of 95%, ultimately resulting in the misassignment of a proportion of the ubiquitination sites; in this case, it was 6.7% (at the isopeptide/peptide level) [66]. The authors have indicated that this strategy can be applied to quantitative analysis using stable isotopic labeling.

6.3.4 SUMOylation Enrichment

Inclusion of enrichment strategies for wild-type SUMOylated proteins or SUMO-isopeptides pre-MS analyses at the protein-level enrichment involves the use of (i) affinity- or epitope-tagged SUMOylated proteins resulting in subsequent isolation and purification of SUMOylated proteins [8]. These enrichment strategies represent a marked improvement in directing analytical-based proteomic analyses toward wild-type SUMOylation; (ii) anti-SUMO antibodies, which have the capability to recognize, isolate, and purify SUMOylated proteins [8, 79]; and (iii) SUMO affinity traps containing the presence of sumo interacting motif (SIM) sequences to capture SUMOylated proteins [8].

A recent enrichment strategy has been developed toward improving analytical-based proteomic analysis toward wild-type SUMOylation. This strategy is termed Protease-Reliant Identification of SUMO modification (PRISM) and has been developed and applied to enable the LC-MS/MS analysis of wild-type SUMO(2) ylation [80]. Briefly, the PRISM workflow involves the use of (a) affinity-tagged SUMO(2) proteins expressed at controlled levels to purify SUMOylated proteins, (b) covalent attachment of the primary amino groups of lysine residues on SUMOylated proteins on to beads via acetylation, resulting in the acetylation and subsequent blocking of free lysine residues not involved in SUMO(2)ylation, thus enabling the concentration of SUMO(2)ylated proteins and subsequent removal of free SUMO(2) proteins that are not involved in the SUMOylation of target proteins, and (c) the utilization of the specificity of SENP2s for SUMO(2)/(3) proteins to specifically remove the SUMO(2) iso-chain from the SUMO(2)ylated proteins resulting in the reintroduction of previously SUMO(2)ylated lysine residues as free lysine residues, subsequently resulting in the generation of peptides rather than isopeptides post-tryptic digestion. Trypsinization of these proteins resulted in the free lysine residues of the peptides enabling the identification of 751 wild-type SUMO(2)ylation sites on 400 proteins on a global scale in HeLa cells [81]. The removal of the SUMO(2) iso-chains from the SUMO(2)ylated proteins using SENP2s presents an attractive analytical advantage during LC-MS/MS under low-energy type CID conditions by eliminating the dominant complex fragmentation pattern associated with the full-length tryptic iso-chain [80] and the need for analysis using specialist bioinformatic software previously used for SUMOylation [34, 35]. However, acetylation of multiple lysine residues can result in the generation of much larger peptides of higher charge state post-tryptic digestion due to multiple tryptic miscleavages, potentially resulting in complex low-energy CID spectra, which may affect the performance of bioinformatic interpretation of the spectra. As alluded to in the COFRADIC approach [66] to the analysis of ubiquitination, acetylation is not 100% efficient in blocking all free lysine residues; therefore, a proportion (determined to be 3% at the peptide level in data generated from the PRISM strategy [81]) of the SUMOylation sites identified could potentially be due to false-positive identifications. However, this enrichment strategy is a vast improvement in directing analytical-based proteomic analyses away from mutagenic approaches [82] and more toward wild-type SUMOylation on a global scale.

Recently, an enrichment strategy was developed at the SUMO-isopeptide level and involved:

1) The use of a wild-type alpha-lytic protease (WaLP)-specific enzyme, which has been shown to have semi-specific C-terminal cleavage toward threonine

residues at a frequency of 30% and generate peptide lengths amenable to good analytical performance [83]. It has been suggested that this enzyme would be used in the generation of a GG iso-chain specifically on SUMO-isopeptides.

2) The use of the K-ε-GG antibody to enrich for the wild-type alpha-lytic-derived (K)-GG iso-chain in the same way it does for the tryptically derived (K)-GG iso-chain [84]. A 30% cleavage frequency is similar to the combined frequency of the atypically tryptic SUMO(2)/(3)-isopeptides detected in the CRA(K)-enabled approach.

The concept of this strategy sounds promising and future presentation and publication of data will be of high interest to the community.

6.4 Concluding Remarks and Recommendations

Wild-type mammalian protein ubiquitination and SUMOylation are highly regulated PTMs, involved in an extensive range of important biological functions both independently and codependently with each other, and a range of other PTMs. Dysregulation of ubiquitination and SUMOylation pathways has been shown to have a substantial impact on the progression of a number of disease states.

MS-based analysis is a key analytical technique in the identification and detection of the isopeptides derived from these proteins that have been targeted by these two PTMs, essentially enabling critical information to be obtained on a high-throughput scale in order to increase the depth of our understanding of their biological roles. However, MS-based analysis is challenging due to the inherent physicochemical nature of each modification. The MS analysis of each PTM presents different challenges, which effectively render the isopeptides they are present on difficult to robustly structurally elucidate and detect by MS-based proteomic analyses. There have, therefore, been significant efforts to develop improved MS-based proteomic strategies.

First, the complexities arising from MS-based and bioinformatic analyses resulting from SUMO-isopeptides containing highly charged full-length tryptic iso-chains and generating highly complex low-energy CID MS/MS spectra was negated by the development of the following strategies to improve the overall analytical amenability (i) by use of alternative low-energy-based fragmentation techniques such as IRMPD, ECD, and high-energy CID to facilitate an improvement in the complexity of their MS/MS spectra; (ii) the MS-based and bioinformatically CRA(K)-enabled analyses of more analytically amenable SUMO-isopeptides derived from atypical tryptic cleavage, under low-energy CID conditions; (iii) the MS-based analysis of analytically amenable SUMO-isopeptides derived from dual-digestion strategies, under low-energy

CID, ETD, and high-energy CID conditions; and (iv) the MS-based analysis of chemically derivatized SUMO-isopeptides derived from atypical tryptic cleavage, in order to facilitate the generation of modification-specific diagnostic product ions providing direct sequence coverage of the QTGG and TGG iso-chains under low-energy CID conditions. Strategies involving (ii)–(iv) demonstrated marked improvements in the generation of analytically amenable SUMO-isopeptides and their subsequent structural elucidation. In particular, one of the chemical modification–based strategies involved in (iv), termed MEDUSA, enabled the simultaneous detection and quantification of Ub-isopeptides in MS mode and also in MS/MS mode based on the generation of the robust QTGG and TGG iso-chain modification-specific diagnostic ions.

Second, the complexities arising from the MS-based and bioinformatic analyses of Ub-isopeptides containing GG iso-chains, predominantly resulting from CID conditions, were negated by the development of the following strategies to enhance the overall MS analysis of Ub-isopeptides: (i) alternative digestion strategies to produce a longer LRGG iso-chain in order to facilitate the generation of a modification-specific diagnostic ion from the iso-chain under low-energy CID conditions; (ii) chemical modification of Ub-isopeptides in order to facilitate the generation of a series of diagnostic ions indicating the presence of GG iso-chain modification under high-energy CID conditions; and (iii) chemical derivatization of Ub-isopeptides in order to facilitate the generation of modification-specific diagnostic a′/b′-type product ions in RUbI and MEDUSA approaches, providing specific sequence coverage of the GG iso-chain under low-energy CID conditions. The strategies involved in (ii) and (iii) were marked improvements in enhancing the analysis of Ub-isopeptide structure and subsequent robust and confident identification. In particular, one chemical modification-based strategy involved in (iii), termed MEDUSA, enabled the simultaneous detection and quantification of Ub-isopeptides in MS mode and also in MS/MS based on the generation of the robust GG iso-chain modification-specific diagnostic ions.

In addition to the improvements made by the development of these methods to the MS-based proteomic analysis of SUMO and Ub-isopeptides, additional progress has been made in the development of effective enrichment strategies for the biological- and chemical-based enrichment of Ub- and SUMO-isopep-tides to assist with improving global MS-based proteomic analyses of protein ubiquitination and SUMOylation. The most effective enrichment strategies include the (i) utilization of monoclonal antibodies to specifically enrich for both Ub- and SUMO-isopeptides with internal lysine residues bearing a GG iso-chain, (ii) the development of a COFRADIC-based strategy to specifically enrich for approach-generated Ub-G-isopeptides, and (iii) a protease-reliant strategy termed PRISM for the analysis of protein SUMOylation.

Recommendations for the improvement of SUMO- and Ub-isopeptide anal-ysis would involve (i) digestion strategies, which are capable of generating a distinguishable single version of a SUMO(1) iso-chain and a SUMO(2/3) iso-chain, which are both analytically amenable to MS-based proteomic analyses; (ii) development of additional monoclonal antibodies that recognize lysine residues bearing both of these iso-chains and/or the chemically derivatized iso-N-terminal forms of these iso-chains to reap the additional benefits of the chemically facilitated diagnostic ions under low-energy CID conditions for both enhanced detection and subsequent quantification in MS/MS mode (this would also apply to Ub-isopeptides); and (iii) development of bioinformatic software to analyze the potentially large list of putative SUMO and Ub-isopeptide candidates that would be generated from postacquisition ion chromatogram extraction of diagnostic ions generated in the RUbI and MEDUSA approaches, should these approaches be applied to MS-based global analyses.

References

1 Hermann J, Lerman LO, Lerman A. Ubiquitin and ubiquitin like proteins in protein regulation. *Circ Res* 2007;**100**:1276–1291.

2 Breitschopf K, Bengal E, Ziv T, Admon A, Ciechanover A. A novel site for ubiquitination: the N-terminal residue, and not internal lysines of MyoD, is essential for conjugation and degradation of the protein. *EMBO J* 1998;**17**:5964–5973.

3 Cadwell K, Coscoy L. Ubiquitination on nonlysine residues by a viral E3 ubiquitin ligase. *Science* 2005;**309**:127–130.

4 Wang X, Herr RA, Chua WJ, Lybarger L, Wiertz EJ, Hansen TH. Ubiquitination of serine, threonine, or lysine residues on the cytoplasmic tail can induce ERAD of MHC-I by viral E3 ligase mK3. *J Cell Biol* 2007;**177**:613–624.

5 Hay RT. SUMO: a history of modification. *Mol Cell* 2005;**18**:1–12.

6 Rodriguez MS, Dargemont C, Hay RT. SUMO-1 conjugation in vivo requires both modification and motif and nuclear targeting. *J Biol Chem* 2001;**276**:12654–12659.

7 Chung TL, Hsiao HH, Yeh YY, Shia HL, Chen YL, Liang PH, Wang AH, Khoo KH, Shoei-Lung Li S. *In vitro* modification of human centromere protein CENP-C fragments by small ubiquitin-like modifier (SUMO) protein: definitive identification of the modification sites by tandem mass spectrometry analysis of the isopeptides. *J Biol Chem* 2004;**279**:39653–39662.

8 Eifler K, Vertegaal ACO. Mapping the SUMOylated landscape. *FEBS J* 2015;**282**:3669–3680.

9 Nakayama KI, Nakayama K. Ubiquitin ligases: cell-cycle control and cancer. *Nat Rev Cancer* 2006;**6**:369–381.

10 Bettermann K, Benesch M, Weis S, Haybaeck J. SUMOylation in carcinogenesis. *Cancer Lett* 2012;**316**:113–125.

11 Nijman SM, Luna-Vargas MP, Velds A, Brummelkamp TR, Dirac AM, Sixma TK, Bernards R. A genomic and functional inventory of deubiquitinating enzymes. *Cell* 2005;**123**:773–786.

12 Hickey CM, Wilson NR, Hochstrasser M. Function and regulation of SUMO proteases. *Nat Rev Mol Cell Biol* 2012;**13**:755–766.

13 Reyes-Turcu FE, Ventii KH, Wilkinson KD. Regulation and cellular roles of ubiquitin-specific deubiquitinating enzymes. *Annu Rev Biochem* 2009;**78**:363–397.

14 Dou H, Huang C, Van-Nguyen T, Lu LS, Yeh ET. SUMOylation and de- SUMOylation in response to DNA damage. *FEBS Lett* 2011;**585**:2891–2896.

15 Behrends C, Harper JW. Constructing and decoding unconventional ubiquitin chains. *Nat Struct Mol Biol* 2011;**18**:520–528.

16 Kulathu Y, Komander D. Atypical ubiquitylation - the unexplored world of polyubiquitin beyond Lys48 and Lys63 linkages. *Nat Rev Mol Cell Biol* 2012;**13**:508–523.

17 Finley D. Recognition and Processing of Ubiquitin-Protein Conjugates by the Proteasome. *Annu Rev Biochem* 2009;**78**:477–513.

18 Baba D, Maita N, Jee JG, Uchimura Y, Saitoh H, Sugasawa K, Hanaoka F, Tochio H, Hiroaki H, Shirakawa M. Crystal structure of thymine DNA glycosylase conjugated to SUMO-1. *Nature* 2005;**435**:979–982.

19 Tatham MH, Geoffroy MC, Shen L, Plechanovova A, Hattersley N, Jaffray EG, Palvimo JJ, Hay RT. RNF4 is a poly-SUMO-specific E3 ubiquitin ligase required for arsenic-induced PML degradation. *Nat Cell Biol* 2008;**10**:538–546.

20 Dikic I, Wakatsuki S, Walters KJ. Ubiquitin-binding domains - from structures to functions. *Nat Rev Mol Cell Biol* 2009;**10**:659–671.

21 Chen ZJJ, Sun LJJ. Nonproteolytic Functions of Ubiquitin in Cell Signaling. *Mol Cell* 2009;**33**:275–286.

22 Muller S, Matunis MJ, Dejean A. Conjugation with the ubiquitin-related modifier SUMO-1 regulates the partitioning of PML within the nucleus. *EMBO J* 1998;**17**:61–70.

23 Ross S, Best JL, Zon LI, Gill G. SUMO-1 modification represses Sp3 transcriptional activation and modulates its subnuclear localization. *Mol Cell* 2002;**10**:831–842.

24 Hoege C, Pfander B, Moldovan GL, Pyrowolakis G, Jentsch S. RAD6-dependent DNA repair is linked to modification of PCNA by ubiquitin and SUMO. *Nature* 2002;**419**:135–141.

25 Papouli E, Chen S, Davies AA, Huttner D, Krejci L, Sung P, Ulrich HD. Crosstalk between SUMO and ubiquitin on PCNA is mediated by recruitment of the helicase Srs2p. *Mol Cell* 2005;**19**:123–133.

26 Moldovan GL, Pfander B, Jentsch S. PCNA, the maestro of the replication fork. *Cell* 2007;**129**:665–679.

27 Hunter T. The age of crosstalk: phosphorylation, ubiquitination, and beyond. *Mol Cell* 2007;**28**:730–738.

28 Wu S-Y, Chiang C-M. Crosstalk between sumoylation and acetylation regulates p53-dependent chromatin transcription and DNA binding. *EMBO J* 2009;**28**:1246–1259.

29 Bologna S, Ferrari S. It takes two to tango: Ubiquitin and SUMO in the DNA damage response. *Front Genet* 2013;**4**:1–18.

30 Chen Z, Lu W. Roles of Ubiquitination and SUMOylation on Prostate Cancer: Mechanisms and Clinical Implications. *Int J Mol Sci* 2015;**16**:4560–4580.

31 Hochstrasser M. Origin and function of ubiquitin like proteins. *Nature* 2009;**2009**(458):422–429.

32 Sarge KD, Park-Sarge OK. Sumoylation and human disease pathogenesis. *Trends Biochem Sci* 2009;**34**:200–205.

33 Matic I, van Hagen M, Schimmel J, Macek B, Ogg SC, Tatham MH, Hay RT, Lamond AI, Mann M, Vertegaal AC. *In vivo* identification of human small ubiquitin-like modifier polymerization sites by high accuracy mass spectrometry and an *in vitro* to *in vivo* strategy. *Mol Cell Proteomics* 2008;**7**:132–144.

34 Jeram SM, Srikumar T, Zhang XD, Anne Eisenhauer H, Rogers R, Pedrioli PG, Matunis M, Raught B. An improved SUMmOn-based methodology for the identification of ubiquitin and ubiquitin-like protein conjugation sites identifies novel ubiquitin-like protein chain linkages. *Proteomics* 2010;**10**:254–265.

35 Hsiao HH, Meulmeester E, Frank BT, Melchior F, Urlaub H. "ChopNSpice," a mass spectrometric approach that allows identification of endogenous small ubiquitin-like modifier-conjugated peptides. *Mol Cell Proteomics* 2009;**8**:2664–2675.

36 Cooper HJ, Tatham MH, Jaffray E, Heath JK, Lam TT, Marshall AG, Hay RT. Fourier transform ion cyclotron resonance mass spectrometry for the analysis of small ubiquitin-like modifier (SUMO) modification: identification of lysines in RanBP2 and SUMO targeted for modification during the E3 autoSUMOylation reaction. *Anal Chem* 2005;**77**:6310–6319.

37 Gardner MW, Smith SI, Ledvina AR, Madsen JA, Coon JJ, Schwartz JC, Stafford GC Jr, Brodbelt JS. Infrared Multiphoton Dissociation of Peptide Cations in a Dual Pressure Linear Ion Trap Mass Spectrometer. *Anal Chem* 2009;**81**:8109–8118.

38 Paizs B, Suhai S. Fragmentation pathways of protonated peptides. *Mass Spectrom Rev* 2005;**24**:508–548.

39 Chicooree N, Griffiths JR, Connolly Y, Tan C-T, Malliri A, Eyers CE, Smith DL. A proteomic approach for the identification of proteotypic SUMOylated isopeptides in simple and complex systems. *Rapid Commun Mass Spectrom* 2013;**27**:127–134.

40 Wells JM, McLuckey SA. Collision-induced dissociation (CID) of peptides and proteins. *Methods Enzymol* 2005;**402**:148–185.

41 Blomster HA, Imanishi SY, Siimes J, Kastu J, Morrice NA, Eriksson JE, Sistonen L. *In vivo* identification of sumoylation sites by a signature tag and cysteine-targeted affinity purification. *J Biol Chem* 2010;**285**:9324–19329.

42 Galisson F, Mahrouche L, Courcelles M, Bonneil E, Meloche S, Chelbi-Alix MK, Thibault P. A novel proteomics approach to identify SUMOylated proteins and their modification sites in human cells. *Mol Cell Proteomics* 2011;**10**:1–15.

43 Perkins DN, Pappin DJC, Creasy DM, Cottrell JS. Probability- based protein identification by searching sequence databases using mass spectrometry data. *Electrophoresis* 1999;**20**:3551–3567.

44 Harrison AG. Peptide Sequence Scrambling Through Cyclisation of b5 Ions. *J Am Soc Mass Spectrom* 2008;**19**:1776–1780.

45 Hohmann L, Sherwood C, Eastham A, Peterson A, Eng JK, Eddes JS, Shteynberg D, Martin DB. Proteomic analyses using *Grifola frondosa* metalloendoprotease Lys-N. *J Proteome Res* 2009;**8**:1415–1422.

46 Dumont Q, Donaldson DL, Griffith WP. Screening method for isopeptides from small ubiquitin-related modifier-conjugated proteins by ion mobility mass spectrometry. *Anal Chem* 2011;**83**:9638–9642.

47 Osula O, Swatkoski S, Cotter RJ. Identification of protein SUMOylation sites by mass spectrometry using combined microwave-assisted aspartic acid cleavage and tryptic digestion. *J Mass Spectrom* 2012;**47**:644–654.

48 Chen PC, Na CH, Peng J. Quantitative proteomics to decipher ubiquitin signaling. *Amino Acids* 2012;**43**:1049–1060.

49 Warren MRE, Parker CE, Mocanu V, Klapper D, Borchers CE. Electrospray ionization tandem mass spectrometry of model peptides reveals diagnostic product ions for protein ubiquitination. *Rapid Commun Mass Spectrom* 2005;**19**:429–437.

50 Sobott F, Watt SJ, Smith J, Edelmann MJ, Kramer HB, Kessler BM. Comparison of CID Versus ETD Based MS/MS Fragmentation for the Analysis of Protein Ubiquitination. *J Am Soc Mass Spectrom* 2009;**20**:1652–1659.

51 Cooper HJ, Heath JK, Jaffray E, Hay RT, Lam TT, Marshall AG. Identification of sites of ubiquitination in proteins: a Fourier transform ion cyclotron resonance mass spectrometry approach. *Anal Chem* 2004;**76**:6982–6988.

52 Wang D, Cotter RJ. Approach for determining protein ubiquitination sites by MALDI-TOF mass spectrometry. *Anal Chem* 2005;**77**:1458–1466.

53 Wang D, Xu W, McGrath SC, Patterson C, Neckers L, Cotter RJ. Direct identification of ubiquitination sites on ubiquitin-conjugated CHIP using MALDI mass spectrometry. *J Proteome Res* 2005;**4**:1554–1560.

54 Chicooree N, Connolly Y, Tan C-T, Malliri A, Li Y, Smith DL, Griffiths JR. Enhanced Detection of Ubiquitin Isopeptides Using Reductive Methylation. *J Am Soc Mass Spectrom* 2013;**24**:421–430.

55 Hsu JL, Huang SY, Shiea JT, Huang WY, Chen SH. Beyond quantitative proteomics: signal enhancement of the a_1 ion as a mass tag for peptide sequencing using dimethyl labeling. *J Proteome Res* 2005;4:101–108.

56 Fu Q, Li L. De Novo Sequencing of Neuropeptides Using Reductive Isotopic Methylation and Investigation of ESI QTOF MS/MS Fragmentation Pattern of Neuropeptides with N-Terminal Dimethylation. *Anal Chem* 2005;77:7783–7795.

57 Hsu JL, Huang SY, Chow NH, Chen SH. Stable isotope dimethyl labeling for quantitative proteomics. *Anal Chem* 2003;75:6843–6852.

58 Chicooree N, Griffiths JR, Connolly Y, Smith DL. Chemically facilitating the generation of diagnostic ions from SUMO(2/3) remnant isopeptides. *Rapid Commun Mass Spectrom* 2013;27:2108–2114.

59 DeSouza LV, Taylor AM, Li W, Minkoff MS, Romaschin AD, Colgan TJ, Siu KWM. Multiple reaction monitoring of mTRAQ-labeled peptides enables absolute quantification of endogenous levels of a potential cancer marker in cancerous and normal endometrial tissues. *J Proteome Res* 2008;7:3525–3534.

60 Griffiths JR, Chicooree N, Connolly Y, Neffling M, Lane CS, Knapman T, Smith DL. Mass spectral enhanced detection of Ubls using SWATH acquisition: MEDUSA-simultaneous quantification of SUMO and ubiquitin-derived isopeptides. *J Am Soc Mass Spectrom* 2014;25:767–777.

61 Vertegaal ACO. Uncovering Ubiquitin and Ubiquitin-like Signaling Networks. *Chem Rev* 2011;111:7923–7940.

62 Tan F, Lu L, Cai Y, Wang J, Xie Y, Wang L, Gong Y, Xu BE, Wu J, Luo Y, Qiang B, Yuan J, Sun X, Peng X. Proteomic analysis of ubiquitinated proteins in normal hepatocyte cell line Chang liver cells. *Proteomics* 2008;8:2885–2896.

63 Lopitz-Otsoa F, Rodriguez-Suarez E, Aillet F, Casado-Vela J, Lang V, Matthiesen R, Elortza F, Rodriguez MS. Integrative analysis of the ubiquitin proteome isolated using Tandem Ubiquitin Binding Entities (TUBEs). *Proteomics* 2012;75:2998–3014.

64 Xu G, Paige JS, Jaffrey SR. Global analysis of lysine ubiquitination by ubiquitin remnant immunoaffinity profiling. *Nat Biotechnol* 2010;28:868–873.

65 Kim W, Bennett EJ, Huttlin EL, Guo A, Li J, Possemato A, Sowa ME, Rad R, Rush J, Comb MJ, Harper JW, Gygi SP. Systematic and quantitative assessment of the ubiquitin-modified proteome. *Mol Cell Proteomics* 2011;44:325–340.

66 Stes E, Laga M, Walton A, Samyn N, Timmerman E, De Smet I, Goormachtig S, Gevaert K. A COFRADIC protocol to study protein ubiquitination. *J Proteome Res* 2014;13:3107–3113.

67 Yoshida Y, Saeki Y, Murakami A, Kawawaki J, Tsuchiya H, Yoshihara H, Shindo M, Tanaka K. A comprehensive method for detecting ubiquitinated substrates using TR-TUBE. *Proc Natl Acad Sci U S A* 2015;112:4630–4635.

68 Emanuele MJ, Elia AE, Xu Q, Thoma CR, Izhar L, Leng Y, Guo A, Chen YN, Rush J, Hsu PW, Yen HC, Elledge SJ. Global Identification of Modular Cullin-RING Ligase Substrates. *Cell* 2011;147:459–474.

69 Kamitani T, Kito K, Nguyen HP, Yeh ET. Characterization of NEDD8, a developmentally down-regulated ubiquitin-like protein. *J Biol Chem* 1997;**272**:28557–28562.

70 Wagner SA, Beli P, Weinert BT, Scholz C, Kelstrup CD, Young C, Nielsen ML, Olsen JV, Brakebusch C, Choudhary C. Proteomic analyses reveal divergent ubiquitylation site patterns in murine tissues. *Mol Cell Proteomics* 2012;**11**:1578–1585.

71 Wagner SA, Beli P, Weinert BT, Nielsen ML, Cox J, Mann M, Choudhary C. A proteome-wide, quantitative survey of *in vivo* ubiquitylation sites reveals widespread regulatory roles. *Mol Cell Proteomics* 2011;**10**:M111 013284. DOI: 10.1074/mcp.M111.013284.

72 Udeshi ND, Mani DR, Eisenhaure T, Mertins P, Jaffe JD, Clauser KR, Hacohen N, Carr SA. Methods for quantification of *in vivo* changes in protein ubiquitination following proteasome and deubiquitinase inhibition. *Mol Cell Proteomics* 2012;**11**:148–159.

73 Udeshi ND, Svinkina T, Mertins P, Kuhn E, Mani DR, Qiao JW, Carr SA. Refined Preparation and Use of Anti-K-ε-GG Antibody Enables Routine Quantification of 10,000s of Ubiquitination Sites in Single Proteomics Experiments. *Mol Cell Proteomics* 2013;**12**:825–831.

74 Udeshi ND, Mertins P, Svinkina T, Carr SA. Large-scale identification of ubiquitination sites by mass spectrometry. *Nat Protoc* 2013;**8**:1950–1960.

75 Iwabuchi M, Sheng H, Thompson J, Wang L, Dubois LG, Gooden D, Moseley M, Paschen W, Yang W. Characterization of the ubiquitin-modified proteome regulated by transient forebrain ischemia. *J Cereb Blood Flow Metab* 2014;**34**:425–432.

76 Thomas SN, Zhang H, Cotter RJ. Application of quantitative proteomics to the integrated analysis of the ubiquitylated and global proteomes of xenograft tumor tissues. *Clin Proteomics* 2015;**12**:1–15. DOI: 10.1186/s12014-015-9086-5.

77 Danielsen JM, Sylvestersen KB, Bekker-Jensen S, Szklarczyk D, Poulsen JW, Horn H, Jensen LJ, Mailand N, Nielsen ML. Mass spectrometric analysis of lysine ubiquitylation reveals promiscuity at site level. *Mol Cell Proteomics* 2011;**10**:M110 003590. DOI: 10.1074/mcp.M110.003590.

78 Peng J, Schwartz D, Elias JE, Thoreen CC, Cheng D, Marsischky G, Roelofs J, Finley D, Gygi SP. A proteomics approach to understanding protein ubiquitination. *Nat Biotechnol* 2003;**8**:921–926.

79 Filosa G, Barabino SM, Bachi A. Proteomics strategies to identify SUMO targets and acceptor sites: a survey of RNA-binding proteins SUMOylation. *Neuromolecular Med* 2013;**15**:661–676.

80 Chicooree N, Unwin RD, Griffiths JR. The application of targeted mass spectrometry-based strategies to the detection and localization of post-translational modifications. *Mass Spectrom Rev* 2014;**34**:595–626.

81 Hendriks IA, D'Souza RC, Chang J-G, Mann M, Vertegaal ACO. System-wide identification of wild-type SUMO-2 conjugation sites. *Nat Commun* 2015;**6**:1–16.

82 Hendriks IA, D'Souza RCJ, Yang B, Verlaan-de Vries M, Mann M, Vertegaal ACO. Uncovering global SUMOylation signaling networks in a site-specific manner. *Nat Struct Mol Biol* 2014;**21**:927–936.

83 Meyer JG, Yang B, Bennett E, Komives EA. A novel comprehensive discovery approach for SUMO modified proteins. *Mol Cell Proteomics* 2014;**13**:S41–S43.

84 Meyer JG, Kim S, Maltby DA, Ghassemian M, Banderia N, Komives EA. Expanding Proteome Coverage with Orthogonal-specificity α-Lytic Proteases. *Mol Cell Proteomics* 2014;**13**:823–835.

7

The Deimination of Arginine to Citrulline

Andrew J. Creese and Helen J. Cooper

School of Biosciences, University of Birmingham,, Birmingham, UK

7.1 Overview of Arginine to Citrulline Conversion: Biological Importance

The nonstandard α-amino acid citrulline was first isolated in 1914 by Koga and Odake [1]. In 1931 the conversion of arginine to citrulline, using alkaline hydrolysis to drive the reaction, was observed by Ackermann [2]. Figure 7.1 shows schematically the conversion of arginine to citrulline; *in vivo*, the reaction is catalyzed by calcium and the peptidylarginine deiminase (PAD) enzymes. At physiological conditions, arginine has a charge of +1, and citrulline is neutral; therefore the conversion of arginine to citrulline in proteins reduces the overall charge of the protein, which will affect the structure and function of the protein [3].

Citrullination has been linked to multiple autoimmune diseases including rheumatoid arthritis (RA) [4, 5], periodontitis [6], multiple sclerosis [7], and Alzheimer's disease [8]. It has also been shown to play crucial roles in cancer [9, 10] and cardiovascular disease [11, 12]. Citrullination is essential in fetal development [13] and helps control gene expression [14]. In 1998, Schellekens et al. [15] demonstrated that patients with RA produced antibodies against citrullinated proteins (anticitrullinated protein antibody (ACPA)). Since then, the study of the modification has grown with over 3500 publications on citrullinated proteins published (1998–2014, Web of Science) as shown in Figure 7.2.

As mentioned earlier, the deimination of arginine to citrulline is catalyzed *in vivo* by the PAD enzymes. In humans there are five PAD enzymes (PAD1–4 and PAD 6) [16]. The family is highly homologous [16]. Over 50% of the amino acid sequence is conserved across the five proteins [16]. PAD3 is probably the least studied and has been localized to hair follicles. It is known to citrullinate

Analysis of Protein Post-Translational Modifications by Mass Spectrometry,
First Edition. Edited by John R. Griffiths and Richard D. Unwin.
© 2017 John Wiley & Sons, Inc. Published 2017 by John Wiley & Sons, Inc.

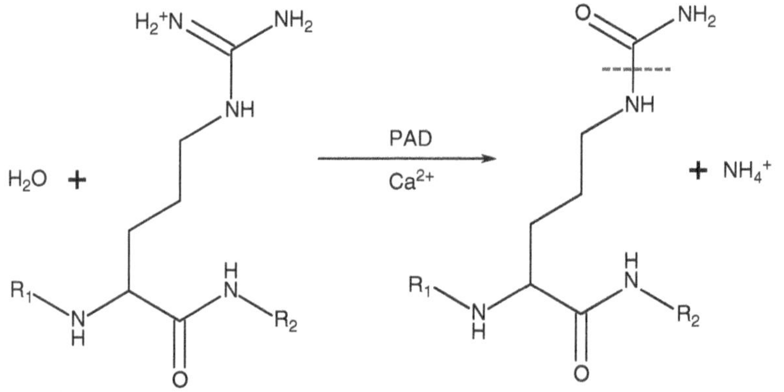

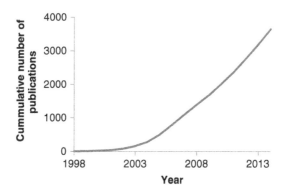

Figure 7.1 The conversion of arginine to citrulline catalyzed by calcium and a peptidylarginine deiminase. The red dashed line indicates cleavage of isocyanic acid during collision-induced dissociation fragmentation.

Figure 7.2 The cumulative number of publications with "citrullinated" in the topic, Web of Science, 08/20/2015.

trichohyalin [17] and filaggrin [18], but there is no literature describing direct involvement in disease.

PAD1 is involved in late-stage epidermal differentiation. The enzyme citrullinates keratin 1 [19], keratin 10 [20], and filaggrin [18]. It has been shown that the citrullination of these three proteins maintains hydration of the stratum corneum [21], which in turn maintains cutaneous barrier function [22]. PAD1 is predominantly expressed in the epidermis and the uterus [20].

Overexpression of PAD2 in mice has been shown to increase citrullination of myelin basic protein (MBP) in brain tissue leading to less stable myelin [23]. Myelin forms a sheath around the axon of neurons and is therefore essential to the development of the nervous system. It has been shown that approximately

20% of MBP is significantly citrullinated (six arginine–citrulline conversions) in healthy white brain matter. In white matter from patients with multiple sclerosis, a greater percent of MBP was heavily citrullinated (45%) [23]. Another study [24] demonstrated that overexpression of PAD2 in transgenic mice led to increased citrullination of MBP and destabilization of myelin. Similar to the results of Moscarello et al. [23], the levels of citrullination increased dramatically when comparing healthy to overexpressing samples. In this case, the increase in citrullination was observed when the mice were 3 months old.

Citrullination of vimentin by PAD2 is involved in the apoptosis of Jurkat T cells [25]. It was observed that activated Jurkat cells expressed greater levels of PAD2 and more citrullinated protein was identified. Overexpression of PAD2 in Jurkat cells showed no difference in apoptosis from cells that did not overexpress PAD2; however, on activation, cell viability decreased significantly. This finding suggests that citrullination has a vital role in cell viability and increased citrullination causes cell death.

PAD2 has been identified in inflamed myofilaments from human hearts. Giles et al. [26] analyzed heart tissue from autopsy samples. Myocardial samples with rheumatic disease were compared to control samples (myocarditis and scleroderma). All five PAD enzymes were identified in the heart tissue. However, PAD2 and PAD4 were the dominant species in the inflamed heart tissue of the rheumatic samples. In a data-independent proteomics study of human heart tissue, Fert-Bober et al. [12] identified significant numbers of citrullinated proteins. They hypothesized that citrullination could regulate muscle contractile protein interactions. PAD2 is ubiquitously expressed; specifically, it has been localized in the brain [24], synovial tissues [27], and muscles [28].

The most studied PAD is PAD4 (over 430 publications to date, 07/20/2015, Web of Science, www.webofscience.com). It is expressed in hematopoietic progenitor cells [29], neutrophils [30], and eosinophils [9] and is associated with inflammatory autoimmune diseases [31]. It has also been observed in carcinoma cells [9, 10]. Of all the PAD enzymes, PAD4 is unique in that it has been localized to the cell nucleus [29]. In the nucleus, PAD4 is known to hypercitrullinate histone, leading to chromatin decondensation and the formation of neutrophil extracellular traps (NETs) [32, 33], an essential part of the immune defense against pathogens.

PAD4 citrullination is considered one of the most important factors in RA. One of the major outcomes of RA is bone loss; this occurs by two mechanisms, inflammation and autoimmunity [34]. It has been shown that ACPAs are specific to RA [35] and the presence of ACPAs generally leads to more destructive bone loss [36]. In RA, increased abundance of citrullinated proteins is linked with increased joint destruction. Hill et al. [37] developed transgenic mice with citrullinated fibrinogen; they found that arthritis was induced in 35% of the mice with citrullinated fibrinogen compared to 0% of mice without citrullinated fibrinogen.

The nuclear localization of PAD4 and the citrullination of histones have been shown recently to be key regulators in gene expression. The tails of histones are unstructured, arginine rich, and heavily post-translationally modified [38]. The modification of arginine by histone methyltransferase results in monomethylation and dimethylation. Methylation of histone can increase and decrease the expression of specific genes [39]. It has recently been observed that PAD4 can demethylate monomethylarginine converting it to citrulline [40] regulating the expression of proteins. The citrullination is nonreversible.

Nakashima et al. [29] have shown that citrullination of histone H3 by PAD4 is involved in the regulation and proliferation of multipotent hematopoietic stem cells. Citrullination of histone H3 affects the expression of c-myc. The regulation of c-myc is essential for normal differentiation of hematopoietic stem cells (source of all blood cells). Decrease in c-myc leads to accumulation of hematopoietic stem cells, resulting in cytopenia (a reduction in the number of blood cells) [41].

PAD4 has been shown to citrullinate antithrombin by Chang et al. [42], who observed that the thrombin inhibitory activity of antithrombin could be hindered by PAD4 citrullination. Thrombin activates coagulation through the formation of fibrin clots, a prominent feature of RA [43]. Increased thrombin activity is also indicative of cancer; it is an activator of angiogenesis, which leads to metastasis [44]. PAD4 was identified as a corepressor of p53 and, together with the histone deacetylase HDAC2, supresses the expression of tumor repressor genes [14, 45].

PAD6 is expressed in the ovaries, oocytes, sperm cells, and the early embryo. It is essential for fertility and embryonic development. Esposita et al. [13] investigated the role of PAD6 in folliculogenesis. By breeding mice deficient in PAD6, they were able to observe the function of the enzyme *in vivo*. It was observed that female PAD6-deficient mice were infertile whereas males were not. On further investigation, it was noted that female mice ovulated and could be fertilized; however the embryo had arrested development at the two-cell stage. This observation is due to the lack of cytoskeletal sheet formation, which the authors attribute to the lack of citrullination by PAD6.

Due to the prevalence of citrullination in disease, there are a variety of PAD inhibitors in development. Wang et al. [46] developed a PAD4 inhibitor to target cancer cells that overexpressed PAD4. They found that the compound YW3-56, which they designed and synthesized, inhibited cancerous cell growth in both human osteosarcoma cells (U2OS) and mouse sarcoma cells (S-180). Wang et al. note that PAD4 may not be a suitable cancer therapy target as it is expressed in multiple cell types and is essential in the innate immune response. Another PAD inhibitor is Cl-amidine [47], which is an irreversible inhibitor of all PAD enzymes. Treatment of mice with collagen-induced arthritis showed a decrease in clinical disease activity and also in the levels of detected citrullinated proteins. Cl-amidine has also been tested on lupus-prone mice [48]. The formation of NETs in lupus-prone mice can result in vascular damage and decreased life expectancy [49]. Treatment with Cl-amidine resulted in less NET formation and protected against other lupus-related damage. Cl-amidine has a lethal effect on mouse embryos [50]. Two-cell mouse embryos cultured in 200 μM Cl-amidine showed complete arrested development at eight cells.

Currently the clinical method for identification of citrullination as part of disease diagnosis is to use synthetic cyclic citrullinated peptides (CCPs) [51]. ACPAs bind to the CCP in an enzyme-linked immunosorbent assay (ELISA), resulting in detection. The body produces ACPAs when proteins are citrullinated. The protein tertiary structure changes sufficiently, such that it is detected as an antigen. ACPAs were first associated with RA in the 1970s when antibody activity against keratin [52] was investigated. It was determined that the keratin was citrullinated. In 2007, a study by Coenen et al. [53] compared the sensitivity and specificity of ACPA detection for RA. From six commercially available antibody tests, it was found that the sensitivity was between 69.6% and 77.5% and the specificity was between 87.8% and 96.4%. A systematic review in 2011 by Taylor et al. [54] highlighted the variability in results from CCP and ACPA tests. Compiling analyses from 56 studies, CCP was found to have 68% sensitivity and 95% specificity. From 29 similar studies of ACPA tests, 60% specificity and 79% sensitivity were calculated.

7.2 Mass Spectrometry-Based Proteomics

It is becoming increasingly clear that citrullination is a widespread protein post-translational modification (PTM) that is essential in both health and disease. To better understand the role citrullination plays, there is a need to identify not only proteins that are citrullinated but also the specific sites of citrullination. The technique best suited for this purpose is mass spectrometry-based proteomics. Proteomics is the large-scale and comprehensive study of a complement of proteins [55]. In recent years mass spectrometry-based proteomics techniques have been developed to identify PTMs [56–58], protein–protein interactions [59], and the complete proteome of whole systems [60, 61]. The term "citrullinome" refers to the large-scale proteomics analysis of citrullinated peptides [62] and was first used by Tutturen et al.

Large-scale proteomics is usually performed on peptide mixtures, termed bottom-up proteomics [63]; a complex protein mixture (e.g., whole cell lysate [64], tissue sample [65], biological fluid [66], etc.) is digested with a protease that cleaves the proteins at specific residues. In most cases the resulting peptide digest is analyzed using C18 reversed-phase (RP) liquid chromatography (LC) coupled with tandem mass spectrometry (LC-MS/MS), though additional prefractionation steps are often included (strong cation-exchange (SCX) chromatography [67] or sodium dodecyl sulfate polyacrylamide gel electrophoresis

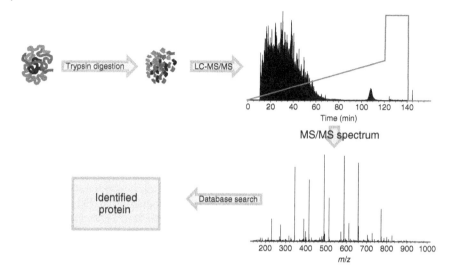

Figure 7.3 A schematic workflow of a basic proteomics experiment. A protein sample is treated with a protease to produce peptides. These peptides are then separated by liquid chromatography and analyzed by tandem mass spectrometry followed by database search identification.

(SDS-PAGE) [68]). Figure 7.3 shows a schematic workflow for a proteomics analysis. During RP LC separation, peptides bind to the C18 chains and are eluted from the column and into the mass spectrometer as the organic content of the mobile phase is increased, that is, peptides that are more hydrophobic elute later than less hydrophobic peptides. The majority of proteomics experiments utilize electrospray ionization (ESI) [69] as the method to transfer peptides from solution to gas phase and create ions. Once in the gas phase, the m/z of the peptide ions is measured. Peptide sequence information is obtained by fragmenting the precursor peptide ion generating product ions (MS/MS).

There are two main fragmentation methods used in proteomics: slow heating and electron-mediated methods. In both cases, the precursor ion of interest is first isolated from all other ions. The fragmentation method predominantly used in proteomics is collision-induced dissociation (CID) [70], which is a slow heating approach. Peptide ions are accelerated and collided with a neutral gas (usually nitrogen, argon, or helium). The collisions result in the conversion of some of the kinetic energy to internal energy. When sufficient energy is internalized, the most labile bond will break. In peptides, this bond is predominantly the $N-C_O$ backbone bond, although there is preferential cleavage of labile PTMs such as serine phosphorylation [71]. The backbone fragment ions are known as b- and y-type ions [72, 73] (b ions contain the N-terminus, and y ions contain the C-terminus; Figure 7.4) and are numbered from the terminus they contain.

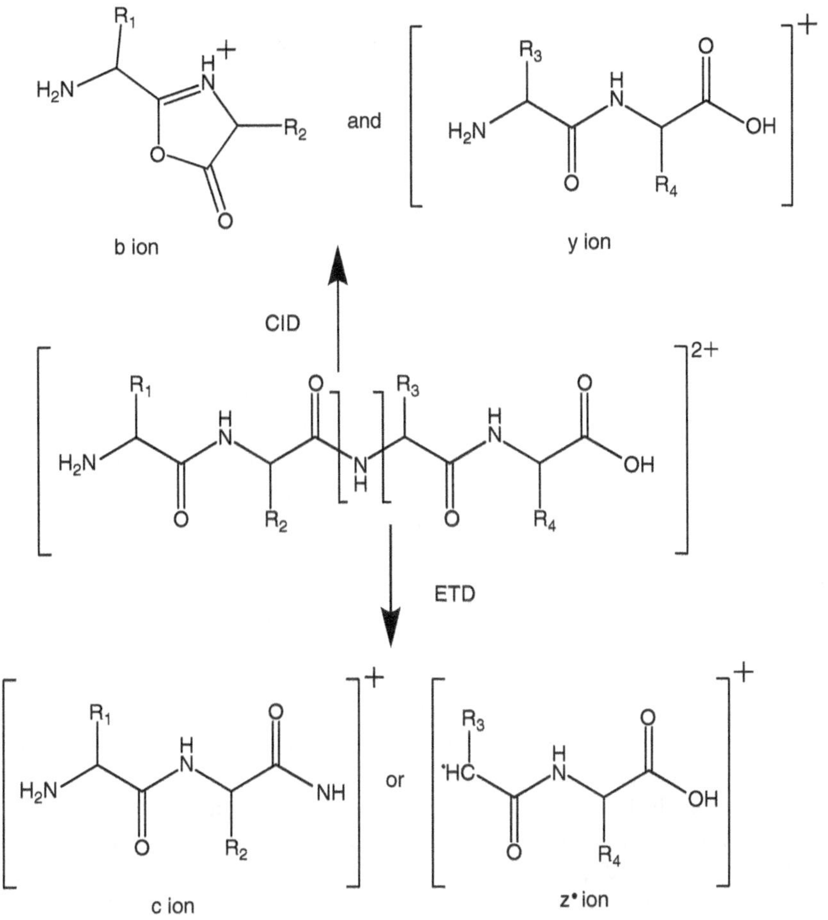

Figure 7.4 A schematic of peptide fragmentation by collision-induced dissociation and electron capture/transfer dissociation. R groups represent amino acid functional groups.

CID is a rapid and efficient process and as such the majority of proteomics experiments make use of this approach. The instrumentation used to generate CID fragments affects both the time for fragmentation and the resulting mass spectrum. Linear ion-trap CID takes between 1 and 100 ms (slow CID), whereas beam-type quadrupole CID (sometimes termed higher-energy collisional dissociation (HCD) or fast CID) takes between 0.5 and 1 ms [74, 75]. Linear ion traps are unable to trap fragment ions with m/z values below approximately 28% of the precursor ion m/z. This "one-third" low mass cutoff can cause problems when trying to identify low mass product ions from glycosylation [76] or mass tags for quantitation [77]. Beam-type CID is unaffected by the one-third rule and can be used to detect low mass product ions [76, 78] and mass tags [79]. One potential drawback of fast CID is that fragmentation often results in additional ions including a- and x-type (fragmentation of the $C-C_O$ bond) and immonium ions, which can interfere with sequence assignment.

There are two main electron-mediated fragmentation methods used in proteomics: electron capture dissociation (ECD) [80] and electron transfer dissociation (ETD) [81]. We will only discuss ETD as ECD requires a highly specialized FT-ICR mass spectrometer. In ETD, analyte (peptide) ions are allowed to react with a radical anion. The radical anion transfers an electron to the multiply charged peptide ion resulting in a charge-reduced radical cation. Transfer of the radical ion induces a fragmentation cascade in the peptide, predominantly cleaving the $N-C\alpha$ bond on the peptide backbone resulting in c- and z-type ions (Figure 7.4). One of the advantages of ETD is the retention of PTMs on the fragment ions [82]. ETD fragmentation is less efficient than CID, and it therefore takes longer to produce mass spectra of sufficient quality for peptide identification. It is possible to improve the fragmentation efficiency in ETD by collisionally activating the ions postfragmentation (supplemental activation) [83]. This process breaks any noncovalent bond holding pairs of fragments together (but no further covalent bonds). The activation time for linear ion-trap CID is usually 10–20 ms, whereas ETD in the same ion trap will often be 80–150 ms [84].

A typical LC-MS/MS analysis of a proteome will result in several thousand MS/MS spectra. It would be impossible to manually assign each of these to a peptide (and subsequently to a protein), and therefore there are multiple search algorithms available that perform this task. There are two main types of spectral analysis: *de novo* sequencing [85] and protein database searching [86]. *De novo* sequencing does not require any prior knowledge of the sample and identifies peptide sequences by matching the difference in mass of ions in the fragmentation spectrum to combinations of amino acids. Combined with the accurate mass of the precursor ion, they are used to identify the most likely amino acid sequence. Protein database search algorithms such as Mascot [86] or SEQUEST [87] work via a multistep process. First, the program produces an *in silico* digest of the proteins within the database using the same enzyme as the experiment. The masses of these *in silico* peptides are then matched against the precursor masses from the proteomics analysis. The top *n* closest mass matches between the *in silico* and the precursor masses are retained. The program then produces a synthetic MS/MS spectrum, which it matches against the experimental MS/MS spectrum. This match and the accuracy of the intact peptide are given a probability score in order to assess whether it is a likely true match or a random match. In both database searching and *de novo* sequencing, it is possible to further inform the search with various parameters. For example, the potential number of missed cleavages in the digestion is often used in searches of samples that are post-translationally modified, since PTMs can hinder full proteolysis of a protein [88]. The mass accuracy of the mass spectrometer is also an important parameter. The closer the experimental and theoretical masses, the higher the score. Another parameter in all search algorithms is the possibility of PTMs on the peptide. The addition of a PTM results in a mass increase or decrease in both the precursor and any fragment ions, which contain the modified residue. This possibility is taken into account in both *de novo* and database searches. The software creates a modified peptide every time the specified amino acid is observed. The mass increase/decrease observed on the fragment ions is used to localize the site of modification.

7.3 Liquid Chromatography and Mass Spectrometry Behavior of Citrullinated Peptides

Like many acidic PTMs, citrullination is known to hinder trypsin cleavage near the site of modification [89]. Unlike phosphorylation and other acidic modifications, which add negative charge near the basic (Arg or Lys) cleavage site, conversion to citrulline, which is neutral, reduces the basicity of the amino acid side chain, decreasing trypsin affinity to the site. It is possible to use other enzymes (Lys-C, Glu-C) without encountering missed cleavages [90]. As the modified arginine is not cleaved, database search results that identify peptides with C-terminal citrullination require manual validation [91]. Many of these sites are reclassified as deamidation of either asparagine or glutamine (exactly the same mass shift as citrullination) after manual validation. The stoichiometry of citrullination is such that, without enrichment, the noncitrullinated form of a peptide is dominant and the concentration of the citrullinated peptides may be outside the dynamic range of the mass spectrometer.

In cases where there are both citrullinated peptides and noncitrullinated counterparts, it has been shown that the citrullinated version of the peptide elutes later from a C18 RP column. This feature can be used to confirm the presence of citrullination. Bennike et al. [89] showed for 24 synthetic peptide sequences (48 peptides total, 24 citrullinated and 24 unmodified) that the retention times of the citrullinated peptides were longer than the noncitrullinated peptides in all but two cases. The maximum shift was observed for a triply citrullinated peptide from the PAD4 protein; a 3.6% increase in retention time was observed. Overall, for the 24 peptides, the average retention time shift was +1.8%. For a 60 min gradient that equates to a retention time shift of +65 s. With modern LC-MS/MS methods, these peptides would not overlap. Two of the peptides in the study by Bennike et al. were positional isomers (NMKEEMARHLcREYQDLLNVK and NMKEEMAcRHLREYQDLLNVK, where cR denotes citrullination site), in which the peptide sequence is identical but the site of modification varies. Interestingly, there was a difference of 0.6% in retention time between these two peptides. The peptide NMKEEMARHLcREYQDLLNVK elutes earlier of the two peptides, suggesting it is more hydrophilic. Like all PTMs, complete conversion of a target site from arginine to citrulline is unlikely. Given the mass shift of +0.984 Da, if the modified peptide does coelute with its unmodified form, with most mass spectrometers, the two would be coisolated for MS/MS fragmentation. (Typical isolation width for MS/MS analysis is m/z 1–3.) Coisolation would complicate the MS/MS spectrum and hinder if not completely negate the successful identification of the peptide and PTM [92]. Because of the propensity for missed cleavages in tryptic digests of biological samples, very few of the citrullinated peptides will elute at the same time as the unmodified counterpeptide. This method is more likely to be of use for Lys-C or Glu-C digests.

As mentioned earlier, citrullination results in an identical mass shift to deamidation of asparagine and glutamine (0.984 Da). Consequently, identification of citrullinated peptides by protein database searches is more difficult. Protease digestion performed in ammonium bicarbonate, Tris-HCl, or triethylammonium bicarbonate results in significant amounts of artificial deamidation of asparagine [93], further complicating the database search. It must also be noted that deamidation of Asn and Gln also results in an increase in RP LC retention time in comparison to the unmodified peptides [94]. The retention time shift observed was similar to those seen for citrullination (~3% increase).

CID [95] of citrullinated peptides has been shown to result in intense peaks corresponding to neutral losses from the precursor ion [96]. The functional group of citrulline contains an $N-C_O$ bond. As mentioned earlier, CID fragmentation predominantly cleaves this bond on the peptide backbone. It is cleaved in citrulline side chain, leading to the loss of isocyanic acid (43 Da, HNCO) (see Figure 7.1). Hao et al. [96] fragmented synthetic citrullinated peptides with CID and observed the neutral loss of isocyanic acid from both the precursor and fragment ions. The neutral loss from the precursor confirms that the peptide is citrullinated and the loss from fragment ions can be used to localize sites of citrullination. Any fragment ion with a neutral loss means that the site of citrullination is located on the remaining portion of the peptide. To test the method, nucleophosmin (NPM) (a molecular chaperone) was analyzed. NPM is known to be citrullinated, but the site of modification was not known. Hao et al. treated recombinant human His6-NPM with PAD4 and digested the modified protein with trypsin. LC-CID-MS/MS resulted in the identification of the peptide SIcRDTPAK. The peptide was identified as both a singly and doubly charged ion, and in both cases a peak corresponding to the expected neutral loss was observed. As there is only one arginine in the peptide, the site can be localized. The unmodified peptide was also identified in an analysis of a non-PAD4-treated sample.

Creese et al. [97] used automatic detection of the peak corresponding to the neutral loss of HNCO from the precursor ion in a CID spectrum to trigger reselection of the same ion and fragmentation by supplemental activation electron transfer dissociation (saETD) [98]. The saETD mass spectra provide additional confidence in the localization of a citrulline modification rather than deamidation of asparagine or glutamine. Creese et al. analyzed a set of four synthetic citrullinated peptides. These peptides were spiked into a commercially available mix of tryptic peptides derived from six proteins and a more complex tryptic digest of human saliva. In the analyses of both the six protein mixture and the saliva digest, all four of the peptides were selected and fragmented by CID. In each case, a peak corresponding to the neutral loss of HNCO was observed, and the precursor ions were subsequently fragmented by saETD. The data were searched using the SEQUEST algorithm [87], and for all but one of the peptides, the ETD mass spectra resulted in higher scores than the corresponding CID mass spectra. It was also noted that analysis of the CID spectra resulted in up to a 1.1% false-positive rate (FPR) (incorrect assignment of citrullination—all peptides identified as citrullinated were manually validated), whereas for all of the saETD analyses the FPR was 0%.

Jin et al. [90] developed a similar method in which (HCD of the precursor ion was triggered by the neutral loss of the isocyanic acid in a CID spectrum. It was noted that HCD fragmentation of citrullinated peptides resulted in a significant reduction in the intensity of the neutral loss peak. In addition, more backbone ions were observed in the HCD spectra. Using the neutral loss-triggered HCD method, Jin et al. were able to characterize previously unlocalized citrullination sites on glial fibrillary acidic protein (GFAP). Two samples were analyzed: a purified GFAP sample analyzed without modification and the same protein treated with PAD2 prior to analysis. The samples were digested with trypsin, Lys-C, and Glu-C. From the PAD2-treated sample, 17 citrullination sites were identified, and 5 sites were identified from the untreated sample. If an identified peptide contained either asparagine or glutamine, the mass spectrum was manually validated. The benefit of both neutral loss methods is the improved confidence that the triggering of the second fragmentation technique gives. Both ETD and HCD give greater confidence in identification (peptide sequence and modification site) than CID; however, observation of the diagnostic peak corresponding to the neutral loss in the CID mass spectra confirms the presence of citrulline.

One method capable of distinguishing citrullination from deamidation utilizes treatment with 2,3-butanedione and antipyrine [99]. The ureido group of citrulline reacts with 2,3-butanedione in the presence of trifluoroacetic acid. This in turn reacts with antipyrine (Figure 7.5), increasing the mass of amino acid residue by 238 Da. It is then possible to detect this modification using UV detection at 464 nm radiation, and modified citrullinated proteins can be identified by Western blot with antimodified citrulline antibodies.

The 2,3-butanedione reaction with the ureido group of citrulline alone can be used to distinguish citrullination from deamidation as shown by De Ceuleneer et al. [91]. The reaction results in a mass increase of 50 Da, easily

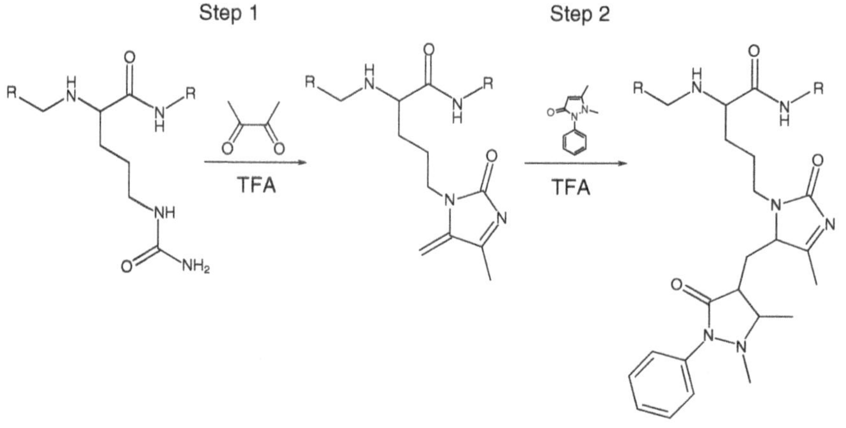

Step 1   Step 2

Figure 7.5 The modification of citrulline using 2,3-butanedione and antipyrine catalyzed by trifluoroacetic acid.

distinguished from deamidated and unmodified peptides by LC-CID-MS/MS analysis. The reaction is performed by mixing citrullinated peptides with 2,3-butanedione in the presence of trifluoroacetic acid. De Ceuleneer et al. observed 95% reaction efficiency when converting a synthetic peptide (STScRSLYASSPG) with a 16 h reaction. The reaction was also performed on the unmodified counterpart peptides to determine the specificity of the reaction. It was found that only the citrullinated peptide was modified. The synthetic citrullinated peptide was spiked into a "complex sample" (cytochrome *c*, Lys-C, and Glu-C digests) and the reaction repeated. The conversion rate was limited to approximately 70% even with an overnight incubation. Human fibrinogen was citrullinated *in vitro* using PAD from rabbit skeletal muscle (PAD enzyme details are not provided) and digested with Glu-C or Lys-C. After analysis by LC-MS/MS and subsequent data analysis with in-house software (msMod searches for the difference in peptide ion signal between the unmodified and modified samples), 15 peptides containing 17 citrullination sites were identified. Of the 15 peptides, only 9 were identified in a Mascot search. It was suggested that the other six peptides were of too low abundance to be selected for fragmentation.

The reaction of citrullinated peptides with 2,3-butanedione and antipyrine is performed with the modifying reagents in large excess. It is therefore necessary for the sample to be purified prior to LC-MS/MS analysis. To remove the excess reagents, Stensland et al. [100] proposed desalting the sample with SCX chromatography followed by C18 RP LC to remove salts. Traditionally SCX is used as an additional dimension in complex proteomics analysis [101]; here it is solely used to remove excess reagent. Using this protocol, they were able to remove over 99% of the excess antipyrine. Stensland et al. used the aforementioned technique to modify citrullinated peptides created by incubating either bovine serum albumin (BSA) or MBP (human) with PAD4. To ensure identification of unmodified counterpeptides, the proteins were digested with Lys-C without incubation with PAD4. An alternating CID and ETD experiment was performed with the peptides separated on a C18 RP column. They observed that CID of peptides with modified citrulline residues resulted in an intense peak at m/z 201.1. This peak was attributed to fragmentation of the exocyclic methyl group to the C4 of the imidazolone moiety and was proposed as a potential trigger for other fragmentation techniques such as product ion-triggered ETD, as used for glycopeptide analysis [76]. The ETD mass spectra of the modified peptides did not yield any fragments of the modification and as such could be used to localize the site of citrullination. LC-CID-MS/MS analysis of citrullinated BSA resulted in the identification of two citrullinated peptides (VPQVSTPTLVEVScRSLGK and LGEYGFQNALIVRYTcRK) whose fragmentation spectra were of sufficient quality to both identify the peptide and localize the site of modification. In both cases, an intense peak at m/z 201.1 was observed. The MBP Lys-C digest was analyzed using an alternating CID–ETD LC-MS/MS analysis. In this analysis, the ETD mass spectra were used to identify and localize the sites of modification, and the CID mass spectra were used to confirm the presence of the m/z 201 signature ion. From this analysis, four peptides were identified containing five sites of citrullination. The sample was also analyzed with a CID-only method from which only two of the four citrullinated peptides were identified. It has been shown by Holm et al. [99] that the use of 2,3-butanedione alone results in multiple additional ions, mass increases including 66, 116, and 162 Da, whereas the reaction with antipyrine only results in a mass increase of 238 Da. Holm et al. observed that the product in the reaction between citrulline and 2,3-butanedione is an intermediate species and the addition of antipyrine reacts with this intermediate, but it is proposed that other nucleophiles could react to produce these unwanted products if antipyrine is not present.

In 2005, Kubota et al. [102] developed a novel labeling method for the identification of citrulline. The method is reliant on *in vitro* citrullination of proteins by one of the PAD enzymes. The reaction is performed in 50% heavy water ($H_2{}^{18}O$). When the PAD enzyme converts arginine to citrulline 50% of the citrulline will contain heavy oxygen. The sample is then digested with trypsin, chymotrypsin, and Glu-C prior to LC-CID-MS/MS analysis. The incorporation of ^{18}O gives the peptide ion a distinct isotope pattern, which can be detected in the mass spectrometer. To ensure all fragment ions shared this isotope distribution, a wider isolation of the ion was used (5 m/z) prior to fragmentation. Kubota et al. used the Mascot search algorithm to identify citrullinated peptides from the data set. Mass spectra assigned to peptides with multiple arginine residues and ambiguous sites of modification were manually analyzed. Fragment ions retaining the citrulline had distinct isotope patterns.

The obvious drawback of this method is the need to citrullinate the proteins *in vitro*, and therefore it cannot be used for the analysis of endogenous citrullination. This method has its uses for purified proteins, which are known to be citrullinated by particular PADs.

7.4 Global Analysis of Citrullination

Prior to efficient enrichment [103, 104], global analysis of citrullination was challenging. Generally, analysis of individual proteins was performed after confirmation of citrullination by Western blot as demonstrated by Bhattacharya et al. [105]. They analyzed healthy and glaucomatous optic nerve tissue; it was observed by Western blot analysis that both PAD2 expression and citrullination levels increased in the glaucomatous optical nerve tissue compared to healthy tissue. They quantified the levels of PAD2 in both the healthy and glaucomatous tissues and observed approximately an eightfold increase in PAD2 in the glaucomatous tissue. The tissue samples were homogenized and the proteins separated by SDS-PAGE, followed by trypsin digestion and LC-MS/MS analysis. Two hundred and fifty proteins were identified, 68 of which were only identified in the glaucomatous samples. Following the proteomics analysis, immunoprecipitation using an anticitrulline antibody was performed on a glaucomatous sample. The immunoprecipitate was also analyzed by LC-MS/MS, and 36 proteins were identified. Several of these proteins had previously been identified as citrullinated including annexin A2 [106], MBP [107], and histone H4 [108]. As discussed earlier, citrullination is known to demethylate monomethylarginine. Bhattacharya et al. performed Western blots using an antibody specific to protein methylarginine. Proteins containing methylarginine were identified in the control groups but not in the glaucomatous samples. This result adds to the evidence that citrullination is involved in the regulation of monomethylation and therefore the regulation of gene expression.

Hermansson et al. [109] analyzed synovial tissue from RA patients. The presence of citrullination was detected by antibody staining of four samples. Though the analysis was of a tissue digest and therefore very complex, the target of the experiment was to identify citrullinated sites on fibrinogen. To achieve this, purified fibrinogen was separately citrullinated using a rabbit PAD enzyme (PAD type not specified) and digested using Lys-C. LC-MS/MS analysis followed by a Mascot search of this digest was then used to identify sites of citrullination. They identified 33% of the arginine residues from the fibrinogen chains (both unmodified and citrullinated). The analysis resulted in a list of citrullinated peptides together with accurate mass and time measurements. In order to use the identified peptides as markers in the *in vitro* study, several criteria had to be met: (i) no asparagine or glutamine residues, (ii) the noncitrullinated peptide had to be identified in the Mascot search with high confidence, and (iii) the citrullinated peptide had to be identified with a Mascot ion score >25 and greater than 95% confidence. Two citrullinated peptides met these criteria: one from the α-fibrinogen (559–575) and one from β-fibrinogen (52–77). For two of the tissue samples where the citrullinated α-fibrinogen peptide was identified, the citrullination occupancy was calculated as 1.4% and 2.5%, respectively (occupancies calculated by dividing the intensity of the citrullinated peptide by the intensity of the unmodified counterpeptide + citrullinated peptide). The citrullinated peptide identified in β-fibrinogen was identified in one of the samples with occupancy of 1.2%.

Bennike et al. [89] analyzed the synovial fluid from a patient with RA. This study was one of the first large-scale analyses of citrullinated peptides from a complex sample. The proteins were digested with trypsin and separated over a 120 min LC gradient. No chemical modifications were employed, and a standard CID fragmentation method was used. As discussed earlier, citrullinated arginine hinders trypsin digestion. Using this feature, they interrogated the results from an LC-MS/MS analysis and database search of a synovial fluid trypsin digest. In total, 58 citrullinated peptides were identified following a protein database search employing the Mascot algorithm. On manual inspection, 14 of the peptides were found to be false positives (24%), 64% of the sites were unambiguously identified (fragment ions for $N–C_O$ bonds either side of the modified arginine), and the remaining 12% of sites were identified by flanking fragment ions (the $N–C_O$ bonds N- and C-terminal to citrulline are not cleaved, but sufficient fragment ions are observed to localize the site of modification). Over 360 proteins were identified from synovial fluid, highlighting the low stoichiometry and specificity of citrullination.

Christophorou et al. [50] measured the differential levels of citrullination in mouse embryonic stem (ES) cells that overexpressed human PAD4 compared to normal mouse ES cells. The quantification was performed by the use of stable isotope labeling by/with amino acids in cell culture (SILAC) [56, 110]. Briefly, in SILAC quantification, one cell population is grown in "light" media (arginine and lysine have ^{12}C or ^{14}N), and another cell population is grown in "medium"/"heavy" media (arginine + $^{13}C_6$ and $^{15}N_4$ and lysine $^{13}C_6$). The isolated proteins are mixed, proteolytically digested, and analyzed by LC-MS/MS. The light and heavy peptides coelute, and extracted ion chromatograms for each pair of peptides can be used to quantify the relative abundance of the peptides. SILAC allows the analysis of three samples at once (light, medium, and heavy). In this experiment, mouse ES cells expressing an empty vector were grown in "light" lysine, and mouse ES cells that overexpressed human PAD4 were grown in "heavy" lysine ($^{13}C_6$). The sample was digested with Lys-C and analyzed by LC-HCD-MS/MS over a 2 h LC gradient. In the four citrullinated HCD spectra presented in this work, the peak corresponding to neutral loss of HNCO, which is observed in CID, was not present. This observation suggests that the modification remains on the arginine, increasing the probability of successfully identifying the site of citrullination. In total, 162 citrullinated peptides were quantified in this first example of large-scale citrullination quantitation by mass spectrometry. Thirteen of the identified citrullinated peptides have localization probabilities of less than 75%. The localization probability used here calculates the chance that a given peak that allows localization was a random match. A localization probability of 75% means that there is a 25% chance that the citrullination is not located to where the search software has assigned it. Given the challenges associated with citrullination site assignment, citrullinated peptide assignments may need to be manually validated to ensure no false sites are reported.

Van Beer et al. [111] were able to confidently identify 149 citrullinated peptides from the synovial fluid of two patients with RA. All citrullinated peptides were manually validated. Synovial fluid from 80 patients was separated into soluble (supernatant) and insoluble (pellet) fractions. The two fractions from each patient were analyzed by Western blotting to identify the two fractions (one soluble and one insoluble) with the greatest levels of citrullinated protein. The soluble and insoluble fractions with the greatest levels of citrullination (from different patients) were depleted for albumin, resulting in two pellet fractions and a supernatant (predominantly albumin) for each sample [112]. The proteins were solubilized and separated by SDS-PAGE. The resulting lanes were dissected into 18 even slices. Each gel slice was analyzed by LC-MS/MS after trypsin digestion using CID fragmentation followed by a Mascot database search. The combined database search of all samples identified 192 proteins with Mascot scores of 30 or greater. Of these, 40 and 45 were identified solely in the soluble and insoluble fractions, respectively. Of the 192 proteins, 53 were identified as citrullinated. Thirty five of these were identified from the soluble fraction. Only six citrullinated proteins were identified in both the patient fractions. Of the modified proteins identified in both fractions, five were identified with multiple citrullination sites in common between the fractions. In β-actin, 11 citrullination sites were identified solely in the supernatant sample. In addition, some peptides identified as citrullinated in the supernatant were only observed as unmodified in the pellet.

Fert-Bober et al. [12] performed a large-scale data-independent (sequence window acquisition of all theoretical fragment ion spectra (SWATH) [113]) quantitative proteomics experiment on human heart tissue from three different groups: ischemic heart disease ($n = 10$), idiopathic cardiomyopathy ($n = 10$), and nonfailing hearts ($n = 10$). No modification of the samples was performed and the citrullinome was compared. SWATH is a two-step process: initially, the sample is analyzed by a traditional data-dependent acquisition (DDA) method. This analysis provides a peptide ion library; this library of accurate mass, time, and fragment measurements is used in step 2. The mass spectrometer rapidly cycles through a set of mass ranges, isolating and fragmenting everything in a specified m/z window (m/z width 25). The mass spectrometer moves to the next mass window and repeats over the whole mass range. The cycle usually takes 2–4 s (instrument, m/z range, and m/z window size dependent). Peptides are therefore fragmented multiple times as they elute from the LC column, allowing for quantitation and identification. Myofilaments were isolated using IN sequencing [114]. Proteins are separated into three fractions based on their solubility at various pH levels. IN sequencing was developed specifically for proteomics analysis of the heart tissue. The heart tissues were fractionated into myofilament- and cytosolic-enriched fractions. To maximize the citrullinome coverage, the samples were treated with a PAD cocktail containing all five PADs. Each sample (both fractions) was analyzed with DDA and data-independent acquisition (DIA) methods using a 2 h 5–35% acetonitrile gradient. To create the SWATH ion library of verified citrullinated peptides, each citrullinated peptide was required to have a partner unmodified peptide with a retention time shift of at least 5 min. For citrullinated peptides, which had unmodified partner peptides with less than 5 min retention time shift or no partner peptide, transitions were manually selected to ensure unambiguous assignment of citrullination, not deamidation. SWATH analysis resulted in the identification and relative quantitation of 304 citrullination sites from 145 proteins. Of the 304 citrullination sites identified, 53 were altered in the heart failure samples compared to the nonfailing controls. A set of citrullinated peptides were identified, which were upregulated in ischemic heart disease and downregulated in idiopathic cardiomyopathy.

7.5 Enrichment Strategies

In vivo citrullination is known to have low stoichiometry [14, 115]. As discussed earlier, proteomics analysis of complex samples including synovial fluid often results in a small number of citrullinated peptides given the number of analysis hours. With the increased interest in large-scale identification of citrullination sites and the low number of citrullinated peptides routinely identified in proteomics experiments, efficient enrichment is essential.

In 2010, Tutturen et al. [104] developed an enrichment technique for citrullinated peptides. Briefly, using a multistep process, sarcosine dimethylacrylamide resin (PL-DMA) is functionalized with ethylenediamine (Figure 7.6). To this, 4-hydroxymethylbenzoic acid is added, followed by bromoacetic acid. Finally, 4-hydroxyphenylglyoxal is added to produce citrulline reactive beads (CRBs). The citrulline residue covalently binds to the CRB, and noncitrullinated peptides are washed from the beads. The bound peptides are released from the beads by washing with NaOH and guanidine. Peptides that were bound to the CRBs have a mass increase of 190.03 Da caused by the glyoxal derivative on the CRB. The technique was initially tested by enriching two synthetic citrullinated peptides (AcRSSVPGVR and SAVQAcRSSVPGVR). The resulting enriched sample was analyzed by MALDI-TOF MS, with both peptides successfully enriched compared to unmodified peptides.

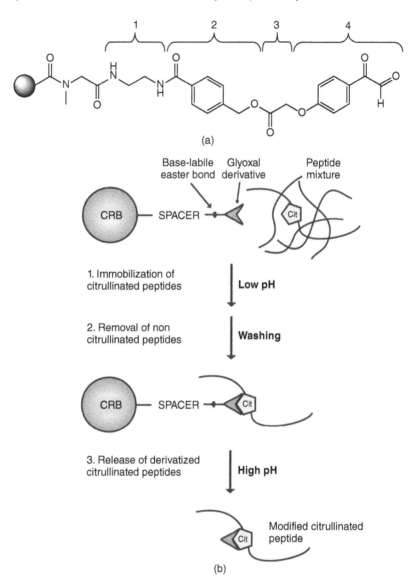

Figure 7.6 A schematic of citrulline reactive bead (CRB) enrichment. (a) The structure of the CRB. (b) Binding citrullinated peptides to the CRB, washing unmodified peptides from the beads, and elution of citrullinated peptides bound to the glyoxal derivative. (c) The structure of the citrulline plus glyoxal derivative. Source: Tutturen, 2010. [104]. Reproduced with permission from Elsevier.

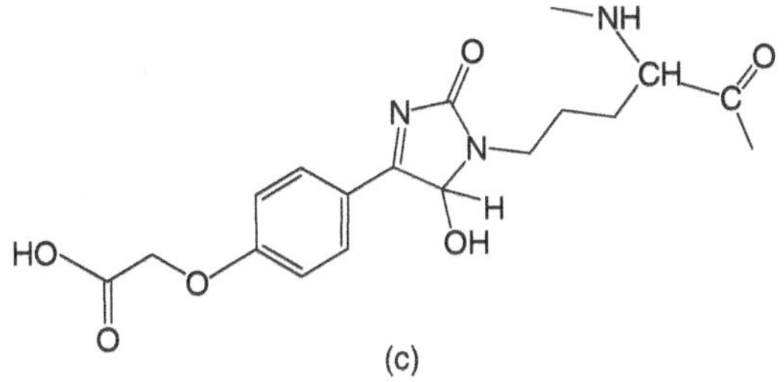

(c)

Figure 7.6 (*Continued*)

To assess the technique further, MBP (human) was treated with PAD4, digested with Lys-C, and enriched. The eluted peptides were analyzed by MALDI-TOF MS. After enrichment, six peaks were observed and identified as five citrullinated peptides (one peptide was identified as both singly and doubly citrullinated; see Figure 7.7). One of the drawbacks of this method is the background signal observed from substances derived from the polymer of the beads (Figure 7.7a).

An alternative enrichment method was developed in 2013 [103] by the same group. Initially, 4-glyoxalbenzoic acid is synthesized by oxidizing 4-acetyl-benzoic acid, which is then reacted with amine-PEG$_2$-biotin to produce biotin-PEG-GBA (BPG). This tag can be enriched using streptavidin beads (Figure 7.8). The authors suggest mixing the biotin tag with the peptide mixture and then using SCX chromatography to separate any unreacted tags from the peptide mixture. The peptide mixture is mixed with the streptavidin beads, and the noncitrullinated peptides are removed. The tagged peptides are eluted from the streptavidin beads using excess biotin. The resulting tagged peptides have a mass addition of 516.4 Da. CID of peptides containing a BPG tag produces an intense fragment ion from the BPG at m/z 270.13, a potential product ion to either trigger additional fragmentation [76] or for use as a marker for modified peptides, giving greater confidence to the citrulline assignment. As in the previous method, MBP was citrullinated with PAD4, digested with Lys-C, and enriched. Analysis was performed by MALDI-TOF. The five citrullinated peptides identified in the previous work were identified; however, the peptide RPSQRHGSK (identified as two singly citrullinated peptides and one doubly citrullinated peptide with CRB enrichment) was only identified with R5 being citrullinated. One additional citrullinated peptide was identified (cRGSGK). This enrichment method appears cleaner than the CRB method as there is no polymer contamination; however the production of the BPG tag requires multiple steps.

Tutturen et al. [62] have applied the BPG enrichment method to a complex synovial fluid sample from a patient with RA to assess the citrullinome. They compared LC-MS/MS analysis of samples with and without enrichment. The supernatant sample was split three ways: one analyzed without purification,

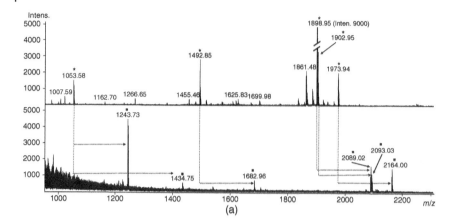

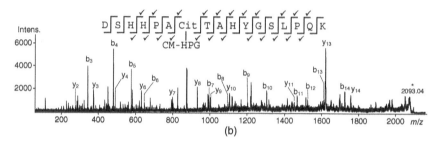

Figure 7.7 The enrichment of citrullinated peptides using the CRB method. (a) Citrulline enrichment of myelin basic protein using citrulline reactive beads: unenriched spectrum (top) and enriched spectrum (bottom). Citrullinated peptides identified before enrichment are marked by asterisks. (b) MALDI-TOF/TOF fragmentation spectrum of a CRB-enriched citrullinated peptide. Complete sequence coverage is observed. Source: Tutturen, 2010. [104]. Reproduced with permission from Elsevier.

one depleted of IgG, and one where IgG and human serum albumin (HSA) were depleted. The synovial pellet was not fractionated. All four samples were digested with Lys-C and half of each digest enriched using the BPG method described earlier. Each sample was analyzed in triplicate; the samples were separated on a 50 cm C18 column over a 295 min gradient. Eluting peptides were fragmented with HCD. Analysis of the supernatant samples without purification or BPG enrichment resulted in confident identification of 119 citrullinated peptides. Analysis of the unpurified supernatant with enrichment revealed 3673 unique monoisotopic masses from fragmentation spectra containing the m/z 270.13 BPG fragment ion (2146 unique masses from the IgG- and HSA-depleted sample). Only 4% of the peptides identified had fragmentation spectra missing the BPG ion. This suggests that the enrichment is very efficient. The database search results of the BPG enrichment samples

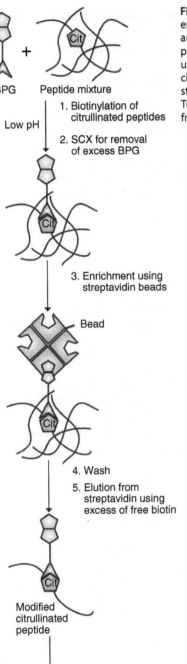

BPG   Peptide mixture

Low pH

1. Biotinylation of citrullinated peptides

2. SCX for removal of excess BPG

3. Enrichment using streptavidin beads

Bead

4. Wash

5. Elution from streptavidin using excess of free biotin

Modified citrullinated peptide

MALDI-TOF MS

Figure 7.8 A schematic of the biotin-based BPG enrichment method. Peptides are mixed with BPG, and unreacted BPG is removed by SCX. The tagged peptides are enriched on streptavidin beads, and unmodified peptides washed off. The modified citrullinated peptides are eluted from the streptavidin and analyzed by MALDI-TOF. Source: Tutturen, 2013. [103]. Reproduced with permission from Elsevier.

were however somewhat disappointing. Only 13 and 10 citrullinated peptides were identified from the supernatant and the depleted supernatant (IgG and HSA), respectively. It was noted that HCD fragmentation of peptides containing the BPG modification resulted in significant signal for m/z 270.13 along with m/z 227.1 and 286.1 (further backbone fragmentation of the BPG tag). The authors suggest that these signals reduced the fragment ion yield for peptide backbone fragments, resulting in poor sequence coverage and low-quality MS/MS spectra. Tutturen et al. acknowledge that there is room to improve the enrichment by modifying the BPG to contain a cleavable site. It is also possible that the use of ETD rather than HCD would not result in fragmentation of the BPG. The two enrichment protocols developed by Tutturen et al. to date have not been applied by other groups, but from the preliminary findings it is hopeful that large-scale citrullinomics is a possibility.

7.6 Bioinformatics

Generally, identification and localization of citrulline are performed using database search algorithms such as Mascot and SEQUEST, as discussed earlier. It should be noted that if the sample was digested with trypsin, the number of allowed missed cleavages should be greater than that used for a normal search.

As mentioned earlier, the detection of citrullinated peptides is complicated by the presence of deamidated asparagine and glutamine residues. Bennike et al. [89] suggest that the presence of missed cleavages on tryptic peptides can be used to increase confidence in assigning modified peptides. Using 24 biologically relevant citrullinated peptides (no C-terminal citrullination sites) and the unmodified counterpart peptides, they assessed the potential for missed cleavages (along with the retention time shift, as described earlier). The 48 peptides were digested with trypsin and analyzed by LC-MS/MS. In all cases, they found that the citrullinated peptides resulted in missed cleavage products, whereas the unmodified counterpart peptides were fully digested. Deamidation does not inhibit trypsin and would behave in the same way as the unmodified peptides. As discussed earlier, they then applied the knowledge that citrullination would result in missed cleavage from trypsin digestion to analyze synovial fluid.

De Ceuleneer et al. [116] describe a method whereby coeluting citrullinated and noncitrullinated peptides can be analyzed to determine the percentage of citrullination ("skewing"). Using the endoprotease Lys-C for the sample digestion will result in citrullinated and noncitrullinated peptides having the same sequence. This method is reliant on the coelution of citrullinated and noncitrullinated peptide, which has been shown to not always be the case by Bennike et al. [89]; however, for those peptides that do coelute, De Ceuleneer et al. [116] provide a template that can be used to determine the percentage "skewing." A large proportion of manual analysis is involved: theoretical isotope distributions need to be manually created for the citrullinated and noncitrullinated forms, and the isotope pattern from the MS data must be manually extracted. With synthetic citrullinated and noncitrullinated peptides, De Ceuleneer et al. observed good linear fit between the calculated and measured percentages ($R^2 > 0.9$). They subsequently calculated the skewing for several citrullinated peptides from synovial fluid.

7.7 Concluding Remarks

Citrullination has been recognized as an important PTM in an increasingly wide range of diseases and other regulatory processes. There are several challenges regarding the detection of citrullinated peptides by mass spectrometry; however, with the advent of efficient enrichment methods like those developed by Tutturen et al. [104], the number of citrullinated peptides it is possible to identify has increased dramatically. The modification of citrulline inherent in these enrichment methods increases the confidence in assignment as the mass shift is no longer the same as that for deamidation. Enrichment of citrullinated peptides is in its infancy but is showing great promise for the improved identification of a PTM, which is being identified as a key regulator in cell biology.

Acknowledgements

AJC and HJC are funded by the Engineering and Physical Sciences Research Council (EPSRC) (EP/L023490/1).

References

1 Koga Y, Ohtake R. Study report of the constituents of squeezed watermelon. *J Tokyo Chem Soc* 1914;**35**:519–528.
2 Ackermann D. The biological decomposition of arginine to citrulline. *Hoppe-Seylers Z Physiol Chem* 1931;**203**:66–69.
3 van Venrooij WJ, Pruijn GJM. Citrullination: a small change for a protein with great consequences for rheumatoid arthritis. *Arthritis Res* 2000;**2**(4):249–251.
4 Vossenaar ER, Nijenhuis S, Helsen MMA, van der Heijden A, Senshu T, van den Berg WB, van Venrooij WJ, Joosten LAB. Citrullination of synovial proteins in murine models of rheumatoid arthritis. *Arthritis Rheum* 2003;**48**(9):2489–2500.
5 Burkhardt H, Sehnert B, Bockermann R, Engstrom A, Kalden JR, Holmdahl R. Humoral immune response to citrullinated collagen type II determinants in early rheumatoid arthritis. *Eur J Immunol* 2005;**35**(5):1643–1652.

6 Liao F, Li ZB, Wang YN, Shi B, Gong ZC, Cheng XR. *Porphyromonas gingivalis* may play an important role in the pathogenesis of periodontitis-associated rheumatoid arthritis. *Med Hypotheses* 2009;**72**(6):732–735.

7 Mastronardi FG, Wood DD, Mei J, Raijmakers R, Tseveleki V, Dosch HM, Probert L, Casaccia-Bonnefil P, Moscarello MA. Increased citrullination of histone H3 in multiple sclerosis brain and animal models of demyelination: A role for tumor necrosis factor-induced peptidylarginine deiminase 4 translocation. *J Neurosci* 2006;**26**(44):11387–11396.

8 Ishigami A, Ohsawa T, Hiratsuka M, Taguchi H, Kobayashi S, Saito Y, Murayama S, Asaga H, Toda T, Kimura N, Maruyama N. Abnormal accumulation of citrullinated proteins catalyzed by peptidylarginine deiminase in hippocampal extracts from patients with Alzheimer's disease. *J Neurosci Res* 2005;**80**(1):120–128.

9 Chang XT, Fang KH. PADI4 and tumourigenesis. *Cancer Cell Int* 2010;**10**(7):1–6.

10 Chang XT, Han JX. Expression of peptidylarginine deiminase type 4 (PAD4) in various tumors. *Mol Carcinog* 2006;**45**(3):183–196.

11 Fert-Bober J, Sokolove J. Proteomics of citrullination in cardiovascular disease. *Proteomics Clin Appl* 2014;**8**(7–8):522–533.

12 Fert-Bober J, Giles JT, Holewinski RJ, Kirk JA, Uhrigshardt H, Crowgey EL, Andrade F, Bingham CO, Park JK, Halushka MK, Kass DA, Bathon JM, Van Eyk JE. Citrullination of myofilament proteins in heart failure. *Cardiovasc Res* 2015;**108**(2):232–42.

13 Esposito G, Vitale AM, Leijten FPJ, Strik AM, Koonen-Reemst AMCB, Yurttas P, Robben TJAA, Coonrod S, Gossen JA. Peptidylarginine deiminase (PAD) 6 is essential for oocyte cytoskeletal sheet formation and female fertility. *Mol Cell Endocrinol* 2007;**273**(1–2):25–31.

14 Li P, Wang D, Yao H, Doret P, Hao G, Shen Q, Qiu H, Zhang X, Wang Y, Chen G, Wang Y. Coordination of PAD4 and HDAC2 in the regulation of p53-target gene expression. *Oncogene* 2010;**29**(21):3153–3162.

15 Schellekens GA, de Jong BAW, van den Hoogen FHJ, van de Putte LBA, van Venrooij WJ. Citrulline is an essential constituent of antigenic determinants recognized by rheumatoid arthritis-specific autoantibodies. *J Clin Invest* 1998;**101**(1):273–281.

16 Vossenaar ER, Zendman AJW, van Venrooij WJ, Pruijn GJM. PAD, a growing family of citrullinating enzymes: genes, features and involvement in disease. *Bioessays* 2003;**25**(11):1106–1118.

17 Rogers G, Winter B, McLaughlan C, Powell B, Nesci T. Peptidylarginine deiminase of the hair follicle: Characterization, localization, and function in keratinizing tissues. *J Invest Dermatol* 1997;**108**(5):700–707.

18 Nachat R, Mechin MC, Takahara H, Chavanas S, Charveron M, Serre G, Simon M. Peptidylarginine deiminase isoforms 1–3 are expressed in the epidermis and involved in the deimination of K1 and filaggrin. *J Invest Dermatol* 2005;**124**(2):384–393.

19 Tsuji Y, Akiyama M, Arita K, Senshu T, Shimizu H. Changing pattern of deiminated proteins in developing human epidermis. *J Invest Dermatol* 2003;**120**(5):817–822.

20 Mechin MC, Sebbag M, Arnaud J, Nachat R, Foulquier C, Adoue V, Coudane F, Duplan H, Schmitt A-M, Chavanas S, Guerrin M, Serre G, Simon M. Update on peptidylarginine deiminases and deimination in skin physiology and severe human diseases. *Int J Cosmet Sci* 2007;**29**(3):147–168.

21 Tarcsa E, Marekov LN, Mei G, Melino G, Lee SC, Steinert PM. Protein unfolding by peptidylarginine deiminase - Substrate specificity and structural relationships of the natural substrates trichohyalin and filaggrin. *J Biol Chem* 1996;**271**(48):30709–30716.

22 Matsui T, Miyamoto K, Kubo A, Kawasaki H, Ebihara T, Hata K, Tanahashi S, Ichinose S, Imoto I, Inazawa J, Kudoh J, Amagai M. SASPase regulates stratum corneum hydration through profilaggrin-to-filaggrin processing. *Embo Mol Med* 2011;**3**(6):320–333.

23 Moscarello MA, Wood DD, Ackerley C, Boulias C. Myelin in multiple-sclerosis is developmentally immature. *J Clin Invest* 1994;**94**(1):146–154.

24 Musse AA, Li Z, Ackerley CA, Bienzle D, Lei H, Poma R, Harauz G, Moscarello MA, Mastronardi FG. Peptidylarginine deiminase 2 (PAD2) overexpression in transgenic mice leads to myelin loss in the central nervous system. *Dis Model Mech* 2008;**1**(4–5):229–240.

25 Hsu PC, Liao YF, Lin CL, Lin WH, Liu GY, Hung HC. Vimentin is involved in peptidylarginine deiminase 2-induced apoptosis of activated Jurkat cells. *Mol Cells* 2014;**37**(5):426–434.

26 Giles JT, Fert-Bober J, Park JK, Bingham CO, Andrade F, Fox-Talbot K, Pappas D, Rosen A, van Eyk J, Bathon JM, Halushka MK. Myocardial citrullination in rheumatoid arthritis: a correlative histopathologic study. *Arthritis Research & Therapy* 2012;**14**(1):1–8.

27 Nakayama-Hamada M, Suzuki A, Kubota K, Takazawa T, Ohsaka M, Kawaida R, Ono M, Kasuya A, Furukawa H, Yamada R, Yamamoto K. Comparison of enzymatic properties between hPADI2 and hPADI4. *Biochem Biophys Res Commun* 2005;**327**(1):192–200.

28 van Beers JJBC, Zendman AJW, Raijmakers R, Stammen-Vogelzangs J, Pruijn GJM. Peptidylarginine deiminase expression and activity in PAD2 knock-out and PAD4-low mice. *Biochimie* 2013;**95**(2):299–308.

29 Nakashima K, Arai S, Suzuki A, Nariai Y, Urano T, Nakayama M, Ohara O, Yamamura K, Yamamoto K, Miyazaki T. PAD4 regulates proliferation of multipotent haematopoietic cells by controlling c-myc expression. *Nat Commun* 2013;**4**:1–8.

30 Li PX, Li M, Lindberg MR, Kennett MJ, Xiong N, Wang YM. PAD4 is essential for antibacterial innate immunity mediated by neutrophil extracellular traps. *J Exp Med* 2010;**207**(9):1853–1862.

31 Yamada R. Peptidylarginine deiminase type 4, anticitrullinated peptide antibodies, and rheumatoid arthritis. *Autoimmun Rev* 2005;**4**(4):201–206.

32 Chowdhury CS, Giaglis S, Walker UA, Buser A, Hahn S, Hasler P. Enhanced neutrophil extracellular trap generation in rheumatoid arthritis: analysis of underlying signal transduction pathways and potential diagnostic utility. *Arthritis Research & Therapy* 2014;**16**(3):1–8.

33 Dwivedi N, Neeli I, Schall N, Wan H, Desiderio DM, Csernok E, Thompson PR, Dali H, Briand JP, Muller S, Radic M. Deimination of linker histones links neutrophil extracellular trap release with autoantibodies in systemic autoimmunity. *FASEB J* 2014;**28**(7):2840–2851.

34 Schett G, Gravallese E. Bone erosion in rheumatoid arthritis: mechanisms, diagnosis and treatment. *Nat Rev Rheumatol* 2012;**8**(11):656–664.

35 Aletaha D, Neogi T, Silman AJ. 2010 Rheumatoid arthritis classification criteria: an American College of Rheumatology/European League Against Rheumatism collaborative initiative (vol 69, pg 1580, 2010). *Ann Rheum Dis* 2010;**69**(10):1892.

36 Mustila A, Korpela M, Haapala AM, Kautiainen H, Laasonen L, Mottonen T, Leirisalo-Repo M, Ilonen J, Jarvenpaa S, Luukkainen R, Hannonen P, Grp F-RS. Anti-citrullinated peptide antibodies and the progression of radiographic joint erosions in patients with early rheumatoid arthritis treated with the FIN-RACo combination and single disease-modifying antirheumatic drug strategies. *Clin Exp Rheumatol* 2011;**29**(3):500–505.

37 Hill JA, Bell DA, Brintnell W, Yue D, Wehrli B, Jevnikar AM, Lee DM, Hueber W, Robinson WH, Cairns E. Arthritis induced by post-translationally modified (citrullinated) fibrinogen in DR4-IE transgenic mice. *J Exp Med* 2008;**205**(4):967–979.

38 Cheung P, Lau P. Epigenetic regulation by histone methylation and histone variants. *Mol Endocrinol* 2005;**19**(3):563–573.

39 Rice JC, Briggs SD, Ueberheide B, Barber CM, Shabanowitz J, Hunt DF, Shinkai Y, Allis CD. Histone methyltransferases direct different degrees of methylation to define distinct chromatin domains. *Mol Cell* 2003;**12**(6):1591–1598.

40 Wang Y, Wysocka J, Sayegh J, Lee YH, Perlin JR, Leonelli L, Sonbuchner LS, McDonald CH, Cook RG, Dou Y, Roeder RG, Clarke S, Stallcup MR, Allis CD, Coonrod SA. Human PAD4 regulates histone arginine methylation levels via demethylimination. *Science* 2004;**306**(5694):279–283.

41 Wilson A, Murphy MJ, Oskarsson T, Kaloulis K, Bettess MD, Oser GM, Pasche AC, Knabenhans C, MacDonald HR, Trumpp A. c-Myc controls the balance between hematopoietic stem cell self-renewal and differentiation. *Genes Dev* 2004;**18**(22):2747–2763.

42 Chang X, Yamada R, Sawada T, Suzuki A, Kochi Y, Yamamoto K. The inhibition of antithrombin by peptidylarginine deiminase 4 may contribute to pathogenesis of rheumatoid arthritis. *Rheumatology* 2005;**44**(3):293–298.

43 Undas A, Gissel M, Kwasny-Krochin B, Gluszko P, Mann KG, Brummel-Ziedins KE. Thrombin generation in rheumatoid arthritis: Dependence on plasma factor composition. *Thromb Haemost* 2010;**104**(2):224–230.

44 Ordonez A, Yelamos J, Pedersen S, Minano A, Conesa-Zamora P, Kristensen SR, Stender MT, Thorlacius-Ussing O, Martinez-Martinez I, Vicente V, Corral J. Increased levels of citrullinated antithrombin in plasma of patients with rheumatoid arthritis and colorectal adenocarcinoma determined by a newly developed ELISA using a specific monoclonal antibody. *Thromb Haemost* 2010;**104**(6):1143–1149.

45 Li PX, Yao HJ, Zhang ZQ, Li M, Luo Y, Thompson PR, Gilmour DS, Wang YM. Regulation of p53 target gene expression by peptidylarginine deiminase 4. *Mol Cell Biol* 2008;**28**(15):4745–4758.

46 Wang YJ, Li PX, Wang S, Hu J, Chen XA, Wu JH, Fisher M, Oshaben K, Zhao N, Gu Y, Wang D, Chen G, Wang YM. Anticancer peptidylarginine deiminase (PAD) inhibitors regulate the autophagy flux and the mammalian target of rapamycin complex 1 activity. *J Biol Chem* 2012;**287**(31):25941–25953.

47 Willis V, Gizinski AM, Banda NK, Causey CP, Knuckley B, Cordova KN, Luo YA, Levitt B, Glogowska M, Chandra P, Kulik L, Robinson WH, Arend WP, Thompson PR, Holers VM. N-alpha-benzoyl-N5-(2-chloro-1-iminoethyl)-L-ornithine amide, a protein arginine deiminase inhibitor, reduces the severity of murine collagen-induced arthritis. *J Immunol* 2011;**186**(7):4396–4404.

48 Knight JS, Subramanian V, O'Dell AA, Yalavarthi S, Zhao WP, Smith CK, Hodgin JB, Thompson PR, Kaplan MJ. Peptidylarginine deiminase inhibition mitigates NET formation and protects against kidney, skin, and vascular disease in lupus-prone MRL/Lpr mice. *Arthritis Rheumatol* 2014;**66**:S281–S281.

49 Becker-Merok A, Nossent JC. Prevalence, predictors and outcome of vascular damage in systemic lupus erythematosus. *Lupus* 2009;**18**(6):508–515.

50 Christophorou MA, Castelo-Branco G, Halley-Stott RP, Oliveira CS, Loos R, Radzisheuskaya A, Mowen KA, Bertone P, Silva JCR, Zernicka-Goetz M, Nielsen ML, Gurdon JB, Kouzarides T. Citrullination regulates pluripotency and histone H1 binding to chromatin. *Nature* 2014;**507**(7490):104–108.

51 Luban S, Li ZG. Citrullinated peptide and its relevance to rheumatoid arthritis: an update. *Int J Rheum Dis* 2010;**13**(4):284–287.

52 Young BJJ, Mallya RK, Leslie RDG, Clark CJM, Hamblin TJ. Anti-keratin antibodies in rheumatoid-arthritis. *Br Med J* 1979;**2**(6182):97–99.

53 Coenen D, Verschueren P, Westhovens R, Bossuyt X. Technical and diagnostic performance of 6 assays for the measurement of citrullinated protein/peptide antibodies in the diagnosis of rheumatoid arthritis. *Clin Chem* 2007;**53**(3):498–504.

54 Taylor P, Gartemann J, Hsieh J, Creeden J. A systematic review of serum biomarkers anti-cyclic citrullinated Peptide and rheumatoid factor as tests for rheumatoid arthritis. *Autoimmune Dis* 2011;**2011**:815038.

55 Wilkins M. Proteomics data mining. *Expert Rev Proteomics* 2009;**6**(6):599–603.

56 Cunningham DL, Sweet SMM, Cooper HJ, Heath JK. Differential phosphoproteomics of fibroblast growth factor signaling: identification of Src family kinase-mediated phosphorylation events. *J Proteome Res* 2010;**9**(5):2317–2328.

57 Srivastava S. Move over proteomics, here comes glycomics. *J Proteome Res* 2008;**7**(5):1799.

58 Svinkina T, Gu HB, Silva JC, Mertins P, Qiao J, Fereshetian S, Jaffe JD, Kuhn E, Udeshi ND, Carr SA. Deep, quantitative coverage of the lysine acetylome using novel anti-acetyl-lysine antibodies and an optimized proteomic workflow. *Mol Cell Proteomics* 2015;**14**(9):2429–2440.

59 Doerr A. Interactomes by mass spectrometry. *Nat Methods* 2012;**9**(11):1043.

60 Picotti P, Clement-Ziza M, Lam H, Campbell DS, Schmidt A, Deutsch EW, Rost H, Sun Z, Rinner O, Reiter L, Shen Q, Michaelson JJ, Frei A, Alberti S, Kusebauch U, Wollscheid B, Moritz RL, Beyer A, Aebersold R. A complete mass-spectrometric map of the yeast proteome applied to quantitative trait analysis. *Nature* 2013;**494**(7436):266–270.

61 Hebert AS, Richards AL, Bailey DJ, Ulbrich A, Coughlin EE, Westphall MS, Coon JJ. The one hour yeast proteome. *Mol Cell Proteomics* 2014;**13**(1):339–347.

62 Tutturen AEV, Fleckenstein B, de Souza GA. Assessing the citrullinome in rheumatoid arthritis synovial fluid with and without enrichment of citrullinated peptides. *J Proteome Res* 2014;**13**(6):2867–2873.

63 Kelleher NL, Lin HY, Valaskovic GA, Aaserud DJ, Fridriksson EK, McLafferty FW. Top down versus bottom up protein characterization by tandem high-resolution mass spectrometry. *J Am Chem Soc* 1999;**121**:806–812.

64 Kocher T, Pichler P, Swart R, Mechtler K. Analysis of protein mixtures from whole-cell extracts by single-run nanoLC-MS/MS using ultralong gradients. *Nat Protoc* 2012;**7**(5):882–890.

65 Sarsby J, Martin NJ, Lalor PF, Bunch J, Cooper HJ. Top-down and bottom-up identification of proteins by liquid extraction surface analysis mass spectrometry of healthy and diseased human liver tissue. *J Am Soc Mass Spectrom* 2014;**25**(11):1953–1961.

66 Amado FML, Vitorino RMP, Domingues PMDN, Lobo MJC, Duarte JAR. Analysis of the human saliva proteome. *Expert Rev Proteomics* 2005;**2**(4):521–539.

67 Creese AJ, Shimwell NJ, Larkins KPB, Heath JK, Cooper HJ. Probing the complementarity of FAIMS and strong cation exchange chromatography in shotgun proteomics. *J Am Soc Mass Spectrom* 2013;**24**(3):431–443.

68 Chevalier F. Highlights on the capacities of "Gel-based" proteomics. *Proteome Science* 2010;**8**:1–10.

69 Fenn JB, Mann M, Meng CK, Wong SF, Whitehouse CM. Electrospray ionization for mass spectrometry of large biomolecules. *Science* 1989;**246**:64–71.

70 Jennings KR. Collision-induced decompositions of aromatic molecular ions. *Int J Mass Spectrom. Ion Phys.* 1968;**1**(4–5):227–235.

71 Sweet SMM, Creese AJ, Cooper HJ. Strategy for the identification of sites of phosphorylation in proteins: Neutral loss triggered electron capture dissociation. *Anal Chem* 2006;**78**(21):7563–7569.

72 Roepstorff P, Fohlman J. Proposal for a common nomenclature for sequence ions in mass spectra of peptides. *Biol Mass Spectrom* 1984;**11**:601.

73 Biemann K. Contributions of mass spectrometry to peptide and protein structure. *Biomed Environ Mass Spectrom* 1988;**16**:99–111.

74 Wells JM, McLuckey SA. Collision-induced dissociation (CID) of peptides and proteins. *Biol Mass Spectrom* 2005;**402**:148–185.

75 Esteban-Fernandez D, El-Khatib AH, Moraleja I, Gomez-Gomez MM, Linscheid MW. Bridging the gap between molecular and elemental mass spectrometry: Higher energy collisional dissociation (HCD) revealing elemental information. *Anal Chem* 2015;**87**(3):1613–1621.

76 Singh C, Zampronio CG, Creese AJ, Cooper HJ. Higher energy collision dissociation (HCD) product ion-triggered electron transfer dissociation (ETD) mass spectrometry for the analysis of N-linked glycoproteins. *J Proteome Res* 2012;**11**(9):4517–4525.

77 Guo T, Gan CS, Zhang H, Zhu Y, Kon OL, Sze SK. Hybridization of pulsed-Q dissociation and collision-activated dissociation in linear ion trap mass spectrometer for iTRAQ quantitation. *J Proteome Res* 2008;**7**(11):4831–4840.

78 Ritchie MA, Gill AC, Deery MJ, Lilley K. Precursor ion scanning for detection and structural characterization of heterogeneous glycopeptide mixtures. *J Am Soc Mass Spectrom* 2002;**13**(9):1065–1077.

79 Wiese S, Reidegeld KA, Meyer HE, Warscheid B. Protein labeling by iTRAQ: A new tool for quantitative mass spectrometry in proteome research. *Proteomics* 2007;**7**(3):340–350.

80 Zubarev RA, Kelleher NL, McLafferty FW. ECD of multiply charged protein cations. A non-ergodic process. *J Am Chem Soc* 1998;**120**:3265–3266.

81 Mikesh LM, Ueberheide B, Chi A, Coon JJ, Syka JEP, Shabanowitz J, Hunt DF. The utility of ETD mass spectrometry in proteomic analysis. *Biochim Biophys Acta Proteins Proteomics* 2006;**1764**(12):1811–1822.

82 Wiesner J, Premsler T, Sickmann A. Application of electron transfer dissociation (ETD) for the analysis of post-translational modifications. *Proteomics* 2008;**8**(21):4466–4483.

83 Swaney DL, McAlister GC, Wirtala M, Schwartz JC, Syka JEP, Coon JJ. Supplemental activation method for high-efficiency electron transfer dissociation of doubly-protonated peptide precursors. *Anal Chem* 2007;**79**(2):477–485.

84 Good DM, Wirtala M, McAlister GC, Coon JJ. Performance characteristics of electron transfer dissociation mass spectrometry. *Mol Cell Proteomics* 2007;**6**(11):1942–1951.

85 Ma B, Zhang KZ, Hendrie C, Liang CZ, Li M, Doherty-Kirby A, Lajoie G. PEAKS: powerful software for peptide de novo sequencing by tandem mass spectrometry. *Rapid Commun Mass Spectrom* 2003;**17**(20):2337–2342.

86 Perkins DN, Pappin DJC, Creasy DM, Cottrell JS. Probability-based protein identification by searching sequence databases using mass spectrometry data. *Electrophoresis* 1999;**20**(18):3551–3567.

87 Eng JK, Mccormack AL, Yates JR. An approach to correlate tandem mass-spectral data of peptides with amino-acid-sequences in a protein database. *J Am Soc Mass Spectrom* 1994;**5**(11):976–989.

88 Siepen JA, Keevil E-J, Knight D, Hubbard SJ. Prediction of missed cleavage sites in tryptic peptides aids protein identification in proteomics. *J Proteome Res* 2007;**6**(1):399–408.

89 Bennike T, Lauridsen KB, Kruse Olesen M, Andersen V, Birkelund S, Stensballe A. Optimizing the identification of citrullinated peptides by mass spectrometry: Utilizing the inability of trypsin to cleave after citrullinated amino acids. *J Proteomics Bioinf* 2013;**6**(12):288–295.

90 Jin ZC, Fu ZM, Yang J, Troncoso J, Everett AD, Van Eyk JE. Identification and characterization of citrulline-modified brain proteins by combining HCD and CID fragmentation. *Proteomics* 2013;**13**(17):2682–2691.

91 De Ceuleneer M, De Wit V, Van Steendam K, Van Nieuwerburgh F, Tilleman K, Deforce D. Modification of citrulline residues with 2,3-butanedione facilitates their detection by liquid chromatography/mass spectrometry. *Rapid Commun Mass Spectrom* 2011;**25**(11):1536–1542.

92 De Ceuleneer M, Van Steendam K, Dhaenens M, Deforce D. In vivo relevance of citrullinated proteins and the challenges in their detection. *Proteomics* 2012;**12**(6):752–760.

93 Hao P, Ren Y, Datta A, Tam JP, Sze SK. Evaluation of the effect of trypsin digestion buffers on artificial deamidation. *J Proteome Res* 2015;**14**(2):1308–1314.

94 Dasari S, Wilmarth PA, Rustvold DL, Riviere MA, Nagalla SR, David LL. Reliable detection of deamidated peptides from lens crystallin proteins using changes in reversed-phase elution times and parent ion masses. *J Proteome Res* 2007;**6**(9):3819–3826.

95 McLaffery FW, Bryce TA. Metastable-ion characteristics: Characterization of isomeric molecules. *Chem Commun* 1967;**23**:1215–1217.

96 Hao G, Wang DC, Gu J, Shen QY, Gross SS, Wang YM. Neutral loss of isocyanic acid in peptide CID spectra: A novel diagnostic marker for mass spectrometric identification of protein citrullination. *J Am Soc Mass Spectrom* 2009;**20**(4):723–727.

97 Creese AJ, Grant MM, Chapple LLC, Cooper HJ. On-line liquid chromatography neutral loss-triggered electron transfer dissociation mass spectrometry for the targeted analysis of citrullinated peptides. *Anal Methods* 2011;**3**(2):259–266.

98 Syka JEP, Coon JJ, Schroeder MJ, Shabanowitz J, Hunt DF. Peptide and protein sequence analysis by electron transfer dissociation mass spectrometry. *Proc Natl Acad Sci* 2004;**101**(26):9528–9533.

99 Holm A, Rise F, Sessler N, Sollid LM, Undheim K, Fleckenstein B. Specific modification of peptide-bound citrulline residues. *Anal Biochem* 2006;**352**(1):68–76.

100 Stensland M, Holm A, Kiehne A, Fleckenstein B. Targeted analysis of protein citrullination using chemical modification and tandem mass spectrometry. *Rapid Commun Mass Spectrom* 2009;**23**(17):2754–2762.

101 Slebos RJC, Brock JWC, Winters NF, Stuart SR, Martinez MA, Li M, Chambers MC, Zimmerman LJ, Ham AJ, Tabb DL, Liebler DC. Evaluation of strong cation exchange versus isoelectric focusing of peptides for multidimensional liquid chromatography-tandem mass spectrometry. *J Proteome Res* 2008;**7**(12):5286–5294.

102 Kubota K, Yoneyama-Takazawa T, Ichikawa K. Determination of sites citrullinated by peptidylarginine deiminase using O-18 stable isotope labeling and mass spectrometry. *Rapid Commun Mass Spectrom* 2005;**19**(5):683–688.

103 Tutturen AEV, Holm A, Fleckenstein B. Specific biotinylation and sensitive enrichment of citrullinated peptides. *Anal Bioanal Chem* 2013;**405**(29):9321–9331.

104 Tutturen AEV, Holm A, Jorgensen M, Stadtmuller P, Rise F, Fleckenstein B. A technique for the specific enrichment of citrulline-containing peptides. *Anal Biochem* 2010;**403**(1–2):43–51.

105 Bhattacharya SK, Crabb JS, Bonilha VL, Gu XR, Takahara H, Crabb JW. Proteomics implicates peptidyl arginine deiminase 2 and optic nerve citrullination in glaucoma pathogenesis. *Invest Ophthalmol Vis Sci* 2006;**47**(6):2508–2514.

106 Ytterberg AJ, Reynisdottir G, Ossipova E, Rutishauser D, Hensvold A, Eklund A, Skold M, Grunewald J, Lundberg K, Malmstrom V, Jakobsson PJ, Zubarev R, Klareskog L, Catrina AI. Identification of shared citrullinated immunological targets in the lungs and joints of patients with rheumatoid arthritis. *Ann Rheum Dis* 2012;**71**:A19–A19.

107 Cao LG, Sun DM, Whitaker JN. Citrullinated myelin basic protein induces experimental autoimmune encephalomyelitis in Lewis rats through a diverse T cell repertoire. *J Neuroimmunol* 1998;**88**(1–2):21–29.

108 Kan R, Jin M, Subramanian V, Causey CP, Thompson PR, Coonrod SA. Potential role for PADI-mediated histone citrullination in preimplantation development. *BMC Developmental Biology* 2012;**12**(19):1–10.

109 Hermansson M, Artemenko K, Ossipova E, Eriksson H, Lengqvist J, Makrygiannakis D, Catrina AI, Nicholas AP, Klareskog L, Savitski M, Zubarev RA, Jakobsson PJ. MS analysis of rheumatoid arthritic synovial tissue identifies specific citrullination sites on fibrinogen. *Proteomics Clin Appl* 2010;**4**(5):511–518.

110 Ong SE, Blagoev B, Kratchmarova I, Kristensen DB, Steen H, Pandey A, Mann M. Stable isotope labelling by amino acids in cell culture, SILAC, as a simple and accurate approach to expression proteomics. *Mol Cell Biol* 2002;**1**(5):376–386.

111 van Beers JJBC, Schwarte CM, Stammen-Vogelzangs J, Oosterink E, Bozic B, Pruijn GJM. The rheumatoid arthritis synovial fluid citrullinome reveals novel citrullinated epitopes in apolipoprotein E, myeloid nuclear differentiation antigen, and ss-actin. *Arthritis Rheum* 2013;**65**(1):69–80.

112 Colantonio DA, Dunkinson C, Bovenkamp DE, Van Eyk JE. Effective removal of albumin from serum. *Proteomics* 2005;**5**(15):3831–3835.

113 Gillet LC, Navarro P, Tate S, Rost H, Selevsek N, Reiter L, Bonner R, Aebersold R. Targeted data extraction of the MS/MS spectra generated by data-independent acquisition: A new concept for consistent and accurate proteome analysis. *Mol Cell Proteomics* 2012;**11**(6):1–17.

114 Kane LA, Neverova I, Van Eyk JE. Subfractionation of heart tissue: the "in sequence" myofilament protein extraction of myocardial tissue. *Methods Mol Biol* 2007;**357**:87–90.

115 Kinloch A, Lundberg K, Wait R, Wegner N, Lim NH, Zendman AJW, Saxne T, Malmstrom V, Venables PJ. Synovial fluid is a site of citrullination of autoantigens in inflammatory arthritis. *Arthritis Rheum* 2008;**58**(8):2287–2295.

116 De Ceuleneer M, Van Steendam K, Maarten D, Elewaut D, Deforce D. Quantification of citrullination by means of skewed isotope distribution pattern. *J Proteome Res* 2012;**11**(11):5245–5251.

8

Glycation of Proteins

Naila Rabbani[1] and Paul J. Thornalley[1,2]

[1] *Warwick Systems Biology Centre, Coventry House, University of Warwick, Coventry, UK*
[2] *Warwick Medical School, Clinical Sciences Research Laboratories, University of Warwick, University Hospital, Coventry, UK*

8.1   Overview of Protein Glycation

Protein glycation is a spontaneous post-translational modification (PTM) of proteins found in physiological systems and food products. It involves the non-enzymatic covalent attachment of a reducing sugar or sugar derivative to a protein [1]. It is a PTM that is often thermally and chemically labile, particularly at high pH and temperature when removed from the physiological setting. Analysis of protein glycation is compromised by the use of heating and high pH in preanalytic processing for mass spectrometric analysis. This is, perhaps, one of the most important points for experts in mass spectrometry new to protein glycation to consider and adapt experimental protocols accordingly for reliable analysis [2]. Initially, protein glycation was thought to be restricted to the modification of amino groups of lysine residue side chains and N-terminal amino acid residues by glucose. In more recent times, glycation of arginine residues by dicarbonyl metabolites has emerged as a major feature of protein glycation in physiological systems and linked to functional impairment [3]. There is also minor involvement of cysteine residues. This chapter mainly focuses on the detection and quantitation of the major early glycation adduct, N_ε-(1-deoxy-D-fructos-1-yl)lysine of fructosyl-lysine (FL), and the major advanced glycation end product (AGE), methylglyoxal (MG)-derived hydroimidazolone MG-H1, and related compounds.

Protein glycation finds application in clinical metabolic monitoring and diagnosis. Glycation of albumin and hemoglobin by glucose produces glycated derivatives, glycated albumin, and glycated hemoglobin, used clinically in the assessment of glycemic control in diabetes over the 3–4 weeks and 6–8 weeks

Analysis of Protein Post-Translational Modifications by Mass Spectrometry,
First Edition. Edited by John R. Griffiths and Richard D. Unwin.
© 2017 John Wiley & Sons, Inc. Published 2017 by John Wiley & Sons, Inc.

prior to sampling [4]. A defined range of glycated hemoglobin, 39–47 mmol/ mol Hb, intermediate between those found in healthy people and patients with diabetes is a diagnostic indicator of impaired glucose tolerance preceding diabetes and is used clinically in diagnostic screening for prediabetes [5]. Formation of AGEs is implicated in aging – including age-related macular degeneration and cataract, obesity, vascular complications of diabetes (nephropathy, retinopathy, and neuropathy), cirrhosis, cardiovascular disease, renal failure, and neurological disorders (Alzheimer's disease and Parkinson's disease) [6–17].

Glycation of proteins occurs by a complex series of sequential and parallel reactions called collectively the Maillard reaction. Many different adducts may be formed – some of which are fluorescent and colored "browning pigments." In the physiological setting, one of the most important saccharides participating in glycation is glucose – forming lysine and N-terminal residue-derived fructosamine derivatives, and one of the most important saccharide derivatives is the reactive dicarbonyl metabolites MG, forming mainly arginine-derived hydroimidazolone derivatives [1].

Glycation adducts are classified into two groups: early-stage glycation adducts and AGEs. Glucose reacts with amino groups of lysine residue side chains and N-terminal amino acid residues to form sequentially a Schiff's base and then, via the Amadori rearrangement, N_ε-(1-deoxy-D-fructos-1-yl)lysine (FL) and $N\alpha$-(1-deoxy-D-fructos-1-yl)amino acid residues – called collectively fructosamines (Figure 8.1). These are early-stage glycation adducts.

Examples of proteins susceptible to fructosamine formation are given in Table 8.1. Collectively these constitute the "fructosamine proteome." Schiff's base adducts are usually a minor component of glucose adducts *in situ*, ca. 10% of the level of FL residues in the steady state. They are also relatively rapidly reversed during sample isolation and processing, whereas fructosamines have much slower reversibility of formation; chemical relaxation times for reversal of Schiff's base and fructosamine formation are *ca.* 2.5 and 38 h at pH 7.4 and 37 °C, respectively [45]. The rate of fructosamine degradation at 37 °C increases markedly above pH 8 through increased reversal of the Amadori rearrangement and oxidative degradation to N_ε-carboxymethyl-lysine (CML) and related $N\alpha$-carboxymethyl amino acids [45, 46]. Accordingly when adducts of early-stage glycation by glucose are detected and quantified, it is typically the fructosamine proteome that is characterized. Fructosamine modification of proteins is usually low, 5–10% modified protein, and may often have only moderate functional effects. This may be related to the respective low and moderate probability of location of N-terminal and lysine residues in the functional domains of proteins [47] and that the fructosamine residue retains the positive charge of the precursor lysyl side chain or N-terminal amino acid residue under physiological conditions. Gene knockout of fructosamine-3-phosphokinase (F3PK), an enzyme that repairs the fructosamine residues of cellular proteins, leads to accumulation of the fructosamine proteome without significant health impairment [48].

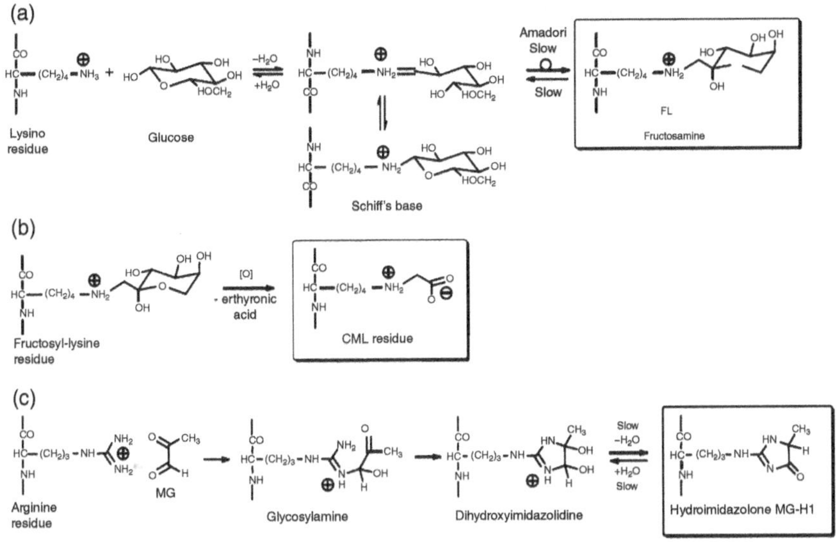

(a)

Lysino residue   Glucose

Schiff's base

Amadori
Slow
Slow

Fructosamine

FL

(b)

Fructosyl-lysine residue

[O]
- erythronic acid

CML residue

(c)

Arginine residue

MG

Glycosylamine   Dihydroxyimidazolidine

Slow
-H₂O
+H₂O
Slow

Hydroimidazolone MG-H1

Figure 8.1 Major protein glycation processes in physiological systems. (a) Early glycation. Formation of the Schiff's base and fructosamine (Amadori product) of lysine residues. (b) Oxidative degradation of fructosyl-lysine to N_ε-carboxymethyl-lysine. Similar processes occur on N-terminal amino acid residues. (c) Glycation of arginine residues by methylglyoxal with the formation of dihydroxyimidazolidine and hydroimidazolone MG-H1 residues. There are related structural isomers and similar adducts formed from glyoxal and 3-deoxyglucosone [1, 18–20].

In later-stage reactions of glycation, Schiff's base and fructosamine adducts degrade to form many stable end-stage adducts or AGEs [1]. Endogenous α-oxoaldehyde metabolites are potent glycating agents and react with proteins to form AGEs directly. Important dicarbonyl glycating agents are glyoxal, MG, and 3-deoxyglucosone (3-DG). Further classification of AGEs relates to the mechanism of AGE formation. "Glycoxidation" refers to glycation processes in which oxidation is involved and the AGEs formed thereby are called "glycoxidation products" [1]. CML and Nα-carboxymethyl amino acids are glycoxidation products. These are often mainly formed by oxidative degradation of fructosamine and hence have the same proteome site-specific coverage as their fructosamine precursor (Figure 8.1b). There are also minor contributions to CML and Nα-carboxymethyl amino acid residue formation from glycation of proteins by glycolaldehyde and glyoxal, which may have a different site-specific distribution [49, 50]. The major AGE quantitatively found in physiological systems is the MG-derived hydroimidazolone [18]. MG reacts predominantly with arginine residues to form sequentially a glycosylamine, dihydroxyimidazolidine, and hydroimidazolone MG-H1 residues (Figure 8.1c). Other structural isomers are also found: MG-H1, MG-H2, and MG-H3 [19];

Table 8.1 Components of the fructosamine proteome.

Species	Protein	Hotspot sites	Extent of modification	Functional impairment	Reference
Human	Apolipoprotein A1	K239	4%	None	[21]
	Apolipoprotein E	K93	Unknown	Impairs heparin binding	[22]
	Bisphosphoglycerate mutase	K158	Unknown	Inactivation	[23]
	CD59	K41	Unknown	Inactivation	[24]
	Complement factor B	K266	Unknown	–	[25]
	Gastric inhibitory polypeptide	Y1	Unknown	Increased insulin release	[26]
	Glucagon-like peptide-1	1H	Unknown	Decreased insulin release	[27]
	Hemoglobin $\alpha_2\beta_2$	α-K61 β-V1 β-K66	5% (α:β, 0.6:1)	Increased oxygen binding in T state	[28–30]
Human	Insulin	β-F1		Decreased activity	[31]
	Microglobulin, β2	I1	Unknown	Aggregation in chronic renal dialysis	[32]
	Serum albumin	D1 K199 K439 K525	10%	Decreased drug binding and leakage through the glomerular filter	[33, 34]
	Superoxide dismutase-1	K122 K128	Unknown	Inactivation	[35]
Bovine	Crystallin, αA	K11 K78	Unknown		[36]
Bovine	Crystallin, αB	K90 K92	Unknown		[36]
	Crystallin, γB	G1 K2	Unknown		[37]
	Glutathione peroxidase-1	K117	Unknown	Inactivated	[38]
	Insulin	α-G1 β-F1 β-K29	Unknown		[39]

Table 8.1 (Continued)

Species	Protein	Hotspot sites	Extent of modification	Functional impairment	Reference
	Major intrinsic peptide	K238 K259	Unknown	Affects membrane permeability	[40]
Bovine	Ribonuclease A	K1 K7 K41	Unknown		[41]
	Serum albumin	K12 K136 K211 K232 K377 K524	10%		[42]
Rat	Collagen I	α1-K434 α2-K453 α2-K479 α2-K924	50–70% 27–33% 24–29% 22–28%	Increased susceptibility to cross-linking	[43]
	Aldo-keto reductase 1A1	K67 K84 K140	18%	Inactivation	[44]

Modifying agent: D-glucose.

isomer MG-H1 is usually dominant *in vivo* [51]. The half-life for reversal of glycosylamine/dihydroxyimidazolidine formation is *ca.* 1.8 days and for reversal of hydroimidazolone is *ca.* 12 days at pH 7.4 and 37 °C [52]. The stability of the hydroimidazolone decreases with increasing pH; the half-life of MG-H1 is 0.87 days at pH 9.4 [19]. Hence dihydroxyimidazolidine and hydroimidazolone residues derived from arginine residues may be detected in mass spectrometric analysis of glycated proteins. Glucose-derived Schiff's base and fructosamines degrade to form glyoxal, MG, and 3-DG, and so dicarbonyl-derived AGEs may be detected in proteins glycated by glucose too [45, 53].

Glycation of proteins by MG is found at levels of 1–5% in most proteins but increases to *ca.* 50% in the human lens of elderly where there is limited protein turnover [20, 51]. It often occurs at functional domain of proteins and leads to protein inactivation and dysfunction. This may be because arginine residues have the highest probability (20%) of any amino acid to be found in a functional domain and there is loss of positive charge on the formation of MG-H1 [3]. Gene knockout of glyoxalase 1 (Glo1), the enzyme that protects against glycation by MG, is embryonically lethal, and increased MG concentration, or dicarbonyl stress, imposed by Glo1 deficiency accelerates the aging process and exacerbates diseases – including cardiovascular disease, diabetes, renal failure, and neurological disorders [6]. Proteins susceptible to MG glycation are called the "dicarbonyl proteome" (Table 8.2).

Glycated proteins undergo proteolysis in physiological systems to release glycated amino acids called "glycation-free adducts." These are found in plasma and other body fluids. They are excreted from the body in urine. Urinary excretion of glycation-free adduct increases from 2- to 15-fold in diabetes and renal failure [8, 61]. Glycation-free adducts in ultrafiltrate of physiological fluids are detected and quantified by stable isotopic dilution analysis liquid chromatography–tandem mass spectrometry (LC-MS/MS) in multiple reaction monitoring (MRM) data acquisition mode. In positive ion mode electrospray tandem mass spectrometry, FL dehydrates in the vapor phase and enters the mass analyzer as the singly charged oxonium ion (M+144) [18, 83]. The LC-MS/MS analysis is extended to quantify total glycation adduct contents of purified proteins and protein extracts of cells and extracellular matrix by prior exhaustive enzymatic hydrolysis [18]. Conventional acid hydrolysis cannot be used because of low analytical recoveries [19].

Table 8.2 Components of the dicarbonyl proteome.

Species	Protein	Hotspot sites	Arg agent	Extent of modification	Functional impairment	Reference
Human	Apolipoprotein A1	R27 R123 R149	MG	*ca.* 1%	R27, increased catabolism; R123, decreased stability; R149, impaired functional activity	[12]
	Apolipoprotein B100	R18	MG		Increased density, proteoglycan binding, and atherogenicity	[11]
	CD4 antigen	R59	*p*HOPhG	Unknown	Binding of HIV pp120	[54]
	Collagen-IV	α1-R390 α2-R889 α2-R1452 α3-1404	MG	*ca.* 5%	Decreased integrin binding	[55, 56]

Table 8.2 (Continued)

Species	Protein	Hotspot sites	Arg agent	Extent of modification	Functional impairment	Reference
Human	Crystallin, αA	R12 R65 R157 R163	MG	Unknown	Increased chaperone activity	[57]
	β-Defensin-2	R22 R23	Glyoxal/MG	Unknown	Decreased antimicrobial activity	[58]
	Fibrin(ogen)	α-R167 α-R199 α-R491 α-R528 β-R149 β-R304	MG	Unknown	Abnormal thrombosis and fibrinolysis	[59]
Human	Heat shock protein-27	R75 R89 R94 R127 R136 R140 R188	MG	Unknown	Enhanced protection against oxidative stress	[60]
	Hemoglobin α₂β₂	α-R31 α-92 α-141 β-R30 β-R 40 β-R104	MG	*ca.* 2.6%	Increased oxygen binding	[61–63]
	HIF1α– coactivator p300	R354	MG	Unknown	Decreased hypoxia response	[64]
	HIV-1 nucleocapsid protein	R7 R10 R32	Kethoxal	Unknown	Binding of RNA stem-loops 2 and 3	[65]

(*Continued*)

Table 8.2 (Continued)

Species	Protein	Hotspot sites	Arg agent	Extent of modification	Functional impairment	Reference
	HLA-DR1	α-50 α-123 β-189	*p*HOPhG	Unknown	Surface ligand binding	[66]
	Insulin	R46	MG	Unknown	Aggregation	[67]
	IgG (monoclonal)	LC-R30	MG	5%	Acidic variant	[68]
	Plasminogen	R504 R530 R561	MG	Unknown	Likely functional changes to cleavage and Lys binding pocket in fibrinolysis	[69]
	Proteasome, 20S subunits	β2-R85 β4-R224 β4-231 β5-123 β5-128	MG	Unknown	Decreased proteasome activity	[70]
Bovine	Ribonuclease A	R10 R39 R85	Glyoxal/ MG	Unknown	Inhibition	[71, 72]
	Seminal plasma protein PDC-109	R57 R64 R104	cHxG	Unknown	Heparin binding	[73]
	Serum albumin	R114 R186 R218 R257 R410 R428	MG	*ca.* 1%	Inhibition of esterase activity, prostaglandin breakdown, and decreased drug binding	[74, 75]
	Ubiquitin	R54 R72 R74	Kethoxal	Unknown	–	[65]
Rabbit	Muscle creatine kinase	R129 R131 R134	PhG	Unknown	Inactivation	[76]

Table 8.2 (Continued)

Species	Protein	Hotspot sites	Arg agent	Extent of modification	Functional impairment	Reference
Mouse	mSin3a corepressor	R925	MG	Unknown	Increased angiopoietin-2 activity	[77]
Chicken	Lysozyme	R5	cHxG	Unknown		[78]
		R73				
		R112				
		R125				
Streptomyces coelicolor	3-Dehydroquinate dehydratase	R23	cHxG	Unknown	Inactivation	[79]
Aspergillus nidulans	3-Dehydroquinate dehydratase	R19	cHxG	Unknown	Inactivation	[79]
Aspergillus sp.	Amadoriase II	R112	*p*HOPhG	Unknown	Inactivation	[80]
		R114				
Sorghum bicolor	Malate dehydrogenase	R87	cHxG	Unknown	Cofactor binding	[81]
		R134			Catalytic mechanism	
		R140			Substrate binding	
		R204			Catalytic mechanism	
Hepatitis C virus	E2 envelope protein	R587		Unknown	Antibody binding	[82]
		R630				
		R651				

Modifying agent key: cHxG, cyclohexanedione; MG, methylglyoxal; *p*HOPhG, *p*-hydroxyphenylglyoxal; PhG, phenylglyoxal. Other: glyoxal and kethoxal. NB: These are all α-oxoaldehydes, but only glyoxal and Mg are physiological glycating agents. Reactivity of sites to other agents is assumed to indicate reactivity of the arginine residue site to dicarbonyl glycation.

8.2 Mass Spectrometry Behavior of Glycated Peptides

The mass increment indicating the detection of FL and other fructosamine-modified peptides is +162 Da. Glycation of intact proteins and large peptide chains has been detected by electrospray positive ion mass spectrometry and matrix-assisted laser desorption ionization time-of-flight (MALDI-TOF) mass spectrometry. Roberts and coworkers detected and quantified fructosamine-modified α- and β-chains of hemoglobin by deconvolution of multiply charged ion series [84], shown in later studies by peptide mapping to reflect fructosamine formation at sites α-K61, β-V1, and β-K66 [28]. Increase in molecular mass of human serum albumin (HSA) glycated by glucose prepared *in vitro* was measured by MALDI-TOF. This revealed that preparations of glycated HSA had a large increase in mass due to the high extent of glycation, dissimilar from the low increase in mass of glycated HSA in plasma samples *in vivo*. For example, HSA from human plasma had a mean mass increment of +243 Da, whereas model glucose-modified albumin prepared *in vitro* had a mean mass increment of +6780 Da [85]. This suggested that the albumin prepared with very high extent of glycation was a poor model for the albumin with minimal extent of glycation found *in vivo*.

For mass spectrometric analysis of glycated peptides, collision-induced dissociation (CID) and higher-energy collisional dissociation (HCD) fragmentation of fructosamine-containing peptides produced characteristic fragment ions of the precursor fructosamines (M+162): by dehydration to an oxonium ion (M+144), further dehydration to a pyrylium ion (M+108), and dehydration and formaldehyde loss to an immonium ion (M+78) [42, 86–88] (Figure 8.2a). Pyrylium and furylium ions are detected in y ion series providing for fructosamine location [42]. In electron transfer dissociation (ETD) fragmentation, abundant and almost complete series of c- and z-type ions were observed, which greatly facilitated the peptide sequencing and fructosamine site location [89].

The FL degradation product and AGE CML were detected at the same sites as fructosamine residues in serum albumin, hemoglobin, and ribonuclease A [42, 90, 91].

MG-derived hydroimidazolone and dihydroxyimidazolidine may be detected in peptides glycated by MG and tryptic peptides of proteins glycated *in vivo*. They have mass increments on arginine residues of +72 Da and +54, respectively. A further minor MG-derived and stable AGE, N_ε-(1-carboxyethyl))lysine (CEL), may be detected as +72 Da on lysine residues [11, 12, 71, 74]. High collision energy fragmentation may dehydrate dihydroxyimidazolidine to hydroimidazolone, and so discrimination is provided by detection of the peptide molecular ion [71]. In analysis of MG-modified lipoproteins, no advantage of ETD over CID in the detection of hydroimidazolone and dihydroxyimidazolidine has been found [11, 12]. Fragmentation of peptides modified by MG-H1 and related isomers gave complete series of b and y ions with mass increment of 54 Da relative to those of unmodified peptide and no neutral losses [11, 12, 71, 74]. A MG-H1-related fragment ion of $m/z = 166.1$ can be observed in the low mass region of the MS/MS spectra, with proposed immonium ion structure (Figure 8.2b). A similar fragment ion of $m/z = 152.1$ can be observed for

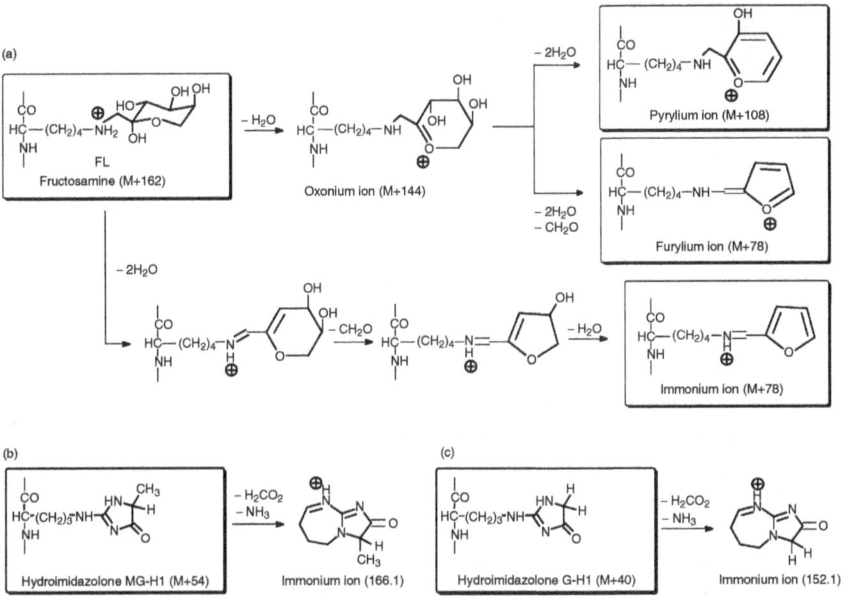

Figure 8.2 Fragmentation of fructosamine and hydroimidazolone glycation adducts. (a) Fragmentation of fructosyl-lysine by CID leading to the formation of oxonium, pyrylium, furylium, and immonium ions. (b) and (c) Fragmentation of hydroimidazolones formed by MG and glyoxal to immonium ions in CID and HCD [42, 86–88].

glyoxal-modified peptides [92] (Figure 8.2c). Hydroimidazolone and dihydrox-yimidazolidine residues are chemically labile AGEs, and conditions of preanalytic processing for proteomics analysis may influence mass spectrometric analysis outcomes. Tryptic digestion methods with prolonged periods of samples incubated at high pH and/or temperature leads to reversal of hydroimidazolone to dihydroxyimidazolidine and deglycation. Alternatively, high pH and temperature may also stimulate dicarbonyl formation [93]. In earlier studies, using N-hydroxysuccinimidyl active ester derivatization of MG-H1 in chromatographic analysis, we found that incubation of MG-H1 in the presence of [^{15}N$_2$]arginine at pH 8.8 for 10 min at 55 °C led to migration of the MG moiety from MG-H1 to [^{15}N$_2$]arginine [19]. Hence, use of high pH and temperature in preanalytic processing may induce migration of the MG moiety between arginine residues and, potentially, also between proteins. Conventional tryptic digestion techniques require modification to minimize the increase of pH and avoid sample heating for peptide mapping and proteomics analysis of MG-modified proteins and related PTMs.

Trypsinization cleavage after lysine and arginine residues is impaired by glycation by glucose and MG, and glycated peptide with missed cleavage at the glycation site is detected [74, 75, 94]. In some cases, cleavage after dicarbonyl glycation of arginine was observed [92].

8.3 Global Analysis of Glycation

The total amount of glycation adducts of different chemically defined structures may be determined by exhaustive enzymatic digestion of protein extracts to hydrolyze protein to their constituent amino acids, followed by stable isotopic dilution analysis LC-MS/MS. This procedure has been automated and is often combined with a screen of protein oxidation and nitration adduct PTMs [2, 83]. Examples of application of this are detection increased FL, CML and MG-H1 in cytoplasmic proteins of the kidney glomeruli, retina and peripheral nerve of rats in experimental diabetes. FL: 0.62–0.96 vs 0.18–0.43 mmol/mol Lys. CML: 0.25–0.46 vs 0.17–0.27 mmol/mol Lys. MG-H1: 4.37–5.98 vs 1.70–2.99 mmol/mol Arg). This provides evidence for increased early glycation and AGE accumulation at tissue sites of development of microvascular complications of diabetes [95]. Genetic and pharmacological prevention of AGE accumulation in Glo1 transgenic mice and high-dose thiamine-treated rats without change in FL with concomitant prevention of microvascular complications provided supporting evidence that AGEs rather than FL cause damage in diabetes [6, 95]. This was translated clinically to show increased FL and MG-H1 in plasma protein of patients with type 1 diabetes with respect to healthy controls (FL: 3.68 ± 0.86 vs. 1.35 ± 0.16 mmol/mol Lys; and MG-H, 0.99 ± 0.27 vs. 0.31 ± 0.20 mmol/mol Arg) [61]. Skin collagen contents of FL and certain AGEs were found to be risk predictors of development of microvascular complications and cardiovascular disease in clinical diabetes [9, 10].

A great challenge for global screening of glycated proteins by limited proteolysis of cellular and extracellular proteomes and mass spectrometric analysis of related peptides is to maximize sequence coverage of proteins in mass spectrometric analysis. Leading research teams performing total proteome analysis report a typical median sequence coverage of *ca.* 20% [96]. A contributory factor to this is the production of short peptides of ambiguous protein origin. This may be improved in MG-modified proteins where MG-H1 formation typically produces missed cleavages with trypsin and Lys-C and longer peptides. A recent computational approach has indicated that with judicious use of proteases the sequence coverage in proteomics analysis may increase to *ca.* 90% [97]. Until this is routinely implemented, it should be remembered that in glycation proteomics only a minor proportion of glycated proteins are likely to be identified.

For the fructosamine proteome, phenylboronate affinity chromatography was used to enrich glycated proteins and glycated tryptic peptides from both human plasma and erythrocyte membranes. Subsequent analysis by ETD-tandem mass spectrometry identified 76 and 31 proteins with fructosamine modification from human plasma and erythrocyte membranes, respectively. The ETD fragmentation mode permitted the identification of a significantly higher number of glycated peptides (88% of all identified peptides) versus CID mode (17% of all identified peptides) when utilizing enrichment on first the protein and then the peptide level [98].

In pilot studies using nanoflow liquid chromatography-Orbitrap Fusion^{TM} mass spectrometry with peptide HCD fragmentation, we analyzed cytosolic protein extracts of human aortic endothelial cells in primary culture. In control samples, cell cytosolic protein had a total MG-H1 residue content of *ca.* 0.4 mmol/mol Arg measured by LC-MS/MS analysis of exhaustive enzymatic digests, while in tryptic digests, Orbitrap Fusion analysis detected 1027 proteins, of which 12 contained MG-H1 or MG-derived dihydroxyimidazolidine residues. After incubation of cytosolic protein extracts with exogenous MG to increase the MG-H1 content *ca.* 20-fold, we then detected total MG-H1 residue content of *ca.* 8 mmol/mol Arg and identified 1366 proteins, of which 344 now contained MG-H1 or MG-derived dihydroxyimidazolidine residues [3]. In a recent report [92], plasma digests were analyzed by nanoflow chromatography-LTQ Orbitrap XL ETD mass spectrometry and tryptic peptides scanned for m/z 152.1 and 166.1 indicative of glyoxal- and MG-modified proteins. Forty-four peptides representing 42 proteins were annotated. Arginine modifications were mostly represented by glyoxal-derived hydroimidazolones (34 peptides/39 sites) and MG-derived dihydroxyimidazolidine (8 peptides/8 sites) and MG-H1 (14 peptides/14 sites). Use of high temperature and pH processing in this study may have compromised the outcome; many glyoxal-modified proteins were detected, whereas LC-MS/MS analysis typically shows very low amounts of glyoxal-derived-AGEs, hydroimidazolone, and $N\omega$-carboxymethylarginine in plasma protein [61].

8.4 Enrichment Strategies

Enrichment methods to facilitate detection and quantitation of PTM proteins are a common strategy in proteomics. A boronate affinity chromatography method has been used for the fructosamine proteome based on the binding of the *cis*-1,2-diol structure of fructosamine-modified proteins, with subsequent release from the boronate affinity matrix with weak acid. Although some enzymatically glycosylated proteins contain *cis*-1,2-diol moieties, steric effects, proximate negatively charged groups, and acetylation limit the retention and interference in this method by glycoproteins [98]. A similar affinity method is used in the routine separation of hemoglobin in clinical chemistry to quantify glycated hemoglobin HbA_{1c} for the assessment of glycemic control in diabetes [99]. Boronate affinity enrichment of protein glycated by glucose was employed in a study of glycated proteins in human plasma and red blood cells, and 7749 unique glycated peptides corresponding to 3742 unique glycated proteins were identified [94].

The dihydroxyimidazolidine residues present in proteins glycated by MG and glyoxal are also a potential interference in boronate affinity chromatography for the enrichment of proteins as they contain a side chain with a *cis*-1,2-diol moiety [100]. This has been exploited to identify proteins with arginine residues activated for reaction with glyoxal derivatives. Reaction of proteins with butane-2,3-dione formed 4,5-dihydroxy-4,5-dimethylimidazolidine residues of proteins containing activated arginine residues. Proteins with such residues on the surface were retained in boronate affinity chromatography [101, 102].

In principle antibodies to fructosamine and hydroimidazolones may be used for immunoaffinity purification and enrichment of proteins glycated by glucose and MG, respectively. Limited specificity of antifructosamine and anti-AGE antibodies, as indicated by poor performance in immunoassay [2], suggests that this is not currently a reliable enrichment strategy. Where immunoaffinity enrichment is employed, it is vital to confirm the presence of a glycation adduct residue in the retained proteins by mass spectrometric analysis.

8.5 Bioinformatics

Bioinformatics tools for protein glycation are, as yet, poorly developed. Mature developed bioinformatics tools will ideally predict sites of protein glycation within the proteome and in combination with experimental studies will assist in surety of location of sites of glycation adducts within proteins. Glycation is a nonenzymatic process, and so selectivity for sites of glycation is determined by the reactivity of the lysine and arginine of N-terminal residue under consideration. This is linked to (i) microscopic pK_a of the residue being modified, (ii) surface exposure of the modification site, and (iii) a proximate conjugate base catalyzing the dehydration step involved in dehydration steps for FL and MG-H1 residue formation (Figure 8.3).

Microscopic pK_a values of lysine, N-terminal, and arginine residues are an important determinant of glycation sites because the rate of glycation is determined by the basicity of the side chain residue and essentially the reaction proceeds through the usually minor proportion of the amino moieties of lysine side chain and N-terminal residues and guanidino side chain moieties of arginine. This has a profound influence on the site of glycation by glucose on N-terminal and lysine residues and on glycation by MG of arginine residues. Microscopic pK_a values may be computed for proteins of known crystal structure – for example, by using the H++ automated system (http://biophysics. cs.vt.edu/H++) [103]. There is marked diversity of pK_a values of Lys and Arg

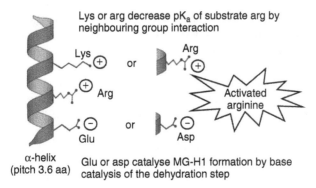

Lys or arg decrease pK_a of substrate arg by neighbouring group interaction

α-helix (pitch 3.6 aa)

Glu or asp catalyse MG-H1 formation by base catalysis of the dehydration step

Figure 8.3 Activation of arginine residues in α-helix domains of proteins by neighboring group interactions with basic and acidic amino acid residues. Source: Rabbani, 2010. [20]. Reproduced with permission from Springer.

residues in proteins. For example, in HSA, microscopic pK_a values of the 59 lysine residues vary from 7.9 to 14.0 and of the 24 arginine residues from 12.2 to 18.6, an expected reactivity range of $>10^6$ (cf. reactivity of N-terminal serine pK_a of 7.9) [104]. The major sites of glycation by glucose in HSA are, in order of reactivity, N-terminal D1, K525, K199, and K439 [33]. This compares to rank order by increasing pK_a value of lysine side chain and N-terminal amino groups of first-equal, sixteenth, third and fourteenth. Low pK_a values are likely driving glycation of D1 and K199. Activating features of K525 and K439 may be deprotonation catalyzed by proximate E520/R521 and E442, respectively. In a study of the hotspot sites of glycation of HSA by MG, three of the five sites with MG-H1 residue formation had the lowest microscopic predicted pK_a values: R218, $pK_a = 12.2$; and R186 and R410, $pK_a = 12.5$. However, the remaining two sites – R114 and R428 with predicted pK_a values of 13.6 and 15.1 – ranked 8th and 14th of 24 in order of increasing microscopic predicted pK_a value. R114 has high surface exposure, which likely also facilitates MG modification. All activated arginine residues have a positively charged R or K residue 3 or 4 residues, further along in the sequence decreasing microscopic pK_a value, and R428 only has a negatively charged residue, E425, preceding in the sequence. A subsequent study confirmed these hotspot sites except for R114 and suggested R257 as a further hotspot modification site, which has a relatively low pK_a (=12.9) [75]. The proximity of a negatively charged, D or E, residue provides a conjugate base to promote the rate-limiting removal of a proton from the protein-glucose Schiff's base and arginyl-dihydroxyimidazolidine precursors of fructosamine and MG-H1 adducts. The combination of proximate cationic and anion side chain residues for lysine and arginine residue activation was initially proposed to explain site specificity of lysine residue glycation by glucose [105] and then applied to MG-H1 formation from arginine [20].

The aforementioned considerations are features relating to the rate of formation of glycation adducts. FL and MG-H1 residues have half-lives of *ca.* 25 and 12 days, respectively [19, 106], which exceed the half-lives of most human proteins (median half-life of 1.9 days [107]). Therefore, for many proteins, the steady-state extent of protein glycation is also influenced by the half-life of the protein. Hence, early studies found that the extent of glycation by glucose of several proteins *in vivo* was linked to the protein half-life [108]. Since glycation leads to protein distortion and misfolding, it is also expected that glycated proteins are targeted for cellular proteolysis and have an unusually decreased half-life. This remains to be determined in robust unfocused proteome dynamics studies. The level of FL and N-terminal fructosamine residues in cellular proteins is also influenced by enzymatic removal and repair by F3PK [109]. F3PK has different specific activity for FL residues in different sites in proteins. The FL residues detected at different sites in proteins are, therefore, a balance of the intrinsic reactivity for glycation and the reactivity of the FL residue site for repair by FP3K – see glycated hemoglobin, for example [110]. There is no known enzymatic mechanism for repair of MG-H1 residues.

An examination of protein motifs for glucose glycation forming FL was made empirically by compiling and combining peptide motifs from published peptide mapping studies. It was found that K and R residues dominate in the N-terminal region and D and E residues dominate in the C-terminal region of FL sites, but no clear motif for FL formation was found [111]. A problem with this approach is inclusion of data where studies have produced highly glycated proteins *in vitro* with markedly greater than 2–3 molar equivalents of modification (which has been common in glycation research [85, 112]). Under these conditions, glycation occurs at many sites – both those favored and those less favored – since the glycation process is so highly driven by high glycated agent concentration. Information on site-specific and selective modification is thereby obscured. In a study of human plasma and red blood cells, detection and filtering for unique peptides with ≥5 spectrum counts gave 361 and 443 unique glycated peptide sequences from native human plasma and red blood cells, respectively. There was only limited evidence to support the hypothesis of N-terminal enrichment of K and R residues and C-terminal enrichment of D and E residues in the sequence motif for hotspot glycation by glucose. Overall, hydrophobic short-chain or uncharged side chain amino acids A, V, L, and S occurred most frequently close to the sites of glycation. E also was highly represented at some amino acid positions close to the glycation site. H, Y, M, W, C, and F were the least frequent amino acid residues close to the sites of glycation [94].

A further consideration for a nonenzymatic PTM such as glycation is to ask, when is the modification of functional importance to the protein substrate? We addressed this question by considering that glycation will have functional effect when it occurs in a domain where the protein interacts with other

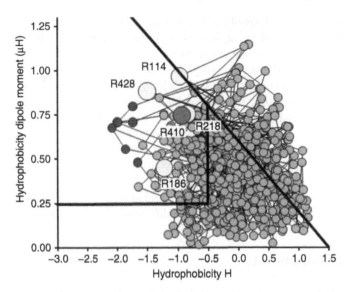

Figure 8.4 Receptor-binding domain (RBD) analysis. RBD plot of human serum albumin prepared by the method of Gallet *et al.* [47] showing sites of hotspot modification by MG. The RBD area is the top truncated trapezium area in the top left of the chart. It was computed with a sequence amino acid interval of 5. Key: gray circles, amino acid residues in human serum albumin; R410 hotspot of MG modification; and R114, R186, R218, and R428, also modified by MG. Other arginine residues within the RBD but are not modified by MG. This figure was originally published in color in Source: Rabbani 2014 [3], Reproduced with permission from Portland Press.

proteins, DNA, or substrates. A sequence-based predictor of such functional domains has been developed – the receptor-binding domain (RBD) analysis [47]. By such analysis, we were able to show that MG modifications of albumin occur at functional sites (Figure 8.4). We have applied the RBD analysis proteome-wide, and this may now be linked to glycation proteomics to identify glycated proteins that are expected to be dysfunctional.

8.6 Concluding Remarks

Protein glycation is a spontaneous PTM of the proteome focused mainly on N-terminal and lysine side chain amino groups by glucose forming fructosamine residues and on guanidino groups of arginine residues forming mainly MG-derived hydroimidazolone residues. The latter appears to be most functionally damaging in physiological systems. Glycation is a relatively labile PTM, and so customized preanalytical processing is required to avoid compromising mass spectrometric analysis outcomes. ETD fragmentation has a clear advantage for the analysis of the fructosamine proteome, whereas CID, HCD, and ETD may be used for arginine-based dihydroxyimidazolidine/hydroimidazolone detection. Robust procedures are available for the detection and quantitation of total glycation adducts in protein extracts, and these may be combined with proteomics analysis for added security of findings – particularly as it is still challenging to achieve high sequence coverage in proteomics experiments. LC-MS/MS quantitative analysis of glycated proteins by MRM of glycated tryptic peptides may provide valuable diagnostic markers where functional impairment is linked to disease, for example, proatherogenicity of MG-modified apolipoprotein B100 in MG-modified low-density lipoprotein [11]. Bioinformatics tools for the analysis of protein glycation sites are in development and will help surety of detection and identification of functionally and physiologically important proteome modifications by glycation.

Acknowledgements

We thank the Biotechnology and Biological Sciences Research Council, European Union FP7 program, and British Heart Foundation for the support for our glycation research.

References

1 Rabbani N, Thornalley PJ. Glycation research in amino acids: A place to call home. *Amino Acids* 2012;**42**:1087–1096.

2 Thornalley PJ, Rabbani N. Detection of oxidized and glycated proteins in clinical samples using mass spectrometry - A user's perspective. *Biochim Biophys Acta* 2014;**1840**:818–829.

3 Rabbani N, Thornalley PJ. Dicarbonyl proteome and genome damage in metabolic and vascular disease. *Biochem Soc Trans* 2014;**42**:425–432.

4 Goldstein DE, Malone JI, Little RR, Nathan D, Lorenz RA, Petersen CM. Tests for glycaemia in diabetes. *Diabetes Care* 1995;**18**:896–909.

5 American-Diabetes-Association. Classification and diagnosis of diabetes. *Diabetes Care* 2015;**38**:S8–S16.

6 Rabbani N, Thornalley PJ. Dicarbonyl stress in cell and tissue dysfunction contributing to ageing and disease. *Biochem Biophys Res Commun* 2015;**458**:221–226.

7 Ahmed N, Thornalley PJ. Advanced glycation endproducts: What is their relevance to diabetic complications? *Diabetes Obes Metab* 2007;**9**:233–245.

8 Agalou S, Ahmed N, Babaei-Jadidi R, Dawnay A, Thornalley PJ. Profound mishandling of protein glycation degradation products in uremia and dialysis. *J Am Soc Nephrol* 2005;**16**:1471–1485.

9 Genuth S, Sun W, Cleary P, Gao X, Sell DR, Lachin J, et al. Skin advanced glycation endproducts (AGEs) glucosepane and methylglyoxal hydroimidazolone are independently associated with long-term microvascular complication progression of type I diabetes. *Diabetes* 2015;**64**:266–278.

10 Hanssen NMJ, Beulens JWJ, van Dieren S, Scheijen JLJM, van der A DL, Spijkerman AMW, et al. Plasma advanced glycation end products are associated with incident cardiovascular events in individuals with type 2 diabetes: A case-cohort study with a median follow-up of 10 years (EPIC-NL). *Diabetes* 2015;**64**:257–265.

11 Rabbani N, Godfrey L, Xue M, Shaheen F, Geoffrion M, Milne R, et al. Conversion of low density lipoprotein to the pro-atherogenic form by methylglyoxal with increased arterial proteoglycan binding and aortal retention. *Diabetes* 2011;**60**:1973–1980.

12 Godfrey L, Yamada-Fowler N, Smith JA, Thornalley PJ, Rabbani N. Arginine-directed glycation and decreased HDL plasma concentration and functionality. *Nutr Diabetes* 2014;**4**:e134.

13 Ahmed N, Ahmed U, Thornalley PJ, Hager K, Fleischer GA, Munch G. Protein glycation, oxidation and nitration marker residues and free adducts of cerebrospinal fluid in Alzheimer's disease and link to cognitive impairment. *J Neurochem* 2004;**92**:255–263.

14 Ahmed N, Thornalley PJ, Luthen R, Haussinger D, Sebekova K, Schinzel R, et al. Processing of protein glycation, oxidation and nitrosation adducts in the liver and the effect of cirrhosis. *J Hepatol* 2004;**41**:913–919.

15 Kurz A, Rabbani N, Walter M, Bonin M, Thornalley PJ, Auburger G, et al. Alpha-synuclein deficiency leads to increased glyoxalase I expression and glycation stress. *Cell Mol Life Sci* 2011;**68**:721–733.

16 Dammann P, Sell DR, Begall S, Strauch C, Monnier VM. Advanced glycation end-products as markers of aging and longevity in the long-lived Ansell's mole-rat (*Fukomys anselli*). *J Gerontol Ser A Biol Sci Med Sci* 2012;**67**:573–583.

17 Uchiki T, Weikel KA, Jiao W, Shang F, Caceres A, Pawlak D, et al. Glycation-altered proteolysis as a pathobiologic mechanism that links dietary glycemic index, aging, and age-related disease (in nondiabetics). *Aging Cell* 2012;**11**:1–13.

18 Thornalley PJ, Battah S, Ahmed N, Karachalias N, Agalou S, Babaei-Jadidi R, et al. Quantitative screening of advanced glycation endproducts in cellular and extracellular proteins by tandem mass spectrometry. *Biochem J* 2003;**375**:581–592.

19 Ahmed N, Argirov OK, Minhas HS, Cordeiro CA, Thornalley PJ. Assay of advanced glycation endproducts (AGEs): Surveying AGEs by chromatographic assay with derivatisation by aminoquinolyl-N-hydroxysuccinimidyl-carbamate and application to Nε-carboxymethyl-lysine- and Nε-(1-carboxyethyl) lysine-modified albumin. *Biochem J* 2002;**364**:1–14.

20 Rabbani N, Thornalley PJ. Methylglyoxal, glyoxalase 1 and the dicarbonyl proteome. *Amino Acids* 2012;**42**:1133–1142.

21 Calvo C, Ulloa N, Campos M, Verdugo C, Ayrault-Jarrier M. The preferential site of non-enzymatic glycation of human apolipoprotein A-I *in vivo*. *Clin Chim Acta* 1993;**217**:193–198.

22 Shuvaev VV, Fujii J, Kawasaki Y, Itoh H, Hamaoka R, Barbier A, et al. Glycation of apolipoprotein E impairs its binding to heparin: Identification of the major glycation site. *Biochim Biophys Acta* 1999;**1454**:296–308.

23 Fujita T, Suzuki K, Tada T, Yoshihara Y, Hamaoka R, Uchida K, et al. Human erythrocyte bisphosphoglycerate mutase: Inactivation by glycation *in vivo* and *in vitro*. *J Biochem* 1998;**124**:1237–1244.

24 Acosta J, Hettinga J, Flückiger R, Krumrei N, Goldfine A, Angarita L, et al. Molecular basis for a link between complement and the vascular complications of diabetes. *Proc Natl Acad Sci U S A* 2000;**97**:5450–5455.

25 Niemann MA, Bhown AS, Miller EJ. The principal site of glycation of human complement factor B. *Biochemical J* 1991;**274**:473–480.

26 O'Harte FPM, Abdel-Wahab YHA, Conlon JM, Flatt PR. Amino terminal glycation of gastric inhibitory polypeptide enhances its insulinotropic action on clonal pancreatic B-cells. *Biochim Biophys Acta* 1998;**1425**:319–327.

27 O'Harte FPM, Abdel-Wahab YHA, Conlon JM, Flatt PR. Glycation of glucagon-like peptide-1(7–36)amide: Characterization and impaired action on rat insulin secreting cells. *Diabetologia* 1998;**41**:1187–1193.

28 Zhang X, Medzihradszky KF, Cunningham J, Lee PDK, Rognerud CL, Ou CN, et al. Characterization of glycated hemoglobin in diabetic patients: Usefulness of electrospray mass spectrometry in monitoring the extent and distribution of glycation. *J Chromatogr B* 2001;**759**:1–15.

29 Coletta M, Amiconi G, Bellelli A, Bertollini A, Čarsky J, Castagnola M, et al. Alteration of T-state binding properties of naturally glycated hemoglobin, HbA1c. *J Mol Biol* 1988;**203**:233–239.

30 Wang SH, Wang TF, Wu CH, Chen SH. In-depth comparative characterization of hemoglobin glycation in normal and diabetic bloods by LC-MSMS. *J Am Soc Mass Spectrom* 2014;**25**:758–766.

31 O'Harte FPM, Højrup P, Barnett CR, Flatt PR. Identification of the site of glycation of human insulin. *Peptides* 1996;**17**:1323–1330.

32 Miyata T, Inagi R, Wada Y, Ueda Y, Iida Y, Takahashi M, et al. Glycation of human á 2 -microglobulin in patients with hemodialysis-associated amyloidosis: Identification of the glycated sites. *Biochemistry* 1994;**33**:12215–12221.

33 Barnaby OS, Cerny RL, Clarke W, Hage DS. Quantitative analysis of glycation patterns in human serum albumin using 16O/18O-labeling and MALDI–TOF MS. *Clin Chim Acta* 2011;**412**:1606–1615.

34 Rabbani N, AntonySunil A, Rossing K, Rossing P, Tarnow L, Parving HH, et al. Effect of Irbesartan treatment on plasma and urinary protein glycation, oxidation and nitration markers in patients with type 2 diabetes and microalbuminuria. *Amino Acids* 2011;**42**:1627–1639.

35 Arai K, Maguchi S, Fujii S, Ishibashi H, Oikawa K, Taniguchi N. Glycation and inactivation of human Cu-Zn-superoxide dismutase - identification of the *in vitro* glycated sites. *J Biol Chem* 1987;**262**:16969–16972.

36 Abraham EC, Cherian M, Smith JB. Site selectivity in the glycation of αA-crystallin and αB-crystallins by glucose. *Biochem Biophys Res Commun* 1994;**201**:1451–1456.

37 Casey EB, Zhao HR, Abraham EC. Role of glycine 1 and lysine 2 in the glycation of bovine g B-crystallin. *J Biol Chem* 1995;**270**:20781–20786.

38 Baldwin JS, Lee L, Leung TK, Muruganandam A, Mutus B. Identification of the site of nonenzymatic glycation of glutathione-peroxidase - rationalization of the glycation-related catalytic alterations on the basis of 3-dimensional protein-structure. *Biochim Biophys Acta* 1995;**1247**:60–64.

39 Guedes S, Vitorino R, Domingues MRM, Amado F, Domingues P. Mass spectrometry characterization of the glycation sites of bovine insulin by tandem mass spectrometry. *J Am Soc Mass Spectrom* 2009;**20**:1319–1326.

40 SwamyMruthinti S, Schey KL. Mass spectroscopic identification of *in vitro* glycated sites of MIP. *Curr Eye Res* 1997;**16**:936–941.

41 Watkins NG, Thorpe SR, Baynes JW. Glycation of amino groups in protein. (Studies on the specificity of modification of RNase by glucose). *J Biol Chem* 1985;**260**:10629–10636.

42 Frolov A, Hoffmann P, Hoffmann R. Fragmentation behavior of glycated peptides derived from D-glucose, D-fructose and D-ribose in tandem mass spectrometry. *J Mass Spectrom* 2006;**41**:1459–1469.

43 Reiser KM, Amigable M, Last JA. Nonenzymatic glycation of type I collagen (the effects of aging on preferential glycation sites. *J Biol Chem* 1992;**267**:24207–24216.

44 Takahashi M, Lu Y-b, Myint T, Fujii J, Wad aY, Taniguchi N. *In vivo* glycation of aldehyde reductase, a major 3-deoxyglucosone reducing enzyme: Identification of glycation sites. *Biochemistry* 1995;**34**:1433–1438.

45 Thornalley PJ, Langborg A, Minhas HS. Formation of glyoxal, methylglyoxal and 3-deoxyglucosone in the glycation of proteins by glucose. *Biochem J* 1999;**344**:109–116.

46 Smith PR, Thornalley PJ. Influence of pH and phosphate ions on the kinetics of enolisation and degradation of fructosamines. Studies with the model fructosamine, Nε-1-deoxy-D-fructos-1-yl hippuryllysine. *Biochem Int* 1992;**28**:429–439.

47 Gallet X, Charloteaux B, Thomas A, Braseur R. A fast method to predict protein interaction sites from sequences. *J Mol Biol* 2000;**302**:917–926.

48 Veiga-Da-Cunha M, Jacquemin P, Delpierre G, Godfraind C, Theate I, Vertommen D, et al. Increased protein glycation in fructosamine 3-kinase-deficient mice. *Biochem J* 2006;**399**:257–264.

49 Wells-Knecht MC, Thorpe SR, Baynes JW. Pathways of formation of glycoxidation products during glycation of collagen. *Biochemistry* 1995;**34**:15134–15141.

50 Anderson MM, Requena JR, Crowley JR, Thorpe SR, Heinecke JW. The myeloperoxidase system of human phagocytes generates Nε-(carboxymethyl) lysine on proteins: A mechanism for producing advanced glycation end products at sites of inflammation. *J Clin Invest* 1999;**104**:103–113.

51 Ahmed N, Thornalley PJ, Dawczynski J, Franke S, Strobel J, Stein G, et al. Methylglyoxal-derived hydroimidazolone advanced glycation endproducts of human lens proteins. *Invest Ophthalmol Visual Sci* 2003;**44**:5287–5292.

52 Lo TWC, Selwood T, Thornalley PJ. Reaction of methylglyoxal with aminoguanidine under physiological conditions and prevention of methylglyoxal binding to plasma proteins. *Biochem Pharmacol* 1994;**48**:1865–1870.

53 Ahmed N, Thornalley PJ. Chromatographic assay of glycation adducts in human serum albumin glycated *in vitro* by derivatisation with aminoquinolyl-N-hydroxysuccinimidyl-carbamate and intrinsic fluorescence. *Biochem J* 2002;**364**:15–24.

54 Hager-Braun C, Tomer KB. Characterization of the tertiary structure of soluble CD4 bound to glycosylated full-length HIVgp120 by chemical modification of arginine residues and mass spectrometric analysis. *Biochemistry* 2002;**41**:1759–1766.

55 Dobler D, Ahmed N, Song LJ, Eboigbodin KE, Thornalley PJ. Increased dicarbonyl metabolism in endothelial cells in hyperglycemia induces anoikis and impairs angiogenesis by RGD and GFOGER motif modification. *Diabetes* 2006;**55**:1961–1969.

56 Pedchenko VK, Chetyrkin SV, Chuang P, Ham AJ, Saleem MA, Mathieson PW, et al. Mechanism of perturbation of integrin-mediated cell-matrix interactions by reactive carbonyl compounds and its implication for pathogenesis of diabetic nephropathy. *Diabetes* 2005;**54**:2952–2960.

57 Gangadhariah MH, Wang BL, Linetsky M, Henning C, Spanneberg R, Glomb MA, et al. Hydroimidazolone modification of human alpha A-crystallin: Effect on the chaperone function and protein refolding ability. *Biochim Biophys Acta* 2010;**1802**:432–441.

58 Kiselar JG, Wang X, Dubyak GR, El Sanadi C, Ghosh SK, Lundberg K, et al. Modification of β-defensin-2 by dicarbonyls methylglyoxal and glyoxal inhibits antibacterial and chemotactic function *in vitro*. *PLoS One* 2015;**10**:e0130533.

59 Lund T, Svindland A, Pepaj M, Jensen AB, Berg JP, Kilhovd B, et al. Fibrin(ogen) may be an important target for methylglyoxal-derived AGE modification in elastic arteries of humans. *Diab Vasc Dis Res* 2011;**8**:284–294.

60 Oya-Ito T, Naito Y, Takagi T, Handa O, Matsui H, Yamada M, et al. Heat-shock protein 27 (Hsp27) as a target of methylglyoxal in gastrointestinal cancer. *Biochim Biophys Acta* 2011;**1812**:769–781.

61 Ahmed N, Babaei-Jadidi R, Howell SK, Beisswenger PJ, Thornalley PJ. Degradation products of proteins damaged by glycation, oxidation and nitration in clinical type 1 diabetes. *Diabetologia* 2005;**48**:1590–1603.

62 Chen Y, Ahmed N, Thornalley PJ. Peptide mapping of human hemoglobin modified minimally by methylglyoxal *in vitro. Ann N Y Acad Sci* 2005;**1043**:905.

63 Gao Y, Wang Y. Site-selective modifications of arginine residues in human hemoglobin induced by methylglyoxal. *Biochemistry* 2006;**45**:15654–15660.

64 Thangarajah H, Yao DC, Chang EI, Shi YB, Jazayeri L, Vial IN, et al. The molecular basis for impaired hypoxia-induced VEGF expression in diabetic tissues. *Proc Natl Acad Sci U S A* 2009;**106**:13505–13510.

65 Akinsiku OT, Yu ET, Fabris D. Mass spectrometric investigation of protein alkylation by the RNA footprinting probe kethoxal. *J Mass Spectrom* 2005;**40**:1372–1381.

66 Carven GJ, Stern LJ. Probing the ligand-induced conformational change in HLA-DR1 by selective chemical modification and mass spectrometric mapping. *Biochemistry* 2005;**44**:13625–13637.

67 Oliveira L, Lages A, Gomes R, Neves H, Familia C, Coelho A, et al. Insulin glycation by methylglyoxal results in native-like aggregation and inhibition of fibril formation. *BMC Biochem* 2011;**12**:41.

68 Chumsae C, Gifford K, Lian W, Liu H, Radziejewski CH, Zhou ZS. Arginine modifications by methylglyoxal: Discovery in a recombinant monoclonal antibody and contribution to acidic species. *Anal Chem* 2013;**85**:11401–11409.

69 Kinsky, O.R., Dicarbonyl protein adduction: Plasminogen as a target and metformin as a scavenging therapeutic in type 2 diabetes. PhD thesis. University of Arizona, Tucson, USA (2014)

70 Queisser MA, Yao D, Geisler S, Hammes HP, Lochnit G, Schleicher ED, et al. Hyperglycemia impairs proteasome function by methylglyoxal. *Diabetes* 2010;**59**:670–678.

71 Brock JWC, Cotham WE, Thorpe SR, Baynes JW, Ames JM. Detection and identification of arginine modifications on methylglyoxal-modified ribonuclease by mass spectrometric analysis. *J Mass Spectrom* 2007;**42**:89–100.

72 Cotham WE, Metz TO, Ferguson PL, Brock JWC, Hinton DJS, Thorpe SR, et al. Proteomic analysis of arginine adducts on glyoxal-modified ribonuclease. *Mol Cell Proteomics* 2004;**3**:1145–1153.

73 Calvete JJ, Campanero-Rhodes MA, Raida M, Sanz L. Characterisation of the conformational and quaternary structure-dependent heparin-binding region of bovine seminal plasma protein PDC-109. *FEBS Lett* 1999;**444**:260–264.

74 Ahmed N, Dobler D, Dean M, Thornalley PJ. Peptide mapping identifies hotspot site of modification in human serum albumin by methylglyoxal involved in ligand binding and esterase activity. *J.Biol Chem* 2005;**280**:5724–5732.

75 Kimzey MJ, Yassine HN, Riepel BM, Tsaprailis G, Monks TJ, Lau SS. New site(s) of methylglyoxal-modified human serum albumin, identified by

multiple reaction monitoring, alter warfarin binding and prostaglandin metabolism. *Chem Biol Interact* 2011;**192**:122–128.

76 Wood TD, Guan Z, Borders CL, Chen LH, Kenyon GL, McLafferty FW. Creatine kinase: Essential arginine residues at the nucleotide binding site identified by chemical modification and high-resolution tandem mass spectrometry. *Proc Natl Acad Sci U S A* 1998;**95**:3362–3365.

77 Yao D, Taguchi T, Matsumura T, Pestell R, Edelstein D, Giardino I, et al. High glucose increases angiopoietin-2 transcription in microvascular endothelial cells through methylglyoxal modification of mSin3A. *J Biol Chem* 2007;**282**:31038–31045.

78 Suckau D, Mak M, Przybylski M. Protein surface topology-probing by selective chemical modification and mass spectrometric peptide mapping. *Proc Natl Acad Sci U S A* 1992;**89**:5630–5634.

79 Krell T, Pitt AR, Coggins JR. The use of electrospray mass spectrometry to identify an essential arginine residue in type II dehydroquinases. *FEBS Lett* 1995;**360**:93–96.

80 Wu X, Chen SG, Petrash JM, Monnier VM. Alteration of substrate selectivity through mutation of two arginine residues in the binding site of amadoriase II from *Aspergillus* sp. *Biochemistry* 2002;**41**:4453–4458.

81 Schepens I, Ruelland E, Miginiac-Maslow M, Le Maréchal P, Decottignies P. The role of active site arginines of sorghum NADP-malate dehydrogenase in thioredoxin-dependent activation and activity. *J Biol Chem* 2000;**275**:35792–35798.

82 Iacob RE, Keck Z, Olson O, Foung SKH, Tomer KB. Structural elucidation of critical residues involved in binding of human monoclonal antibodies to hepatitis C virus E2 envelope glycoprotein. *Biochim Biophys Acta* 2008;**1784**:530–542.

83 Rabbani N, Shaheen F, Anwar A, Masania J, Thornalley PJ. Assay of methylglyoxal-derived protein and nucleotide AGEs. *Biochem Soc Trans* 2014;**42**:511–517.

84 Roberts NB, Amara AB, Morris M, Green BN. Long-term evaluation of electrospray ionization mass spectrometric analysis of glycated hemoglobin. *Clin Chem* 2001;**47**:316–321.

85 Thornalley PJ, Argirova M, Ahmed N, Mann VM, Argirov OK, Dawnay A. Mass spectrometric monitoring of albumin in uraemia. *Kidney Int* 2000;**58**:2228–2234.

86 Jerić I, Versluis C, Horvat Š, Heck AJR. Tracing glycoprotein structures: Electrospray ionization tandem mass spectrometric analysis of sugar–peptide adducts. *J Mass Spectrom* 2002;**37**:803–811.

87 Horvat Š, Jakas A. Peptide and amino acid glycation: New insights into the Maillard reaction. *J Pept Sci* 2004;**10**:119–137.

88 Mennella C, Visciano M, Napolitano A, Del Castillo MD, Fogliano V. Glycation of lysine-containing dipeptides. *J Pept Sci* 2006;**12**:291–296.

89 Zhang QB, Frolov A, Tang N, Hoffmann R, van de Goor T, Metz TO, et al. Application of electron transfer dissociation mass spectrometry in analyses of non-enzymatically glycated peptides. *Rapid Commun Mass Spectrom* 2007;**21**:661–666.

90 Cai J, Hurst HE. Identification and quantitation of N-(carboxymethyl)valine adducts in hemoglobin by gas chromatography/mass spectrometry. *J. Mass Spectrom.* 1999;**34**:537–543.

91 Brock JWC, Hinton DJS, Cotham WE, Metz TO, Thorpe SR, Baynes JW, et al. Proteomic analysis of the site specificity of glycation and carboxymethylation of ribonuclease. *J Proteome Res* 2003;**2**:506–513.

92 Schmidt R, Böhme D, Singer D, Frolov A. Specific tandem mass spectrometric detection of AGE-modified arginine residues in peptides. *J Mass Spectrom* 2015;**50**:613–624.

93 Rabbani N, Thornalley PJ. Measurement of methylglyoxal by stable isotopic dilution analysis LC-MS/MS with corroborative prediction in physiological samples. *Nat Protoc* 2014;**9**:1969–1979.

94 Zhang Q, Monroe ME, Schepmoes AA, Clauss TRW, Gritsenko MA, Meng D, et al. Comprehensive identification of glycated peptides and their glycation motifs in plasma and erythrocytes of control and diabetic subjects. *J Proteome Res* 2011;**10**:3076–3088.

95 Karachalias N, Babaei-Jadidi R, Rabbani N, Thornalley PJ. Increased protein damage in renal glomeruli, retina, nerve, plasma and urine and its prevention by thiamine and benfotiamine therapy in a rat model of diabetes. *Diabetologia* 2010;**53**:1506–1516.

96 Nagaraj N, Kulak NA, Cox J, Neuhauser N, Mayr K, Hoerning O, et al. System-wide perturbation analysis with nearly complete coverage of the yeast proteome by single-shot ultra HPLC runs on a bench Top orbitrap. *Mol Cell Proteomics* 2012;**11**:11.

97 Meyer JG. In silico proteome cleavage reveals iterative digestion strategy for high sequence coverage. *ISRN Comput Biol* 2014;**2014**:7.

98 Zhang Q, Tang N, Brock JWC, Mottaz HM, Ames JM, Baynes JW, et al. Enrichment and analysis of nonenzymatically glycated peptides: Boronate affinity chromatography coupled with electron-transfer dissociation mass spectrometry. *J Proteome Res* 2007;**6**:2323–2330.

99 Goodall I. HbA1c standardisation destination--global IFCC Standardisation. How, why, where and when--a tortuous pathway from kit manufacturers, via inter-laboratory lyophilized and whole blood comparisons to designated national comparison schemes. *Clin Biochem Rev* 2005;**26**:5–19.

100 Lo TWC, Westwood ME, McLellan AC, Selwood T, Thornalley PJ. Binding and modification of proteins by methylglyoxal under physiological conditions. A kinetic and mechanistic study with Nα-acetylarginine, Nα-acetylcysteine, Nα-acetyl-lysine, and bovine serum albumin. *J Biol Chem* 1994;**269**:32299–32305.

101 Leitner A, Lindner W. Functional probing of arginine residues in proteins using mass spectrometry and an arginine-specific covalent tagging concept. *Anal Chem* 2005;**77**:4481–4488.

102 Leitner A, Amon S, Rizzi A, Lindner W. Use of the arginine-specific butanedione/phenylboronic acid tag for analysis of peptides and protein digests using matrix-assisted laser desorption/ionization mass spectrometry. *Rapid Commun Mass Spectrom* 2007;**21**:1321–1330.

103 Anandakrishnan R, Aguilar B, Onufriev AV. H++ 3.0: Automating pK prediction and the preparation of biomolecular structures for atomistic molecular modeling and simulations. *Nucleic Acids Res* 2012;**40**:W537–W541.

104 Harris R, Patel SU, Sadler PJ, Viles JH. Observation of albumin resonances in proton nuclear magnetic resonance spectra of human blood plasma: N-terminal assignments aided by use of modified recombinant albumin. *Analyst* 1996;**121**:913–922.

105 Venkatraman J, Aggarwal K, Balaram P. Helical peptide models for protein glycation: Proximity effects in catalysis of the Amadori rearrangement. *Chem Biol* 2001;**8**:611–625.

106 Smith PR, Thornalley PJ. Mechanism of the degradation of non-enzymatically glycated proteins under physiological conditions. (Studies with the model fructosamine, N-(1-deoxy-D-fructose-1-yl)hippuryl-lysine). *Eur J Biochem* 1992;**210**:729–739.

107 Schwanhausser B, Busse D, Li N, Dittmar G, Schuchhardt J, Wolf J, et al. Global quantification of mammalian gene expression control. *Nature* 2011;**473**:337–342.

108 Schleicher E, Wieland OH. Kinetic analysis of glycation as a tool for assessing the half-life of proteins. *Biochim Biophys Acta* 1986;**884**:199–205.

109 Delpierre G, Rider MH, Collard F, Stroobant V, Vanstapel F, Santos H, et al. Identification, cloning, and heterologous expression of a mammalian fructosamine-3-kinase. *Diabetes* 2000;**49**:1627–1634.

110 Delpierre G, Vertommen D, Communi D, Rider MH, Van Schaftingen E. Identification of fructosamine residues deglycated by fructosamine-3-kinase in human hemoglobin. *J Biol Chem* 2004;**279**:27613–27620.

111 Johansen MB, Kiemer L, Brunak S. Analysis and prediction of mammalian protein glycation. *Glycobiology* 2006;**16**:844–853.

112 Westwood ME, Thornalley PJ. Molecular characteristics of methylglyoxal-modified bovine and human serum albumins. Comparison with glucose-derived advanced glycation endproduct-modified serum albumins. *J Prot Chem* 1995;**14**:359–372.

9

Biological Significance and Analysis of Tyrosine Sulfation

Éva Klement[1], Éva Hunyadi-Gulyás[1] and Katalin F. Medzihradszky[1,2]

[1] *Laboratory of Proteomics Research, Institute of Biochemistry, Biological Research Centre of the Hungarian Academy of Sciences, Szeged, Hungary*
[2] *Department of Pharmaceutical Chemistry, University of California San Francisco, San Francisco, USA*

9.1 Overview of Protein Sulfation

Classical analysis of proteins and peptides using a wide array of enzymatic digestions, simple sample fractionation methods, and organic chemistry reactions yielded quite sophisticated results: not only the amino acid sequences were deciphered but also a wide variety of post-translational modifications (PTMs) were identified. O-sulfation of tyrosine residues is one of these PTMs. (Note: Sometimes, it is mistakenly referred to as sulfonation although that indicates a carbon–sulfur linkage.) The presence of sulfated Tyr in different proteins and polypeptides has been reported from the 1950s [1–3]. It has been demonstrated that the modification can be removed by acid hydrolysis [1] and was later shown that arylsulfatases recognize and remove the sulfate linked to Tyr residues [4]. With the introduction of [35S]-labeling, gradually it became obvious that the modification is linked to the secretory pathway and the derivatization must occur in the *trans*-Golgi network [5–8]. The first proteomics-style Tyr-sulfation experiment was also performed using radioactive labeling and classical analytical techniques resulting in the detection of fibrinogen, α-fetoprotein, and fibronectin with sulfated Tyr residues [9].

Approximately 30 years after the first reports on Tyr sulfation, the two protein tyrosine sulfotransferase (PTST) enzymes performing the modification, using 3′-phosphoadenosine-5′-phosphosulfate (PAPS) as the sulfo-donor, were identified [10, 11]. Tyr sulfotransferases have also been described in plants [12, 13] and, recently, in a Gram-negative bacterium [14].

Numerous articles attest to the biological importance of Tyr sulfation: it has been shown that the sulfation of Factor V and Factor VIII is important for

Analysis of Protein Post-Translational Modifications by Mass Spectrometry,
First Edition. Edited by John R. Griffiths and Richard D. Unwin.
© 2017 John Wiley & Sons, Inc. Published 2017 by John Wiley & Sons, Inc.

optimal activity and thrombin cleavage [15–17]; sulfation has been implicated in protein–protein interaction [18–21]; and modified Tyr residues play important role(s) in ligand–receptor interactions [22–25]. Tyrosine sulfation modulates the activity of cytokine receptors [26], facilitates HIV entry in the cell [27, 28], and participates in both autoimmune processes and inflammatory responses [29].

Besides the modification of Tyr residues, the sulfation of Ser, Thr, and Cys side chains has also been reported [30–32]. The origin and biological role of these modifications will not be discussed in this chapter; however, it should be noted that they display the same chromatographic and mass spectrometry behavior as their Tyr-modified counterparts.

In addition, we would like to draw attention to the fact that the sulfation of the aforementioned residues, including Tyr, can be achieved by exposing the proteins/digests to certain chemicals, for example, thiosulfates, during sample preparation. Artifactual peptide sulfation during silver staining has been reported [33].

While working on this manuscript we extracted all human UniProt entries that were assigned as sulfated. The search yielded only 50 entries. All assignments based on "sequence analysis," "similarity" or "curated," were removed. Supporting references were scrutinized, and as a result reliable human sulfation data included in the UniProt database as of May 2015 are listed in Table 9.1.

More than half of these sites were identified by mutation and radioactive labeling. Mass spectrometry analyses were performed on purified proteins, and frequently a combination of different tools had to be applied. Perhaps the best example for a dedicated study is the sulfation analysis of extracellular leucine-rich repeat proteins when different proteolytic enzymes, MALDI and ESI, positive and negative ionization were used [40].

9.2 Mass Spectrometry Behavior of Sulfated Peptides

Tyrosine sulfation introduces a strongly acidic modification to the peptide. One would expect that this modification negatively influences the mass spectrometry response in positive ion mode and lowers the number of charges deposited on the modified peptide by electrospray ionization. From this angle, sulfopeptides resemble phosphopeptides. In addition, both modifications increase the peptide mass by nominally 80 Da: the mass difference between these isobaric structures is only 9 mmu (the exact additive masses of the modifications are 79.9663 Da and 79.9568 Da for phosphate and sulfate, respectively). What makes the sulfopeptides unique is that this modification is more fragile than phosphorylation, and a partial neutral loss of SO_3 is regularly observed upon mass measurement of the intact peptide (Figure 9.1, inset),

Table 9.1 Reliably assigned human sulfo-Tyr sites from UniProt May 2015.

Entry	Gene names	Position of the modification	References
P08697	SERPINF2 AAP PLI	484	[34]
P19021	PAM	961	[35]
Q16581	C3AR1 AZ3B C3R1 HNFAG09	174, 184, 318	[36]
P21730	C5AR1 C5AR C5R1	11, 14	[22]
P06307	CCK	97	[3]
P51681	CCR5 CMKBR5	3	[27]
P0C0L4	C4A CO4 CPAMD2	1417, 1420, 1422	[37]
P0C0L5	C4B CO4 CPAMD3; C4B_2	1417, 1420, 1422	[37]
P49682	CXCR3 GPR9	27, 29	[25]
P61073	CXCR4	7, 12, 21	[28]
P00451	F8 F8C	365, 737, 738, 742, 1683, 1699	[16, 38]
P02679	FGG PRO2061	444, 448	[39]
Q06828	FMOD FM SLRR2E	20, 38, 39, 45, 47, 53, 55	[40]
P01350	GAST GAS	87, partial	[2]
P07359	GP1BA	292, 294, 295	[41]
P15515	HTN1 HIS1	46, 49, 53, 55	[42]
P51884	LUM LDC SLRR2D	20, 21, 23, 30	[43]
Q99983	OMD SLRR2C UNQ190/PRO216	22, 25, 31, 39, 51, 58, 77	[40]
Q8NBP7	PCSK9 NARC1 PSEC0052	38	[44]
Q9NZ53	PODXL2 UNQ1861/PRO3742	97, 118	[24]
Q14242	SELPLG	46, 48, 51	[45]
P21815	IBSP BNSP	313 or 314	[46]
P07477	PRSS1 TRP1 TRY1 TRYP1	154	[47]
P07478	PRSS2 TRY2 TRYP2	154	[47]
P04004	VTN	75, 78, 282, 417, 420	[43]

while gas-phase desulfation is 100% upon collisional activation (Figure 9.1). This fragmentation step is favored so much that in ion trap collision-induced dissociation (CID), that is, under resonance-activation conditions, usually the only fragment observed is the product of desulfation (data not shown).

Mass spectrometry analysis of a sulfopeptide has been described as early as 1987: an α2-antiplasmin peptide was analyzed using fast atom bombardment

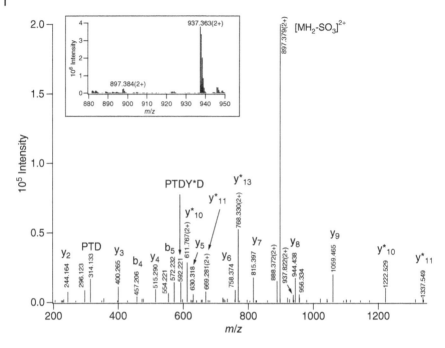

Figure 9.1 Beam-type CID (higher-energy C-trap dissociation (HCD) in the Thermo nomenclature) spectrum of *m/z* 937.3632(2+) corresponding to <QFPTDY(sulfo) DEGQDDRPK of bovine fibrinopeptide B. <Q stands for pyroglutamic acid. The inset shows the MS survey data, indicating *some* gas-phase desulfation. CID fragments labeled with an asterisk underwent gas-phase desulfation. These data were acquired on an LTQ-Orbitrap Velos instrument in LC/MS/MS analysis of a fraction from a bovine serum tryptic digest. The temperature of the heated capillary was 160 °C. The instrument was tuned as usual, acquisition parameters were not altered for the analysis of sulfopeptides. The normalized collision energy for the HCD data acquisition was set to 35%.

(FAB) ionization in positive ion mode, and the detection of the intact peptide, its sodium adduct, and the sulfate loss yielding an abundant unmodified peptide ion have been reported [34]. Later, it became obvious that gas-phase sulfate loss (−80 Da) is a characteristic feature of sulfopeptides: in positive mode, the modification may completely disappear, but desulfation may occur even in negative ion mode [48]. The extent of this gas-phase desulfation is influenced by the ionization method: under normal MALDI conditions, sulfopeptides require analysis to be performed in negative mode, while the gentler atmospheric pressure MALDI permits sulfopeptide detection even in positive mode [49]. We postulate that the same results could be achieved with regular MALDI-TOF instruments using 2,6-dihydroxyacetophenone as the matrix [50] as for equally fragile phospho-His-containing peptides [51]. Larger polypeptides may retain the modification in positive linear mode even in conventional matrices [42].

Unfortunately, even the gentler electrospray ionization may trigger abundant SO_3 losses [31, 38, 49]. MS acquisition conditions have to be adjusted to prevent/minimize this phenomenon in the mass measurement experiments. For example, lowering the temperature of the heated capillary in certain mass spectrometers significantly eliminates gas-phase desulfation and, thus, increases the sulfopeptide signal [38]. Unfortunately, while the sulfate loss may be minimized in the MS surveys, collisional activation leads to complete elimination of the modification. Thus, the peptide fragments allow the amino acid sequence assignment; however, the site localization is impossible from CID data [31, 38]. At the same time, Wolfender et al. presented high-energy CID data in negative mode where the diagnostic fragment ions for sulfation (m/z 80) and for sulfo-Tyr (m/z 214) were detected [49].

An obvious solution for assigning fragile modifications to precise amino acids is the application of electron-transfer dissociation (ETD). Indeed, Mikesh et al. demonstrated that the method may work for sulfopeptides, albeit the synthetic peptide analyzed was modified on a Ser residue [52]. However, more frequently than not, sulfation works against forming a precursor ion of sufficient charge density. One can expect that most tryptic sulfopeptides will be detected as doubly charged ions. Doubly charged precursor ions may produce sufficient information for modification site assignment as illustrated by Figure 9.2. At the same time, in most cases – unless a peptide features "reasonable" charge density – the resulting ETD spectra will not be informative enough even for peptide identification [53], with longer or multiply modified sulfopeptides most likely fitting this category.

Since sulfopeptides form negative ions with relatively high efficiency, especially with multiple modifications, some research groups turned to this direction when searching for alternative solutions. Hersberger and Håkansson described sulfopeptide fragmentation in negative ion CID, electron detachment dissociation (EDD), negative ETD (NETD), and negative ion mode ECD (niECD) [54]. In EDD, radical, charge-reduced ions are formed using a high-energy electron beam hitting multiply charged negative ions. Fragmentation occurs along the peptide backbone, between the α-carbons and the carbonyl groups, resulting in a• and x-ion formation, and, crucially, leaving the side chains intact. In the negative variation of ETD (NETD), the multiply charged anion transfers an electron to an acceptor molecule, eventually leading to the same type of fragmentation as in EDD. These methods suffer from the same problem as ETD: producing high charge density negative ions is more difficult than achieving the same in positive mode. Thus, these methods do not offer real solution. On the other hand, niECD produces *charge-increased* precursor ions via triggering electron capture by irradiating peptide anions with lower-energy electrons. In these multiply charged peptide radicals, the fragmentation occurs between the amino group and the α-carbon, producing primarily c′ and z• ions, while retaining the sulfate modification (Figure 9.3). While this method may yield more extensive sequence cover-

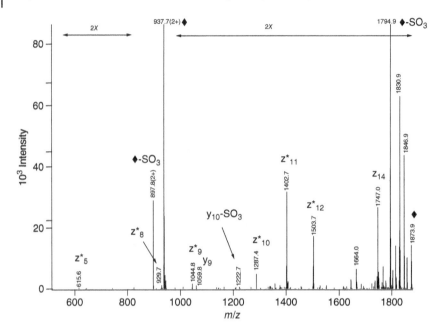

Figure 9.2 ETD spectrum of m/z 937.3632(2+) corresponding to <QFPTDY(sulfo) DEGQDDRPK of bovine fibrinopeptide B. <Q stands for pyroglutamic acid. ◆ indicates the precursor selected or its charge-reduced state. Asterisks indicate z + 1 fragments frequently formed from doubly charged precursor ions. This spectrum was acquired following the HCD acquisition shown in Figure 9.1. The ETD analysis was performed in the linear trap with an isolation window of 3 Da, a trigger intensity of 2000, and an activation time of 100 ms. One microscan was taken, and the AGC setting was 10^4. The supplemental collision energy was set at 15%.

age for sulfopeptides than ETD, due to its low sensitivity and efficiency, its utility as a truly viable alternative is questionable.

Using another alternative technique, Robinson et al. demonstrated that 193 nm ultraviolet photodissociation (UVPD) generates an informative MS/MS spectrum from sulfopeptide anions. The results are similar to ETD fragmentation in that while sulfate loss is regularly detected from the precursor and charge-reduced precursor ions, the fragments usually retain the modification (Figure 9.4) [55]. Unfortunately, this method is not sensitive enough for routine LC–MS/MS analysis either.

Chemical derivatization has been used for improving/controlling MS/MS fragmentation as well as for enabling the site assignment of other elusive PTMs. Yu et al. used a clever "trick" for finding sulfated residues in purified proteins: the unmodified Tyr residues were acetylated using sulfosuccinimidyl acetate (S-NHSAc) in the presence of imidazole at pH 7.0 [43]. This way even

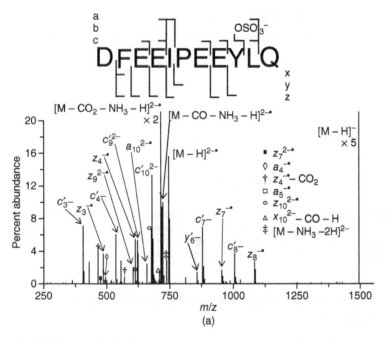

Figure 9.3 NiECD data of sulfo-hirudin, the precursor ion was singly charged. Source: Hersberger, 2012. [54]. Reproduced with permission from American Chemical Society.

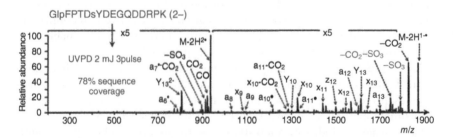

Figure 9.4 Negative UVPD mass spectrum (three pulses at 2 mJ) of bovine fibrinopeptide B, (2-) from the average of 18 MS/MS scans acquired over 12 s. SO_3 neutral loss products are annotated in light grey. Glp stands for pyroglutamic acid. Source: Robinson, 2014 [55]. Reproduced with permission from Springer.

the gas-phase desulfation does not prevent the accurate site assignments since detected tyrosine residues without the modification are assumed to be sulfated (Figure 9.5).

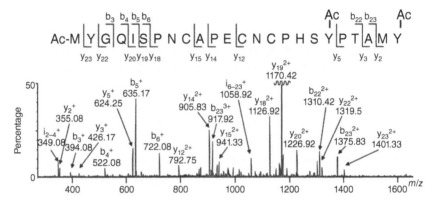

Figure 9.5 Identification of the site of sulfation in lumican peptide [29]Met–Tyr[52]. The unmodified Tyr residues as well as the N-terminus were derivatized with sulfosuccinimidyl acetate. Thus, the only "unmodified" Tyr represents the modification site. Source: Yu, 2007. [43], Reproduced with permission from Nature Publishing Group.

9.3 Enrichment Strategies and Global Analysis of Sulfation

Post-translationally modified peptides represent a small percentage of the digested proteomes, even when the level of modification within specific proteins is relatively high. Thus, for systems biology studies, peptides bearing a particular PTM have to be selectively isolated or at least enriched. In addition, diagnostic mass spectrometric features that enable selective detection/more reliable assignment of the modification of interest are extremely valuable even in the analysis of mixtures rich in specifically modified sequences.

To date, no large-scale, high-throughput sulfation studies have been performed. This can be explained by the mass spectrometry behavior of sulfopeptides as presented earlier. In addition, no reliable and sufficiently selective enrichment method has been established for sulfated sequences yet.

Large-scale PTM identification studies are frequently aided by specific antibodies. The generation of sulfo-Tyr recognizing antibodies has been reported [56, 57], and the latter group also used their antibody for the enrichment of modified proteins [57]. A sulfo-Tyr-specific antibody is also marketed by multiple vendors (Abcam, Millipore). However, no large-scale sulfotyrosine studies using any of these antibodies have been reported. Thus, we cannot draw any conclusions about their specificity and binding affinity.

Phosphopeptides are often enriched based on their strong affinity to a variety of metal ions. Immobilized metal ion affinity chromatography (IMAC) with Ga(III) was reported to enrich sulfopeptides from the skin secretion of a frog, however, with *very* modest selectivity [58]. The utility of a weak anion exchanger at a close to neutral pH has been demonstrated on model sulfopeptides and bovine fibrinogen [59]. Following trypsin digestion, C-terminal basic residues were removed by carboxypeptidase B to aid the isolation of sulfated peptides. While this treatment enables the enrichment of the peptides (unless there are additional basic residues within the sequence), it is also counterproductive for the following mass spectrometric analysis. Other chromatographic fractionation methods, such as electrostatic repulsion hydrophilic interaction chromatography (ERLIC) [60, 61] and strong cation exchange (SCX) separations [62], were relatively successful for the enrichment of phosphopeptides. These techniques also take advantage of the acidity of the modifying group, but at a low enough pH where all basic residues are protonated, and only the strong acidic group bears a negative charge. These techniques may also be tested for the enrichment of the more acidic sulfopeptides; however, the elimination of the interfering phosphopeptides by alkaline phosphatase treatment would be recommended.

Even with a certain degree of selective enrichment, finding post-translationally modified peptides in such a complex mixture as, for example, a plasma protein digest is not a simple task to tackle. The similarity of phospho- and sulfopeptides further complicates the matter. Theoretically, sulfopeptides could be identified by accurate mass measurements. However, one usually has to rely only on the mass accuracy of the precursor ion since none of the CID-generated fragment ions retain the modification. Thus, the mass error must not be higher than 2 ppm in order to differentiate between phospho- and sulfopeptides up to 2500 Da.

The characteristic 80 Da loss may be used to our advantage. For example, one can use neutral loss analysis to identify the sulfopeptides [38, 63], or alternatively the diagnostic fragment ion, SO_3^- (*m/z* 80), may be used either in precursor ion scans in negative mode or in a "ping-pong style" acquisition setup, where in-source fragmentation is monitored in negative mode, while the full scan mass measurements and data-dependent MS/MS acquisitions are performed in positive mode [64]. Obviously, the latter approach will indicate which MS surveys contain sulfopeptides, but will not directly identify them.

Conventionally, large-scale proteomic analyses are performed in positive ion mode. The generally used activation methods are ion trap CID (resonance activation), beam-type CID (also sometimes referred to as higher-energy C-trap dissociation (HCD)), and ETD. During database searches, phosphorylation is frequently considered as variable modification, while sulfation is usually not. However, even if it were included, unless very good mass accuracy is specified for the precursor ion (see preceding text), both modifications may fit within the permitted mass error. Based on our experience, most search engines identify sulfopeptides as phosphorylated sequences and even assign the residue modified. As far as we know, only ProteinProspector permits the computer "to admit" that there is no sufficient information to distinguish between positional isomers and displays this automatically in the search results [65]. ProteinProspector also permits searching for fragile modifications as neutral losses; that is, there is no obligatory site localization [66].

The difficulty of sulfopeptide identification is beautifully illustrated by a study initiated by the ABRF Proteome Informatics Research Group [67]. Seventy synthetic, PTM-bearing peptides, among them seven sulfopeptides, were mixed with a yeast proteome tryptic digest. High-quality beam-type CID data were acquired, and identical peak lists and a database were sent out to the study participants. As it turned out, sulfopeptides represented the biggest challenge – "Lack of knowledge of how this modification behaves in CID (either by software or by user) led to many people either completely missing these peptides or reporting them as phosphopeptides instead" [67].

9.4 Sulfation Site Predictions

Protein or rather peptide identification from mass spectrometry data has become straightforward, and for "obscure" species the process is aided by homology and related sequences and/or transcriptome data. However, PTMs cannot be predicted from genome sequences.

Identification of consensus motifs, that is, structural details that are recognized by the modifying enzyme(s), may aid the analytical process. For example, the most promising cleavage method can be selected for generating the "best" modified peptide for chromatographic separation and mass spectrometric characterization. Similarly, if the motif to be considered can be specified, the search space can be properly restricted; for example, Byonic (www.proteinmetrics.com) or ProteinProspector (prospector.ucsf.edu) offer this option for N-glycosylation. Thus, a database search with the MS/MS data becomes significantly faster, and the resulting identifications much more reliable. Narrowing down the options and predicting which sites can be modified has been a desire whenever a new PTM has been reported.

As far as Tyr sulfation is concerned, the first such studies investigated the primary structure of proteins, the amino acid sequences around the sulfated Tyr residues. Several physical attributes were tested, and the presence of acidic residues seemed to be the most distinctive feature for Tyr sulfation [37, 68]. It soon became obvious that considering only the adjacent residues does not provide sufficient information for reliable modification prediction. A wider net was cast, with the assumption that secondary structure may influence the target recognition: a ±5 amino acid-wide sequence stretch was investigated using a position-specific scoring matrix (PSSM) [69]. Despite the fact that no unambiguous motif could be identified by this approach, conclusions were drawn about the potential Tyr sulfation in biologically significant transmembrane receptors [70] and viral proteins [71].

Table 9.2 Prediction softwares tested with a few proteins with reliable site assignments.

Protein	Sites assigned	Sulfinator (cutoff $E = 55$) (default value)	sulfosites (prediction sensitivity: 90%)	PredSulSite (cutoff probability = 0.5)
Vitronectin (P04004)	75, 78, 282, 417, 420	75, 78, 282, 417, 420	54, 75, 78, 239	75, 78
Histatin 1 (P15515)	46, 49, 53, 55	46	49	46, 49, 53
Osteomodulin (Q99983)	22, 25, 31, 39, 51, 58, 77	8, 39	31, 39, 51	39, 416, 417
Lumican (P51884)	20, 21, 23, 30	None	23	None
Fibromodulin (Q06828)	20, 38, 39, 45, 47, 53, 55	None	39	20
Bone sialoglycoprotein (P21815)	313 or 314	259, 263, 265, 271, 278, 290, 293, 297, 299	265, 305	313

Still sticking to the analysis of the linear amino acid sequence using four different hidden Markov models, a new prediction tool was established that is available as part of ExPASy's Proteomics Tool package (http://www.expasy.org/tools/sulfinator/) [72]. We have tested the predictive power of this tool with six proteins with reliably assigned modification sites: vitronectin, fibromodulin, osteomodulin, histatin 1, lumican, and bone sialoglycoprotein (see Table 9.2). Sulfinator was able to identify all the assigned sites for vitronectin, but did not find any potential sites in fibromodulin and lumican; produced 10 candidates in bone sialoglycoprotein, but the assigned site was not among them; predicted 2 sites for osteomodulin, 1 was correct; and indicated a single (assigned) site for histatin 1. The ExPASy toolbox features an additional piece of software, sulfosites (http://sulfosite.mbc.nctu.edu.tw/) [73]. This prediction tool was developed using support vector machine (SVM) learning, and both the linear sequence and the secondary structure around the sulfated sites were considered. This predictor was also tested and missed the marks approximately as much as Sulfinator (Table 9.2). The newest prediction website – http://bioinfo.ncu.edu.cn/inquiries_PredSulSite.aspx – has been developed along the same lines, using SVM and considering both the secondary and primary structures of the protein [74]. It did not perform significantly better than the previous two when tested with the same proteins (Table 9.2).

9.5 Summary

Obviously, the existing prediction methods are not reliable and must not be trusted. Perhaps searching for a single consensus motif is futile, since two enzymes are responsible for Tyr sulfation, and they may have different preferences. Or perhaps, the tertiary structure of the protein modified is also a determinant factor. We do not know, and without acquiring more, reliable information one cannot tell.

Unfortunately, Tyr sulfation presently does not belong to those PTMs that can be studied in a high-throughput manner. First, a selective and sensitive enrichment method has to be developed. Then, the resulting mixtures could be used to develop/fine-tune mass spectrometric workflows for efficient and reliable sulfopeptide detection and characterization, including unambiguous site assignments. We think, presently, HCD/ETD analysis or chemical derivatization and HCD analysis seem to be the most sensitive solutions, compatible with nano-HPLC fractionation. But we predict that the future most certainly will offer better alternatives, and then our knowledge on protein sulfation and its biological function will further expand.

Acknowledgement

KFM was supported by the NIH grant NIGMS 8P41GM103481, the Howard Hughes Medical Institute, and the Miriam and Sheldon Adelson Medical Research Foundation (to the Bioorganic Biomedical Mass Spectrometry Resource at UCSF, director A.L. Burlingame). All authors were also supported by the Hungarian Scientific Research Fund Grant OTKA #105611. The Janos Bolyai Fellowship to EK is gratefully acknowledged.

References

1 Bettelheim FR. Tyrosine-O-sulfate in a peptide from fibrinogen. *J Am Chem Soc* 1954;**76**:2838–2839.

2 Gregory H, Hardy PM, Jones DS, Kenner GW, Sheppard RC. Antral hormone gastrin. Structure of gastrin. *Nature* 1964;**204**:931–933.

3 Mutt V, Jorpes JE. Structure of porcine cholecystokinin-pancreozymin. 1. Cleavage with thrombin and with trypsin. *Eur J Biochem* 1968;**6**:156–162.

4 Fluharty AL, Stevens RL, Goldstein EB, Kihara H. The activity of arylsulfatase A and B on tyrosine O-sulfates. *Biochim Biophys Acta* 1979;**566**:321–326.

5 Hille A, Rosa P, Huttner WB. Tyrosine sulfation: a post-translational modification of proteins destined for secretion? *FEBS Lett* 1984;**177**:129–134.

6 Baeuerle PA, Huttner WB. Tyrosine sulfation is a *trans*-Golgi-specific protein modification. *J Cell Biol* 1987;**105**:2655–2664.

7 Hille A, Braulke T, von Figura K, Huttner WB. Occurrence of tyrosine sulfate in proteins--a balance sheet. 1. Secretory and lysosomal proteins. *Eur J Biochem* 1990;**188**:577–586.

8 Hille A, Huttner WB. Occurrence of tyrosine sulfate in proteins--a balance sheet. 2. Membrane proteins. *Eur J Biochem* 1990;**188**:587–596.

9 Liu MC, Yu S, Sy J, Redman CM, Lipmann F. Tyrosine sulfation of proteins from the human hepatoma cell line HepG2. *Proc Natl Acad Sci U S A* 1985;**82**:7160–7164.

10 Ouyang YB, Lane WS, Moore KL. Tyrosylprotein sulfotransferase: purification and molecular cloning of an enzyme that catalyzes tyrosine O-sulfation, a common post-translational modification of eukaryotic proteins. *Proc Natl Acad Sci U S A* 1998;**95**:2896–2901.

11 Beisswanger R, Corbeil D, Vannier C, Thiele C, Dohrmann U, Kellner R, Ashman K, Niehrs C, Huttner WB. Existence of distinct tyrosylprotein sulfotransferase genes: molecular characterization of tyrosylprotein sulfotransferase-2. *Proc Natl Acad Sci U S A* 1998;**95**:11134–11139.

12 Hanai H, Nakayama D, Yang H, Matsubayashi Y, Hirota Y, Sakagami Y. Existence of a plant tyrosylprotein sulfotransferase: novel plant enzyme catalyzing tyrosine O-sulfation of preprophytosulfokine variants in vitro. *FEBS Lett* 2000;**470**:97–101.

13 Komori R, Amano Y, Ogawa-Ohnishi M, Matsubayashi Y. Identification of tyrosylprotein sulfotransferase in Arabidopsis. *Proc Natl Acad Sci U S A* 2009;**106**:15067–15072.

14 Han SW, Lee SW, Bahar O, Schwessinger B, Robinson MR, Shaw JB, Madsen JA, Brodbelt JS, Ronald PC. Tyrosine sulfation in a Gram-negative bacterium. *Nat Commun* 2012;**3**:1153.

15 Pittman DD, Tomkinson KN, Michnick D, Selighsohn U, Kaufman RJ. Post-translational sulfation of factor V is required for efficient thrombin cleavage and activation and for full procoagulant activity. *Biochemistry* 1994;**33**:6952–6959.

16 Pittman DD, Wang JH, Kaufman RJ. Identification and functional importance of tyrosine sulfate residues within recombinant factor VIII. *Biochemistry* 1992;**31**:3315–3325.

17 Michnick DA, Pittman DD, Wise RJ, Kaufman RJ. Identification of individual tyrosine sulfation sites within factor VIII required for optimal activity and efficient thrombin cleavage. *J Biol Chem* 1994;**269**:20095–20102.

18 Leyte A, van Schijndel HB, Niehrs C, Huttner WB, Verbeet MP, Mertens K, van Mourik JA. Sulfation of Tyr1680 of human blood coagulation factor VIII is essential for the interaction of factor VIII with von Willebrand factor. *J Biol Chem* 1991;**266**:740–746.

19 Wilkins PP, Moore KL, McEver RP, Cummings RD. Tyrosine sulfation of P-selectin glycoprotein ligand-1 is required for high affinity binding to P-selectin. *J Biol Chem* 1995;**270**:22677–22680.

20 Hortin GL, Farries TC, Graham JP, Atkinson JP. Sulfation of tyrosine residues increases activity of the fourth component of complement. *Proc Natl Acad Sci U S A* 1989;**86**:1338–1342.

21 Marchese P, Murata M, Mazzucato M, Pradella P, De Marco L, Ware J, Ruggeri ZM. Identification of three tyrosine residues of glycoprotein Ib alpha with distinct roles in von Willebrand factor and alpha-thrombin binding. *J Biol Chem* 1995;**270**:9571–9578.

22 Farzan M, Schnitzler CE, Vasilieva N, Leung D, Kuhn J, Gerard C, Gerard NP, Choe H. Sulfated tyrosines contribute to the formation of the C5a docking site of the human C5a anaphylatoxin receptor. *J Exp Med* 2001;**193**:1059–1066.

23 Costagliola S, Panneels V, Bonomi M, Koch J, Many MC, Smits G, Vassart G. Tyrosine sulfation is required for agonist recognition by glycoprotein hormone receptors. *EMBO J* 2002;**21**:504–513.

24 Fieger CB, Sassetti CM, Rosen SD. Endoglycan, a member of the CD34 family, functions as an L-selectin ligand through modification with tyrosine sulfation and sialyl Lewis x. *J Biol Chem* 2003;**278**:27390–27398.

25 Gao JM, Xiang RL, Jiang L, Li WH, Feng QP, Guo ZJ, Sun Q, Zeng ZP, Fang FD. Sulfated tyrosines 27 and 29 in the N-terminus of human CXCR3 participate in binding native IP-10. *Acta Pharmacol Sin* 2009;**30**:193–201.

26 Ludeman JP, Stone MJ. The structural role of receptor tyrosine sulfation in chemokine recognition. *Br J Pharmacol* 2014;**171**:1167–1179.

27 Farzan M, Mirzabekov T, Kolchinsky P, Wyatt R, Cayabyab M, Gerard NP, Gerard C, Sodroski J, Choe H. Tyrosine sulfation of the amino terminus of CCR5 facilitates HIV-1 entry. *Cell* 1999;**96**:667–676.

28 Farzan M, Babcock GJ, Vasilieva N, Wright PL, Kiprilov E, Mirzabekov T, Choe H. The role of post-translational modifications of the CXCR4 amino terminus in stromal-derived factor 1 alpha association and HIV-1 entry. *J Biol Chem* 2002;**277**:29484–29489.

29 Hsu W, Rosenquist GL, Ansari AA, Gershwin ME. Autoimmunity and tyrosine sulfation. *Autoimmun Rev* 2005;**4**:429–435.

30 Lim A, Prokaeva T, McComb ME, Connors LH, Skinner M, Costello CE. Identification of S-sulfonation and S-thiolation of a novel transthyretin Phe33Cys variant from a patient diagnosed with familial transthyretin amyloidosis. *Protein Sci* 2003;**12**:1775–1785.

31 Medzihradszky KF, Darula Z, Perlson E, Fainzilber M, Chalkley RJ, Ball H, Greenbaum D, Bogyo M, Tyson DR, Bradshaw RA, Burlingame AL. O-sulfonation of serine and threonine: mass spectrometric detection and characterization of a new post-translational modification in diverse proteins throughout the eukaryotes. *Mol Cell Proteomics* 2004;**3**:429–440.

32 Dave KA, Whelan F, Bindloss C, Furness SG, Chapman-Smith A, Whitelaw ML, Gorman JJ. Sulfonation and phosphorylation of regions of the dioxin receptor susceptible to methionine modifications. *Mol Cell Proteomics* 2009;**8**:706–719.

33 Gharib M, Marcantonio M, Lehmann SG, Courcelles M, Meloche S, Verreault A, Thibault P. Artifactual sulfation of silver-stained proteins: implications for the assignment of phosphorylation and sulfation sites. *Mol Cell Proteomics* 2009;**8**:506–518.

34 Hortin G, Fok KF, Toren PC, Strauss AW. Sulfation of a tyrosine residue in the plasmin-binding domain of alpha 2-antiplasmin. *J Biol Chem* 1987;**262**: 3082–3085.

35 Yun HY, Keutmann HT, Eipper BA. Alternative splicing governs sulfation of tyrosine or oligosaccharide on peptidylglycine alpha-amidating monooxygenase. *J Biol Chem* 1994;**269**:10946–10955.

36 Gao J, Choe H, Bota D, Wright PL, Gerard C, Gerard NP. Sulfation of tyrosine 174 in the human C3a receptor is essential for binding of C3a anaphylatoxin. *J Biol Chem* 2003;**278**:37902–37908.

37 Hortin G, Folz R, Gordon JI, Strauss AW. Characterization of sites of tyrosine sulfation in proteins and criteria for predicting their occurrence. *Biochem Biophys Res Commun* 1986;**141**:326–333.

38 Severs JC, Carnine M, Eguizabal H, Mock KK. Characterization of tyrosine sulfate residues in antihemophilic recombinant factor VIII by liquid chromatography electrospray ionization tandem mass spectrometry and amino acid analysis. *Rapid Commun Mass Spectrom* 1999;**13**:1016–1023.

39 Meh DA, Siebenlist KR, Brennan SO, Holyst T, Mosesson MW. The amino acid sequence in fibrin responsible for high affinity thrombin binding. *Thromb Haemost* 2001;**85**:470–474.

40 Onnerfjord P, Heathfield TF, Heinegård D. Identification of tyrosine sulfation in extracellular leucine-rich repeat proteins using mass spectrometry. *J Biol Chem* 2004;**279**:26–33.

41 Dong JF, Li CQ, López JA. Tyrosine sulfation of the glycoprotein Ib-IX complex: identification of sulfated residues and effect on ligand binding. *Biochemistry* 1994;**33**:13946–13953.

42 Cabras T, Fanali C, Monteiro JA, Amado F, Inzitari R, Desiderio C, Scarano E, Giardina B, Castagnola M, Messana I. Tyrosine polysulfation of human salivary histatin 1. A post-translational modification specific of the submandibular gland. *J Proteome Res* 2007;**6**:2472–2480.

43 Yu Y, Hoffhines AJ, Moore KL, Leary JA. Determination of the sites of tyrosine O-sulfation in peptides and proteins. *Nat Methods* 2007;**4**:583–588.

44 Benjannet S, Rhainds D, Hamelin J, Nassoury N, Seidah NG. The proprotein convertase (PC) PCSK9 is inactivated by furin and/or PC5/6A: functional consequences of natural mutations and post-translational modifications. *J Biol Chem* 2006;**281**:30561–3072.

45 Pouyani T, Seed B. PSGL-1 recognition of P-selectin is controlled by a tyrosine sulfation consensus at the PSGL-1 amino terminus. *Cell* 1995;**83**:333–343.

46 Zaia J, Boynton R, Heinegård D, Barry F. Post-translational modifications to human bone sialoprotein determined by mass spectrometry. *Biochemistry* 2001;**40**:12983–12991.

47 Sahin-Toth M, Kukor Z, Nemoda Z. Human cationic trypsinogen is sulfated on Tyr154. *FEBS J* 2006;**273**:5044–5050.

48 Yagami T, Kitagawa K, Futaki S. Liquid secondary-ion mass spectrometry of peptides containing multiple tyrosine-O-sulfates. *Rapid Commun Mass Spectrom* 1995;**9**:1335–1341.

49 Wolfender JL, Chu F, Ball H, Wolfender F, Fainzilber M, Baldwin MA, Burlingame AL. Identification of tyrosine sulfation in *Conus pennaceus* conotoxins alpha-PnIA and alpha-PnIB: further investigation of labile sulfo- and phosphopeptides by electrospray, matrix-assisted laser desorption/ ionization (MALDI) and atmospheric pressure MALDI mass spectrometry. *J Mass Spectrom* 1999;**34**:447–454.

50 Gorman JJ, Ferguson BL, Nguyen TB. Use of 2,6-dihydroxyacetophenone for analysis of fragile peptides, disulphide bonding and small proteins by matrix-assisted laser desorption/ionization. *Rapid Commun Mass Spectrom* 1996;**10**:529–536.

51 Medzihradszky KF, Phillipps NJ, Senderowicz L, Wang P, Turck CW. Synthesis and characterization of histidine-phosphorylated peptides. *Protein Sci* 1997;**6**:1405–1411.

52 Mikesh LM, Ueberheide B, Chi A, Coon JJ, Syka JE, Shabanowitz J, Hunt DF. The utility of ETD mass spectrometry in proteomic analysis. *Biochim Biophys Acta* 2006;**1764**:1811–1822.

53 Good DM, Wirtala M, McAlister GC, Coon JJ. Performance characteristics of electron transfer dissociation mass spectrometry. *Mol Cell Proteomics* 2007;**6**:1942–1951.

54 Hersberger KE, Håkansson K. Characterization of O-sulfopeptides by negative ion mode tandem mass spectrometry: superior performance of negative ion electron capture dissociation. *Anal Chem* 2012;**84**:6370–6377.

55 Robinson MR, Moore KL, Brodbelt JS. Direct identification of tyrosine sulfation by using ultraviolet photodissociation mass spectrometry. *J Am Soc Mass Spectrom* 2014;**25**:1461–1471.

56 Kehoe JW, Velappan N, Walbolt M, Rasmussen J, King D, Lou J, Knopp K, Pavlik P, Marks JD, Bertozzi CR, Bradbury AR. Using phage display to select antibodies recognizing post-translational modifications independently of sequence context. *Mol Cell Proteomics* 2006;**5**:2350–2363.

57 Hoffhines AJ, Damoc E, Bridges KG, Leary JA, Moore KL. Detection and purification of tyrosine-sulfated proteins using a novel anti-sulfotyrosine monoclonal antibody. *J Biol Chem* 2006;**281**:37877–37887.

58 Demesa Balderrama G, Meneses EP, Hernández Orihuela L, Villa Hernández O, Castro Franco R, Pando Robles V, Ferreira Batista CV. Analysis of sulfated peptides from the skin secretion of the *Pachymedusa dacnicolor* frog using IMAC-Ga enrichment and high-resolution mass spectrometry. *Rapid Commun Mass Spectrom* 2011;**25**:1017–1027.

59 Amano Y, Shinohara H, Sakagami Y, Matsubayashi Y. Ion-selective enrichment of tyrosine-sulfated peptides from complex protein digests. *Anal Biochem* 2005;**346**:124–131.

60 Alpert AJ. Electrostatic repulsion hydrophilic interaction chromatography for isocratic separation of charged solutes and selective isolation of phosphopeptides. *Anal Chem* 2008;**80**:62–76.

61 Alpert AJ, Hudecz O, Mechtler K. Anion-exchange chromatography of phosphopeptides: weak anion exchange versus strong anion exchange and anion-exchange chromatography versus electrostatic repulsion-hydrophilic interaction chromatography. *Anal Chem* 2015;**87**:4704–4711.

62 Villén J, Gygi SP. The SCX/IMAC enrichment approach for global phosphorylation analysis by mass spectrometry. *Nat Protoc* 2008;**3**:1630–1638.

63 Salek M, Costagliola S, Lehmann WD. Protein tyrosine-O-sulfation analysis by exhaustive product ion scanning with minimum collision offset in a NanoESI Q-TOF tandem mass spectrometer. *Anal Chem* 2004;**76**:5136–5142.

64 Carr SA, Annan RS, Huddleston MJ. Mapping post-translational modifications of proteins by MS-based selective detection: application to phosphoproteomics. *Meth Enzymol* 2005;**405**:82–115.

65 Baker PR, Trinidad JC, Chalkley RJ. Modification site localization scoring integrated into a search engine. *Mol Cell Proteomics* 2011;**10**:M111.008078.

66 Chalkley RJ, Baker PR, Medzihradszky KF, Lynn AJ, Burlingame AL. In-depth analysis of tandem mass spectrometry data from disparate instrument types. *Mol Cell Proteomics* 2008;**7**:2386–2398.

67 Chalkley RJ, Bandeira N, Chambers MC, Clauser KR, Cottrell JS, Deutsch EW, Kapp EA, Lam HH, McDonald WH, Neubert TA, Sun RX. Proteome informatics research group (iPRG)_2012: a study on detecting modified peptides in a complex mixture. *Mol Cell Proteomics* 2014;**13**:360–371.

68 Rosenquist GL, Nicholas HB Jr. Analysis of sequence requirements for protein tyrosine sulfation. *Protein Sci* 1993;**2**:215–222.

69 Nicholas HB Jr, Chan SS, Rosenquist GL. Reevaluation of the determinants of tyrosine sulfation. *Endocrine* 1999;**11**:285–292.

70 Yu KM, Liu J, Moy R, Lin HC, Nicholas HB Jr, Rosenquist GL. Prediction of tyrosine sulfation in seven-transmembrane peptide receptors. *Endocrine* 2002;**19**:333–338.

71 Lin HC, Tsai K, Chang BL, Liu J, Young M, Hsu W, Louie S, Nicholas HB Jr, Rosenquist GL. Prediction of tyrosine sulfation sites in animal viruses. *Biochem Biophys Res Commun* 2003;**312**:1154–1158.

72 Monigatti F, Gasteiger E, Bairoch A, Jung E. The Sulfinator: predicting tyrosine sulfation sites in protein sequences. *Bioinformatics* 2002;**18**: 769–770.

73 Chang WC, Lee TY, Shien DM, Hsu JB, Horng JT, Hsu PC, Wang TY, Huang HD, Pan RL. Incorporating support vector machine for identifying protein tyrosine sulfation sites. *J Comput Chem* 2009;**30**:2526–2537.

74 Huang SY, Shi SP, Qiu JD, Sun XY, Suo SB, Liang RP. PredSulSite: prediction of protein tyrosine sulfation sites with multiple features and analysis. *Anal Biochem* 2012;**428**:16–23.

10

The Application of Mass Spectrometry for the Characterization of Monoclonal Antibody-Based Therapeutics

Rosie Upton[1], Kamila J. Pacholarz[1], David Firth[2], Sian Estdale[2] and Perdita E. Barran[1]

[1] *Michael Barber Centre for Collaborative Mass Spectrometry, Manchester Institute of Biotechnology, University of Manchester, Manchester, UK*
[2] *Covance Laboratories Ltd., Harrogate, UK*

10.1   Introduction

Monoclonal antibody (mAb) therapeutics currently represent the fastest-growing class of biopharmaceuticals [1]. With the introduction of several stand-alone immunoglobulin G (IgG) therapies, fragment crystallizable (Fc)-fusion proteins/peptides [2, 3], fragment antigen-binding (Fab) fragments [4], bispecific antibodies (bsAbs) [2, 5, 6], antibody–drug conjugates (ADCs) [7, 8], radioimmunoconjugates [9, 10], as well generic versions known as biosimilars, it is clear that mAbs are of significant interest to the pharmaceutical industry [11], to healthcare providers [12], as well as to potential consumers [13, 14].

Analogous to the development of small-molecule drugs, structural and analytical characterization of biologics is critically required at many points in the pipeline to ensure comparability to the original product [15]. The current limitations in probing the inherent complexities of many biologics, and understanding their potential impact in a safety or clinical setting, presents a substantial challenge. Even very minor alterations to the structure of a biologic can have unintended clinical consequences. mAbs possess specific structural features that affect their function [16]. As a biologic is developed and manufactured, all levels of structure from the primary sequence and post-translational modifications (PTMs) to the quaternary fold must be shown to be highly similar (see ICH Q5E). Effective clinical application relies on the consistency of the formulated product at the molecular level, and hence they must be characterized using sensitive, orthogonal analytical techniques. Effective characterization that provides relevant quantifiable or qualitative information regarding

Analysis of Protein Post-Translational Modifications by Mass Spectrometry,
First Edition. Edited by John R. Griffiths and Richard D. Unwin.
© 2017 John Wiley & Sons, Inc. Published 2017 by John Wiley & Sons, Inc.

the intended clinical outcome (or mechanism of action) can greatly enhance understanding and derisking of manufacturing changes and also prevent unnecessary nonclinical or clinical studies.

As patents for many mAb-based therapeutics are expiring, introduction of generic versions – the so-called biosimilars or biobetters – presents even more regulatory challenges. Clinicians are wary of the interchangeability of biosimilars and rightly have questions regarding the transferral and extrapolation of clinical trial data obtained for a particular disease toward different diseases that have not been through the same vigorous testing [17]. Developing confidence in regulatory procedures and acceptance criteria for biosimilars will be enabled by evidence proving consistency of the products, which relies on robust analytical characterization, pharmacokinetic (PK) and pharmacodynamic (PD) analyses, and *in vivo* efficacy testing [18].

The rapidly evolving market for mAb-based therapeutics demands comprehensive, analytical characterization of these complex and flexible biological structures. One such technique that has shown dominance in this field is mass spectrometry (MS) [19, 20], including liquid chromatography–mass spectrometry (LC-MS), native MS, and recently developed ion mobility–mass spectrometry (IM-MS) and hydrogen/deuterium exchange mass spectrometry (HDX-MS). This review concentrates on the use of these techniques for the characterization of intact mAbs and ADCs while introducing the concept of biosimilars and their characterization requirements. We first introduce the common structural features of mAbs, with a focus on IgG. We then highlight how the structures can be altered during expression with focus on variations in N-glycosylation patterns, which can affect the mode of action. Certain modifications are deliberately introduced in the case of ADCs, and these will be described, as will biosimilars. We describe common MS approaches for examining mAb structure and present some more advanced examples of the use of MS to provide structural and functional insights.

10.1.1 Antibody Structure

mAbs, based upon the immunoglobulin (Ig) "Y" shape, are tetrameric glycoprotein therapeutics with two HC and two LC constituents. These four chains form the basis of the homoheterodimer structure and are linked via a flexible stretch of polypeptide chain referred to as the hinge region [21]. In mammals there are five different isoforms consisting of one or more replicas of the "Y"-shaped unit: gamma (IgG), mu (IgM), alpha (IgA), delta (IgD), and epsilon (IgE) [22]. Therapeutic templates however have been based primarily upon the IgG1 class (though therapeutics based on IgG2 and IgG4 are also used) as it is considered to be the most active within the immune system, resulting from its ability to efficiently engage with cytotoxic cell receptors (FcγR) and its ability to destroy target cells through activation of the complement [23]. The IgG form also accounts for 75% of total serum Ig [24]. A schematic representation of an IgG1 molecule is shown in Figure 10.1. The heavy chain (HC) and light chain (LC) are made up of different regions that each contain approximately 110 amino acids, which fold into discrete compact regions referred to as domains [25]. Each of the four chains has a variable domain (V_H and V_L) at the N-terminus, which is responsible for antigen binding. The nomenclature of the Ig isoforms is based upon the antigen binding specificity that is determined by the amino acid sequence in these variable domains. The remaining domains are referred to as "constant" as their amino acid composition is highly conserved across the different Ig isoforms. The antibody molecule can be further divided into two functionally distinct regions: the Fab and Fc subunits. The Fab corresponds to the two identical antibody "arms," which consist of the entire LC coupled with V_H and C_H1, and is hence responsible for antigen recognition and binding [26]. The Fc interacts with cell surface receptors and complement proteins [27].

The IgGs are split into additional subclasses (IgG1-4), which differ from one another most significantly with the location and quantity of disulfide bonds

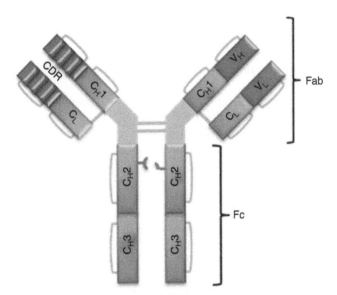

Figure 10.1 The immunoglobulin G antibody structure consists of a variable region in the heavy and light chains (V_H and V_L, respectively), a constant region in light chain (C_L), and three constant regions in the heavy chain (C_H1, C_H2, and C_H3). The complementarity determining region (CDR) consists of three loops connecting the variable domains at which antigen binding and selectivity occur. The Fab region recognizes and binds antigens, while the Fc region interacts with effector cells. N-glycosylation in IgG1 mAbs, such as trastuzumab, occurs in the Fc C_H2 domain.

within the hinge region [28–30]. In each case a single disulfide bridge connects the HC and LC, with two disulfide bonds connecting the two HC strands in IgG1 and IgG4, four in IgG2, and eleven in IgG3 at the flexible hinge region [30, 31]. Contributing to the overall 3D structure of the mAbs are 12 additional intramolecular disulfide bridges (common to all IgGs) and PTMs such as N-glycosylation, O-glycosylation, galactosylation, and mannosylation, to name a few – all of which need to be characterized during the drug development process [32]. It is the N-termini of the variable domains that facilitate antigen binding through six sites in the complementarity determining region (CDR): three from variable HC and three from variable LC (V_H and V_L, respectively). The CDR expresses loops formed from β-pleated sheets with antigen binding affinities that are based upon sequence structure in the variable domains [33], whereas the constant domains host the most abundant PTMs [34]. The majority of approved therapeutic mAbs are produced using Chinese hamster ovary (CHO) cells [35, 36]; however other expression platforms such as *E. coli* [37] and mouse myeloma cells [38] have also been employed. The majority of approved mAbs for therapeutic use are subclasses or derivatives of humanized IgG1, although IgG2 and IgG4 therapeutics have also been reported [39, 40]. IgG3 structures, however, are not currently used in therapeutics as they have been reported to have a significantly shorter half-life and faster clearance rates than the other isotypes [21, 38, 40].

10.1.2 N-Linked Glycosylation

All IgG mAbs [41] contain the Asn^{297}-X-Ser/Thr-Y sequon (exact sequence of amino acids) in the constant C_H2 domain, which is the prerequisite for N-glycosylation, where X is any amino acid besides proline [42, 43]. This sequon is common to the IgG C_H2 domain but is also observed in the variable domains of the Fab region in around 20% of IgGs [44]. As a consequence, the most commonly encountered PTMs in IgG mAbs are N-linked glycans, although the glycosylation on the two HC within a single IgG is not necessarily identical and in some cases they are not glycosylated at all [32].

The degree of glycosylation can have a high impact upon the efficacy, toxicology, and structure of the mAb therapeutic, most commonly by altering the binding affinity to the target receptor [45, 46]. The main glycoforms observed in IgGs expressed in CHO cells are G0 and its fucosylated counterpart, G0F, along with G1F, G2F, and the high-mannose glycan Man5 [21, 47]. Each of the glycans named possesses the same trimannosyl core; G denotes the number of galactose units connected to the two additional GlcNAc units, and for high-mannose species the number denotes the total number of mannose moieties connected to the chitobiose core (GlcNAc–GlcNAc). F denotes the presence of core fucosylation (see Figure 10.2). It has been reported on several occasions, for example, that the absence of the fucose residue at the core of the *N*-glycans

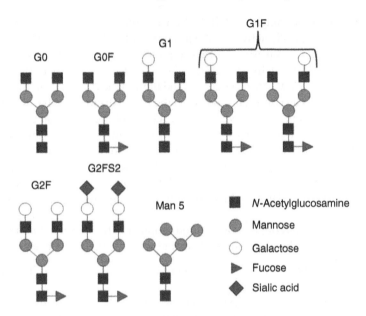

Figure 10.2 Common N-glycans present in CHO produced IgGs. N-glycans are post-translational modifications that add to the Asn-X-Ser/Thr sequons of IgG heavy chains. Variations of the aforementioned also occur. G and the sequential number indicate the number of galactose units attached to the GlcNAc residues. F denotes the presence of fucose within the chitobiose core (GlcNAc–GlcNAc).

of the Fc constant region (referred to as Fc glycan core afucosylation) has a large impact upon binding to protein receptors [43, 44, 48–51]. In trastuzumab (Herceptin®, Genentech) that specifically targets the human epidermal growth factor receptor type 2 (HER2) that is overexpressed in 20–30% of breast carcinomas [52], a high level of afucosylation increases the antibody-dependent cell-mediated cytotoxicity (ADCC) of the mAb. The increase in binding affinity of the therapeutic protein toward the FcγIII receptor (FcγRIII) on the effector function immune system cells is achieved by the reduction in steric hindrance surrounding the N-glycosylated C_H2 domain of the Fc region [53]. With the binding of FcγR, macrophages and natural killer (NK) cells are recruited, hence stimulating the required immune response and activating the ADCC pathway [39, 54].

10.1.3 Antibody-Drug Conjugates

The advantages of mAb specificity have also been exploited in the development of ADCs, which have begun to show great potential, specifically as alternative cancer treatments. ADCs are a class of anticancer biopharmaceutical drug composed of a single mAb linked via a stable chemical linker to a varying number of biologically active cytotoxic small-molecule drugs [8, 55]. The high binding specificity of mAbs and their targeted receptors enables ADCs to be precise and effective delivery systems of the cytotoxins directly to the cancerous cells. *In vivo*, ADCs are recognized and bind to the receptors on the surface of the targeted cell. This then can lead to the internalization of the ADC where lysosomal digestion occurs, enabling the release of the cytotoxic drug, which consequently kills the target cancer cell [56]. Currently two ADCs are approved by the Food and Drug Administration (FDA): brentuximab vedotin (Adcetris® [Seattle Genetics, Inc.]) and ado-trastuzumab emtansine (Kadcyla® [Genentech, Inc.]). But with over 45 still involved in clinical trials and over 160 within preclinical trials [57], ADCs are likely to dominate a new generation of oncology therapeutics.

There are various conjugation methods applied to the covalent linking of the cytotoxic drug to the mAb, currently including three main strategies:

1) Acylation of lysines [58]. Attachment of the cytotoxic drugs occurs through the ε-amine group of lysine.
2) Alkylation of genetically engineered cysteines [59]. Attachment to the thiol side chain of cysteine. The primary sequence of the antibody is mutated by the introduction of unpaired cysteine residues into the HC where they will not form inter- or intrachain disulfide bridges. Unpaired cysteine residues are then conjugated with the cytotoxic drug.
3) Alkylation of cysteine residues generated by the reduction of existing interchain disulfide bonds [60]. The interchain disulfide bridges are partially reduced prior to the conjugation reaction. This results in a mixture of ADCs with drug loading ranging from zero to eight as shown in Figure 10.3.

One of the most important properties of an ADC is the average number of drug molecules attached [61, 62]. This determines the amount of cytotoxic drug being delivered to the tumor cells (the "payload"), which in turn affects the potency, safety, and efficacy of the ADC [63]. The drug component is typically a hydrophobic small drug molecule conjugated to exposed regions of the mAb. For cysteine-linked drugs, conjugates with zero, two, four, six, and eight cytotoxic payloads are observed typically resulting in an average DAR of four.

10.1.4 Biosimilars

Biosimilars are "copies" of the original large-molecule drug. Due to the large market share of approved biologics, there has been much interest in generating biosimilars with a view to accessing market share as soon as patent expiry or loss of exclusivity occurs [38]. With biosimilars becoming increasingly important, there is high demand for effective and efficient characterization techniques from both the biosimilar manufacturers and the governing bodies, who

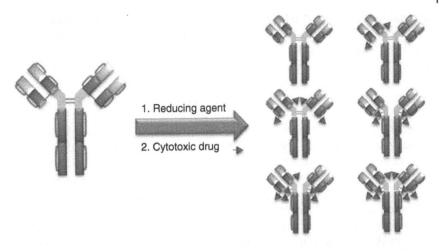

Figure 10.3 mAbs are partially reduced to free cysteine residues from interchain disulfide bonds. The cytotoxic drug and linker are then conjugated via alkylation, with the drug loading ranging from zero to eight.

are concerned with ensuring that a product is as similar as possible to the original biologic. The US FDA and the European Medicines Agency (EMA) both acknowledge that each candidate biopharmaceutical requires individual review [64]. The innovator manufacturing processes and conditions are not made available to competitor companies seeking to produce biosimilars, meaning that no biosimilar is identical to that of the branded, reference therapeutic [46]. A proportion of dissimilarity is also introduced with the use of mammalian cells to produce the biosimilar protein; the DNA sequence is identified and inserted into a vector and then into the mammalian cells for replication. However, given that no two mammalian cells are identical, inter- and intra-batch variability is to be expected, particularly with respect to glycosylation [46, 65–67]. These variations are also present within the original mAb therapeutics due to the use of heterogeneous mammalian cells, and therefore realistic acceptance criteria limits need to be defined for the biosimilars, in line with the individual innovator drugs and their mechanism of action.

According to the EMA "Guideline on similar biological medicinal products containing monoclonal antibodies – non-clinical and clinical issues," which became effective in 2012, the definition for a biosimilar is a biological medicinal product that has similar molecular and biological activity in terms of posology, efficacy, quality, safety, and administration. As a minimum, any biosimilar mAb candidate must contain an identical amino acid sequence, disulfide bond arrangement, and hence tertiary structure as the reference product. Deviations in folding can adversely affect antigen binding and immunogenicity [68, 69]. Due to the nature of mAb manufacturing, variability is inevitable, and PTMs can have significant or little effect upon the therapeutic mechanism of action [67]. Any deviations, however, highlighted during bio-similar comparability testing require justification using sound scientific rationale, which will then help deduce the extent of both nonclinical and clinical *in vivo* and *in vitro* studies required to demonstrate biosimilarity. The requirements for testing are therefore generated on a case-by-case basis while adhering to guidelines [67]. The term biosimilar does not apply to next-generation mAbs, for example, glycoengineered mAbs designed to improve efficacy or potency as they are structurally and/or functionally altered. The world's first biosimilar mAb (infliximab, trade names Remsima® and Inflectra®, manufactured by Celltrion and Hospira, respectively) was approved by the EMA in September 2010 [70] and released to the market in February 2015 following patent expiration of the original. The FDA announced their first biosimilar approval, Zarxio® [Novartis AG], in March 2015, although this was not a mAb. With the approval of Remsima® and Inflectra®, it seems likely that competitor companies will acknowledge the potential of biosimilar mAb development, and therefore a quick, informative technique such as MS will offer data that can derisk biosimilar development programs, limiting investment in unsuitable candidates.

10.2 Mass Spectrometry Solutions to Characterizing Monoclonal Antibodies

Preparing mAb samples for MS analysis varies significantly depending upon whether the sample is enzymatically digested prior to use (bottom up and middle up) or examined intact (top down). In all cases it is usual to start with a purified sample. For intact (top-down) analyses, sample cleanup usually in the form of desalting is all that is required prior to analysis, hence reducing sample preparation and analysis times (see Section 2.3 for details on native (intact) sample preparation and conditions). Reducing agents such as dithiothreitol (DTT) and cleavage-specific enzymes such as immunoglobulin-degrading enzyme (from *Streptococcus pyogenes*) (IdeS), endoglycosidases, and PNGase F can be used to generate antibody subunit fragments between 25 and 50 kDa in size, with and without glycosylation (see Section 3.1 for more details). Middle-up approaches and MS analysis of smaller antibody fragments typically offer improved data quality and mass accuracy over intact analyses without significantly increasing the data analysis time. Liquid chromatography–tandem MS (LC-MS/MS) is frequently used for peptide mapping of mAbs to obtain structural characterization including glycan profiles for each individual glycosylation site [71].

A specialized form of bottom-up analysis is employed in HDX-MS utilizing an online pepsin column to digest the intact molecule, analyzing the resulting peptides to yield site-specific information. Examples of each type of analysis are detailed throughout the rest of this chapter.

10.2.1 Hyphenated Mass Spectrometry (X-MS) Techniques to Study Glycosylation Profiles

Different techniques such as hydrophobic interaction liquid chromatography (HILIC) [72], reversed-phase liquid chromatography (RPLC) [73, 74], size-exclusion chromatography (SEC) [75] and ion-exchange chromatography (IEC), as well as capillary electrophoresis (CE) have been applied to separate mAbs, their derivatives, and released glycans. The hydrophobic nature of antibodies is well suited to hydrophobic stationary phases used for RPLC, allowing selective desorption of individual subunits of peptides with the introduction of increased organic proportions within the mobile phase. Glycan moieties are hydrophilic, however, and therefore are not retained well using RPLC conditions [76]. Typical released glycan analyses utilize HILIC methods that offer better retention.

Diepold et al. [77] combined incubation at elevated temperatures and proteolytic peptide mapping, followed by quantitative LC-MS to simultaneously induce, identify, and quantify Asp isomerization and Asn deamidation. mAbs were first denatured with acidic conditions, reduced using DTT, and then transferred into a trypsin-based digestion buffer before being separated using RPLC with MS detection. Both isomerization and deamidation have been reported to impact the *in vivo* biological activity and the *in vitro* stability of any mAb product, and therefore characterization of these events is critical to the efficacy and safety of the therapeutic.

In 2008, Damen et al. executed the first LC-MS-based quantification of trastuzumab [78]. An advantage of using LC-MS over the typical enzyme-linked immunosorbent assay (ELISA) method is the detection of structural changes that do not just affect the binding properties. LC-MS also provides insight into the degradation patterns of a given mAb and helps identify the PTMs. The results were reported to be in good agreement with the results obtained via UV spectrophotometry and HPLC–UV analyses, and although they had lower sensitivity over ELISA methods, they gained specificity with LC-MS.

In 2011, Gilar et al. applied a HILIC-MS method to characterize the glycosylation sites of a mAb therapeutic [79]. Following tryptic digestion of trastuzumab, RPLC isolated the glycopeptides according to their hydrophobicity, and HILIC was then used to resolve glycoforms. In some HILIC-based analyses, quantification and identification can be performed without MS using UV detection; however, some minor glycoforms do not give sufficient UV responses. For therapeutic mAbs where glycosylation can be highly influential upon efficacy and safety, quantification and identification of these glycans are critical, which is where MS can play an important role.

For released glycan and glycopeptide analysis, HILIC is highly selective [80]. Retention within the silica-based HILIC column is dominated by hydrogen bonding; however, with ionic stationary phases, retention can also be governed by a combination of ionic and dipole–dipole interactions too. With the use of an HILIC column, polar analytes are retained and eluted with mobile phases of higher organic content [81], enabling good LC separation and improved ESI efficiency. Released *N*-glycan analysis typically involves the deglycosylation of the mAb with a peptide *N*-glycosidase, such as PNGase F [82]. This enzyme cleaves the glycan as a glycosylamine, converting asparagine to aspartic acid in the process [83]. MS and in particular LC-MS can be employed for the determination of the mass of antibody postdeglycosylation, which will benefit from increased signal intensity for each charge state as the ion current is no longer shared among the different glycoforms [84]. Analyzing the released glycans can be performed by MS with or without derivatization. 2-Aminobenzamide (2-AB) labeling is commonly used to enable fluorescence detection, but sample preparation and purification can be laborious. There are alternative options, with recent variations including a basic functional group for improved ESI in an HILIC-FLR-MS experiment [85]. Figure 10.4 describes a series of possible approaches for analyzing mAbs that ultimately enable the characterization of the oligosaccharides present.

The microheterogeneity of glycans and the presence of multiple isomers have been investigated with CE coupled to MS (CE-MS), as together this technique can offer high separation efficiencies alongside high mass resolution and mass accuracy [76]. Ma and Nashabeh were the first to demonstrate the use of CE for the analysis and monitoring of mAb N-linked glycans throughout a manufacturing process [87]. Gennaro et al. coupled this with MS to offer an online CE-LIF-MS method capable of identifying minor peaks, unidentified with previous methods [88]. The accurate mass measurements were capable of identifying CE peaks corresponding to important ADCC, regulating afucosylated glycan moieties along with other typical glycans observed for mAb therapeutics.

IEC and SEC methods are often used to monitor the quality and stability of mAb products during all stages of manufacture and handling. It is often the minor peaks present in IEC and SEC separations however that are critical to understanding any changes or degradations. Alvarez et al. therefore developed a two-dimensional LC strategy combining SEC with RP trap cartridges and an MS system [89]. The traps served to collect, concentrate, and desalt IEC or SEC fractions of interest to enable quantitative analysis and separation of poorly resolved and low-level peaks. The incorporation of a disulfide bond DTT reduction step also enabled chain-specific information to be obtained. This 2D LC-MS format made it possible to resolve and identify coeluting fractions while saving time and conserving sample.

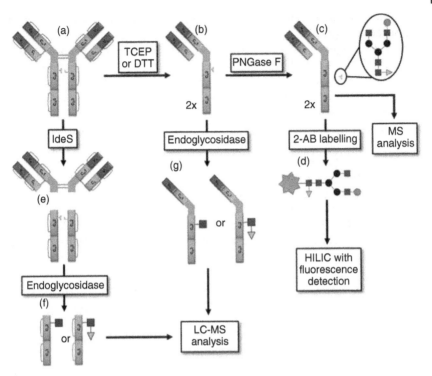

Figure 10.4 A workflow demonstrating the typical strategies for mAb and glycan characterization. Starting with an intact glycosylated mAb (a), reduction of the disulfide bonds separates the two heavy chains and releases light chain (b). Addition of PNGase F cleaves the glycans completely (c), which can be analyzed directly using MS [72, 74], or the free glycan can be fluorescently labeled (d) to enable fluorescence [72]. Alternatively (a) can be digested using IdeS, which cleaves specifically below the hinge to yield Fab and glycosylated Fc subunits (e). Endoglycosidase digestions alongside IdeS can be used to simplify Fc [86] (f) or HC [53] (f) glycan variants into groups with and without core fucose.

10.2.2 Hydrogen/Deuterium Exchange Mass Spectrometry (HDX-MS) to Characterize Monoclonal Antibody Structure

HDX is a protein labeling technique whereby accessible protons within the protein structure are replaced with deuterium through sample dilution in a deuterated buffer [90]. The exchange with deuterium occurs for backbone peptide bond hydrogens and amino acid side chain hydrogens, the rate of which depends upon their structural connection and/or their location [91]. The rate of exchange for different labile hydrogen atoms is also highly dependent upon pH [92]. This pH dependence has been exploited in order to prevent deuterium back-exchange and migration; a pH of approximately 2.6, in conjunction with reduced temperatures (0 °C), is employed to allow sufficient time for MS measurements [93].

This technique was successfully coupled with a mass spectrometer (HDX-MS) in 1991 by Katta and Chait [94]. The exchanged deuterons each have a relative mass increase of one mass unit, which can be easily detected using a mass spectrometer [19]. The information obtainable from an HDX mass measurement experiment includes insights into conformational dynamics by observing the difference in proton exchange rates between the protein backbone and the aqueous solvent [95]; analysis of higher-order structure and dynamics of protein-based therapeutics [90, 96, 97] such as mAbs, both on their own and when interacting with target proteins or antigens [21]; and insights into the degree of glycosylation and the effects of the glycans upon structure [95]. HDX-MS can also be used to locate conformational change in the tertiary fold [98], currently achievable at the peptide level [99]. In addition, paratopes of mAbs and epitopes of antigens can be mapped [100, 101] enabling complementary pairings to be highlighted and proposed during new drug discovery [19]. Figure 10.5 shows the basic workflow of a typical HDX-MS setup.

The information-rich nature and sensitivity of HDX-MS have positioned it as an emerging technique for the analysis of biologics. Houde et al. [102] performed a direct comparison between a glycosylated IgG1 with the deglycosylated equivalent. With use of limited proteolysis to generate Fab and Fc fragments, it was demonstrated that comprehensive deglycosylation affected the Fc C_H2 domain significantly; however, the remainder of the protein was unaffected. This highlights how changes in glycosylation affect structures critical for Fc receptor binding and hence gives insights into the potential efficacy of an IgG1 mAb.

In 2013 Rose et al. [103] used HDX-MS to characterize the allosteric changes induced in the presence of a single point C_H3 domain mutation. This was shown to have a knock-on effect upon the connected C_H2 domains also. In

Figure 10.5 Workflow for a typical HDX-MS setup for the conformational dynamics analysis of monoclonal antibodies and other proteins. The mAb is combined with an equilibration buffer and then exposed to a deuterated labeling buffer for set time intervals. Experiments where no labeling occurs (t_o) are used as a control to determine the rate of deuterium uptake. Following the specified time period, the labeled mAb solution is quenched using harsh conditions: 0 °C and pH 2.6. The quench buffer also contains reducing and denaturing reagents to reduce the disulfide bonds and aid pepsin digestion. The peptide fragments are then separated and analyzed via LC-MS.

order to improve the clinical efficacy of mAb therapeutics, mutations could be introduced into the C_H3 domains to engineer C_H2 domain N-glycosylation, such as afucosylation for enhanced ADCC.

Over recent years, Zhang et al. have investigated mAb aggregation mechanisms [104] and the impact of chemical modification [96] and have developed a platform for automated data processing [105] all using HDX-MS. The aggregation mechanisms were demonstrated using two opposing extreme conditions: freeze–thaw cycles and heat denaturing [104]. Their experiments were performed on the therapeutic mAb bevacizumab, and interestingly the aggregates that formed upon exposure to multiple freeze–thaw cycles consisted of several "native" structural monomers, with little difference observed between the LC and HC fragments of the intact mAb and the freeze–thaw-induced aggregates. This demonstrated the highly stable nature of the mAb toward freezing stress. However, significant disordering under thermal stress was observed, as confirmed by the significant increase in HDX within the LC of the Fab region during aggregation. Three peptides within the CDR became increasingly solvent protected, relative to the native state upon aggregation, implying that these regions were responsible for some of the intermolecular interactions within bevacizumab aggregates.

Aggregation of mAbs has also been explored by Iacob et al. [106] where monomers were compared to their naturally occurring corresponding dimers. Subtle changes in deuterium uptake were observed for one of the samples in the C_H2 domain and close to the hinge region. Conclusions reported, based upon the HDX-MS data and other biophysical techniques, that the dimerization of the mAb (with changes to its structural dynamics) was due to an alternative dimerization pathway: domain swapping at the hinge region. For the mAb with no recorded exchange rate differences, it was proposed that the dimerization pathway predominantly involved C_H2 amino acid side chains where the exchange rates were not measurable via HDX-MS.

Zhang et al. also used HDX-MS to detect local chemical modifications within IgG1 mAb therapeutics such as asparagine deamidation, methionine oxidation, and aspartic acid isomerization [96], in order to study their conformational impact. These modifications can ultimately affect the safety and efficacy of mAb therapeutics through changes in thermal stability, antigen binding capabilities, and Fc effector functions. For methionine oxidation, impact was observed close to the modification site and in some cases also affected the residues linking the C_H2 and C_H3 domains. Through the comparison of a glycosylated mAb and an aglycosylated mAb, the authors demonstrated the importance of N-glycans upon the effect of methionine oxidation; significantly greater conformational changes were observed for the deglycosylated mAb. This supported the idea that N-glycans increase the structural stability of the C_H2 domain. Following on from this, it was demonstrated that aglycosylated mAbs also have decreased thermal stability and significantly increased levels of aggregation as compared to glycosylated mAbs. It was postulated that the effects upon the deglycosylated mAb conformation, following methionine oxidation, could be the structural requirements for changes toward thermal stability and aggregation. The effects of aspartic acid isomerization and asparagine deamidation upon the local mAb conformation were found to be minimal unless sufficient accumulation of succinimide intermediate was present. All of their HDX-MS observations were confirmed using structural modeling.

A combination of IM-MS and HDX-MS analyses has been reported by Edgeworth et al. detailing the conformational and dynamic properties of IgG Fc-engineered variants with reduced thermal stability [107]. Engineered to be active for reduced immune system recruitment and to increase serum half-life meant the samples exhibited increased thermodynamic destabilization. Using these techniques, the authors were able to show that the mutations effected a local destabilization within the C_H2 domain (HDX-MS) without affecting the global conformation of the Fc region (IM-MS). Effected HDX-MS has shown that drug binding to form an ADC does not drastically affect the conformation of the stand-alone mAb [98]. The authors compared interchain cysteine-linked IgG1 ADCs with the equivalent drug-free mAbs and observed 90% similarity between their HDX kinetics. The minor differences observed were a slight increase in the structural dynamics of the Fc region of the ADCs. Upon further testing, these differences were found to be attributable to the absence of intact interchain disulfide bonds and not the presence of the attached drug. Similarly, HDX-MS has been incorporated into comparability testing workflows between biosimilar candidates and originator products and revealed no difference between the exchange rate dynamics [108].

For a mAb therapeutic where a secondary effector function is part of the mechanism of action, receptor binding *in vivo* followed by induction of an immune response is required. As discussed previously, N-linked glycosylation in the Fc domain can affect the binding affinity between IgG and Fc receptors (FcR) and hence ADCC potency. Using HDX-MS, Jensen et al. [109] were able to show that analogous N-linked glycosylation present on neonatal FcRs (FcRn) did not affect IgG binding, implying that binding affinities are due solely to IgG glycosylation. Further HDX-MS experiments also revealed the expected conformational role for the Fab region in IgG1 mAbs with respect to receptor binding. Significant protection from HDX was observed for specific regions within the Fab arms upon receptor binding, indicating the presence of conformational interactions between the receptor, Fc and Fab regions. Generation of a 3D homology model for a full-length IgG1 in complex with FcRn confirmed the feasibility of interactions between the Fab and bound receptor.

Despite HDX-MS being well established, some limitations remain: one issue being the back-exchange of deuterium, as mentioned previously, and the effects that the quenching conditions may have upon protein structure, which could confound the results. This effect needs to be carefully considered for different protein systems [95]. In addition, for antibody–antigen interactions when the binding stoichiometry is very heterogenic, the effect of binding on each component may be difficult to determine. The complexity of the analysis and in particular the data processing has held HDX-MS back from routine use [110]. With the involvement of so many steps – deuterated incubation, proteolytic digestion, chromatographic separation, and MS identification, there is the requirement for an experienced analyst and comprehensive data analysis packages. In summary it is a powerful technique, but more akin to NMR in terms of the length of time needed to solve a protein structure, relative to an experiment based on MS alone.

10.2.3 Native Mass Spectrometry and the Use of IM-MS to Probe Monoclonal Antibody Structure

Native MS enables retention of key structural attributes such as noncovalent interactions within the tertiary and quaternary structures of large protein complexes and can be usefully applied to determine mass and stoichiometric information [111]. The basics of native MS involve using electrospray ionization (ESI) or more usually nanoelectrospray ionization (nESI) with the analyte dissolved in an aqueous volatile buffer, commonly ammonium acetate or ammonium bicarbonate. Most mAb samples are not presented in these buffer conditions, so desalting spin cartridges may be used to buffer exchange the samples prior to analysis. These can also serve to help remove unwanted salts such as sodium and potassium from the solution, which would otherwise form adducts with the sample during ionization, leading to peak broadening [84]. The ionization process is made as soft as possible with the use of lower spray potentials, lower source temperature, and reduced or no desolvation gas flow and by throttling the vacuum in the first desolvation stages of the mass spectrometer [112, 113]. Under these experimental conditions, noncovalent interactions can be retained, as evidenced by stoichiometry data and the preservation of protein–protein or protein–ligand interactions. The charge states observed with native ESI are generally lower, and the charge distributions are commonly narrower in comparison with analytical ESI, meaning data presents with lower charge, z (and higher mass-to-charge ratio, m/z). This can be used to probe mAb–antigen interactions, to examine heterogeneity and glycan content in the product, and most powerfully to check for formulation or storage-dependent aggregation [56, 95, 111, 114].

Native MS approaches coupled to IM-MS are hybrid experiments that directly probe the structures of biological macromolecules, such as mAbs, and their complexes as well as providing mass and stoichiometry [20, 115, 116]. Ion mobility separates ions by pulsing them through a cell filled with an inert buffer gas in the presence of an electric field. The time it takes for an ion to traverse the cell is based on the charge, size, and shape of an ion and its interactions with the buffer gas at a given temperature, resulting in a value termed the rotationally averaged collision cross section (CCS). When coupled to a mass spectrometer, ions are now separated according to two parameters, the mobility (K) and m/z. It is therefore possible to distinguish coincident m/z species based on their oligomeric order and protein conformations. IM-MS for intact protein analysis started in the mid-1990s led by a few groups who used homemade instruments, for example, Jarrold and Clemmer et al. [117, 118] and also following the development of the first commercial traveling wave IM-MS (TWIMS) instrument by Bateman and coworkers in 2006 [119, 120].

Several types of ion mobility–mass spectrometers are available; however, antibody characterization to date has predominantly been reported using TWIMS [20, 116, 121, 122]. More recently use of drift time ion mobility–mass spectrometry (DT IM-MS) has also begun to show promise in this field [115, 123]. For DT IM-MS, CCS can be extrapolated directly from the measured arrival time distributions via the Mason–Schamp equation [115]. When entering the drift tube, ions experience a force imposed by the electric field and collisions with neutral gas molecules. If the applied field intensity $(E = V/L)$ (where L is the length of the drift cell) is sufficiently low and the gas number density (N) sufficiently high, ions will move with a constant velocity called a drift velocity $(v_d = L/t_d)$. This velocity is proportional to the electric field intensity, with the ion mobility (K) being defined as the proportionality constant between v_d and E:

$$v_d = KE \tag{10.1}$$

$$t_d = \frac{L^2}{KV} \tag{10.2}$$

From this experimental measurement of K, it is possible to obtain the experimental ^{DT}CCS of an ion of mass M in the presence of a buffer gas of mass m using

$$K \approx \frac{3}{16} \frac{q}{N} \left(\frac{1}{M} + \frac{1}{m} \right)^{1/2} \left(\frac{2\pi}{k_B T} \right)^{1/2} \frac{12}{\Omega_D T} \tag{10.3}$$

For TWIMS, ^{TW}CCS can be obtained from the arrival time distributions of each ion indirectly using a calibration method [124, 125].

Application of IM-MS to mAbs has been used to solve a number of analytical challenges, from disulfide heterogeneity detection [122] to quantitative analysis of the folding status of the protein [126] as well as to determine the CCS of intact IgG [21, 115, 116, 127]. Currently IM-MS provides structural information at the global level whereby different conformational families can be resolved. For site-specific information, complementary techniques such as top-down MS and fast photochemical oxidation of proteins (FPOP) can be applied to obtain information for conformational changes at the regional and residue levels, respectively [128].

As for HDX-MS, information can also be obtained using IM-MS regarding the induced conformational changes upon the binding of the epitope and paratope of the antigen and intact mAb, respectively [38]. mAbs are inherently flexible, existing in many conformations, in equilibrium, with only subtle differences between them. These subtle changes are distinguishable by IM-MS, whereas traditional techniques are only capable of resolving large shifts in structure [127]. Pritchard et al. discussed the importance of knowing which different conformers are present when looking at the interaction of antigen proteins with antibodies. Different conformers may bind in significantly different ways, and hence performing an immunoassay without taking into account different conformational families and their relative abundances could potentially lead to an inaccurate conclusion upon biological activity [127].

As previously mentioned, the biological activity of mAbs can be significantly affected by PTMs, such as N-glycosylation [45, 46]. These PTMs were probed by Damen et al. [20] who exploited the three-stage IM cell of the SYNAPT HDMS instrument to interrogate the N-glycosylation profiles of different trastuzumab batches. Glycopeptides fragmented within the trap cell were then analyzed by TWIMS. With the transfer cell voltage set low, no further fragmentation was induced, providing MS/MS-style spectra to demonstrate the cleavage of glycosidic linkages. The authors then increased the trap cell voltage to form second-generation fragment ions due to peptide backbone cleavage, comparable to MS^3. This type of analysis was quick and informative, provides intact mass, and could be expanded upon in the future to further characterize the molecular heterogeneity of mAbs. Recent work by Ruotolo et al. used collision-induced unfolding (CIU) followed by IM-MS to demonstrate how varying levels of glycosylation affect the stability and/or conformation and hence unfolding behavior of mAbs; no glycosylation led to unfolding at lower collision energies compared with fully glycosylated equivalents [129]. Through use of collision energy ramps, unfolding was induced, and heat map plots were generated to give visual representation of the unfolding patterns. This method also enabled different IgG isoforms to be distinguished based on their differences in disulfide bond quantity and/or patterns of disulfide bonding.

In 2010, Bagal et al. used TWIMS to resolve structural isoforms of IgG2 mAbs that differed only in their disulfide bond heterogeneity [122]. Analysis of two different IgG2 mAbs showed the presence of 2–3 conformers. A deglycosylated version of the IgG2 mAb also maintained the multiple gas-phase conformers. These findings, alongside structural modeling, provided strong evidence toward the presence of IgG2 disulfide bond heterogeneity. The classical disulfide connectivity within IgG2 of four interchain linkages is referred to as IgG2-A; however, two additional isoforms are also known [130–133]. IgG2-B refers to the disulfide bond arrangement when one bond is formed between a

C_H1 domain cysteine instead of a hinge region cysteine on one side of the molecule, whereas if both sides are bound in this way then the isoform present is IgG2-A/B. The work by Bagal et al. demonstrated the ability of IM-MS to identify and resolve these different isoforms of IgG2. More recently an IgG2 isoform with similar disulfide structure but different biophysical and biochemical properties to IgG2-A was discovered [134, 135]. Reports now refer to the original as IgG2-A1 and the new isoform as IgG2-A2 [136]. These isoforms, along with IgG2-B, have been studied with HDX-MS by Zhang et al. [136]. Their results demonstrated that the IgG2-B isoform was a more compact global structure with regions of increased solvent protection just above the hinge region in comparison with the A1 isoform. This suggests that the two Fab arms in the B isoform are pulled closer together due to the nonclassical disulfide bond arrangement and that intramolecular contacts, in the form of Fab–Fab interactions, are present. These results were in line with thermal stability data also.

Typically IM experiments for mAb characterization are performed using the TWIMS setup; however, Pacholarz et al. have demonstrated how DT IM-MS can be used to explore the gas-phase dynamics of intact IgG1 and IgG4 mAbs [115]. The authors observed an increase in both the IgG isoform CCS values with higher charge state; as expected, however a larger increase was observed for the IgG4 mAb. Due to the linking of the IgG4 LC being positioned further away from the center of mass than in the IgG1 isoform, it was speculated that this would allow for increased flexibility of the Fab arms, hence a larger CCS. This finding was contradictory to those observed for solution-based IgGs 2 and 4 and therefore raises interesting questions for the effects of desolvation upon flexible proteins, with MS holding the key.

The flexibility of IgG Fab arms has led to the development of bsAbs whereby two different IgG half molecules (one LC and one HC) recombine during a Fab-arm exchange (FAE) process to form a chimeric bsAb. These hybrid proteins are of particular interest as they have the potential to simultaneously bind two different antigens. The monitoring of the FAE process and of bsAb formation was performed using TWIMS for the first time by Debaene et al. [116]. Two humanized IgG4 antibodies were analyzed separately and then combined with the addition of a mild reducing agent to facilitate bsAb formation for CCS determination. The SYNAPT G2 IM-MS instrument was capable of differentiating between the three different glycosylated mAbs despite molecular weights that differed by only ~0.5%. In addition native MS was used to monitor the formation of bsAb as a function of time, demonstrating how MS can be used for kinetic measurements as well. With the scope for being able to identify intramolecular differences between structural isoforms and for recognizing structural changes induced upon antigen binding, IM-MS could become increasingly popular within the field of mAb therapeutic development.

More recently, Pacholarz et al. have implemented variable temperature IM-MS (VT-IM-MS) to probe thermal stability of therapeutic mAbs and their fragments [123]. Enhancing the stability of biologics is aimed to reduce aggregation and improve production consistency. Protein engineering is used to shift the mAb away from an aggregation-prone state by increasing the thermodynamic stability of the native fold, which may in turn alter conformational flexibility [137]. VT-IM-MS has been used to study the unfolding of monomeric proteins and more recently for the unfolding and dissociation of large protein complexes [138–141]. Pacholarz et al. used VT-IM-MS to analyze mAbs and observe changes in the conformations of isolated proteins as a function of temperature from 300 to 550 K. The temperature at which the maximum structural collapse occurs correlates remarkably well with loss of quaternary structure and the solution-based melting temperature (Tm). Diversity in the extent of collapse and subsequent unfolding are rationalized by differences in the hinge flexibility and strength of noncovalent interactions at the C_H3 domain interface among IgG subclasses. VT-IM-MS provides insights into the structural thermodynamics of mAbs and was presented as a promising tool for thermal stability studies for proteins of therapeutic interest.

10.3 Advanced Applications

10.3.1 Quantifying Glycosylation

Analysis of large polypeptides from enzymatically digested intact antibodies has been reported to enable the quantification of N-linked glycosylation and in particular the levels of afucosylation [113]. These middle-up approaches are summarized in the workflow in Figure 10.4.

By incorporating endoglycosidases into an analysis, the oligosaccharide population can be simplified into two groups: endoglycosidases cleave between the two GlcNAc residues of the chitobiose core, thus generating mAbs or fragments with either GlcNAc or GlcNAc + Fuc moieties still attached. LC-MS analysis of reduced mAbs enzymatically digested with EndoF2 and EndoH has been reported, which enabled an estimation of both fucosylation and afucosylation [142]. A similar approach was employed by Goetze et al. [86]. However instead of reduction, the immunodegrading enzyme (IdeS) was used to generate Fab and glycosylated Fc/2 subunits prior to analysis. The IdeS cleavage site is located in the hinge region (−PELLG/G− in IgG1) [143, 144] and simplifies the complexity of the analysis by concentrating on the glycosylated Fc/2 fragment. Determination of afucosylation and high mannose levels have also been demonstrated using an IdeS and endoglycosidase strategy by Firth et al. (paper submitted) [145] via the application of EndoS2. EndoS has a lower affinity for high mannose compared with EndoS2, and therefore a sequential application of the two enabled the quantification of the high-mannose species based upon the change in relative MS signals observed.

10.3.2 Antibody-Drug Conjugates

Hydrophobic interaction chromatography (HIC) can be used for the determination of drug-to-antibody ratio (DAR) and to monitor drug distribution for cysteine-linked ADCs; molecules with more drug conjugated exhibit increased hydrophobicity [59, 146]. DAR calculations can also be performed following the reduction of the ADC and the analysis of the composite LC and HC by RPLC with UV detection [147, 148]. MS, however, is currently emerging as a new method of choice for ADC characterization and evaluation of DAR due to the additional structural information offered [21, 56, 84, 149, 150].

In the case of lysine- or engineered cysteine-conjugated ADCs (types 1 and 2, respectively), the interchain disulfide bonds between the HC and LC of the antibody remain intact. This means that DAR can be determined using LC-MS methods employing organic mobile phases [63, 151]. ADCs conjugated at the interchain cysteine residues produce a mixture of noncovalent mAb tetramers (two LC and two HC with a variable number of drug molecules attached) after reduction of interchain disulfide bonds; hence the application of more classic LC-MS analytical strategies would result in the dissociation of this noncovalent ADC. The organic solvents used disrupt the noncovalent nature of the ADC structure producing conjugated LC and HC fragments, making it hard to define average DAR [152]. Careful sample preparation needs to be employed for the analysis of interchain cysteine-linked ADCs, especially if using denaturing conditions, to avoid ADC denaturation, which in turn could affect average DAR calculations.

There are a few different methods reported in the literature that employ native MS for the intact mass and DAR analyses of these cysteine-linked ADCs [56, 153, 154]. In 2012, Valliere-Douglass et al. [153] presented the first method for the rapid determination of intact mass and DAR for interchain cysteine-conjugated ADCs. Following ESI-MS analysis of deglycosylated ADCs, the authors observed some dissociation of the noncovalent ADCs into conjugated LC and HC but reported levels sufficiently low enough to not impact the relative levels of the intact drug-loaded species. The results acquired by MS were similar to those obtained from the orthogonal HIC method.

In 2013, Chen et al. [56] reported an MS method engaging limited enzymatic digestion, nESI, and native MS for the direct determination of intact mass and calculation of the average DAR of cysteine-linked ADCs. Utilizing an nESI source provided improved ionization, sensitivity, and increased confidence in DAR determination. The cytotoxic drug monomethyl auristatin E (MMAE) is reported to have high hydrophobicity. This meant that high-drug-load species were underestimated due to reduced affinity for protons causing inefficient ionization. To minimize this ion suppression and equalize ionization efficiency among species, proteolytic drug removal was induced. The hydrophobic moiety was selectively cleaved, using cysteine protease, from the ADC while leaving the linker attached indicative of the original drug load. The DAR values obtained following enzymatic digestion were more comparable with those determined by HIC methods.

Hengel et al. [154] have described a customized sample preparation method for cysteine-linked ADCs from an *in vivo* source using native SEC LC-MS for accurate drug load distribution determination. The results confirmed that higher-drug-loaded species were undetectable shortly after dosing; however, after 30 min the predominant species was the ADC with four conjugated drug molecules. The observed shift was likely due to clearance of the higher-loaded species and/or deconjugation. The MS dimension of the analysis was also capable of quantifying the *in vivo* increase in ADC mass heterogeneity. Overall this method provides useful PK and stability insights into the *in vivo* changes in drug load distribution and ADC structure.

Shortly afterward, Debaene et al. [55] developed a semiquantitative method for the determination of average DAR and DAR distributions based on native IM-MS data. The authors utilized TWIMS to conformationally characterize and separate each drug-loaded species and observed constant drift time/CCS shifts between consecutive DAR variants. This indicated that no conformational changes were induced upon drug binding and that all shifts were attributable to changes in mass. By plotting the intensities for each drift peak for the different drug binding stoichiometries against charge state, the areas under each curve were found to be representative of the relative abundances of the different ADC species present. This enabled the calculation of an average DAR value with excellent agreement toward HIC and native MS obtained data as well as accurate DAR distribution profiling.

An Orbitrap Exactive® mass spectrometer with extended mass range (EMR) [Thermo/Finnigan] enabled the Heck group to isolate native ADCs as part of tandem MS/MS experiments for localizing bound drug moieties [155]. The cysteine-linked ADC brentuximab vedotin was studied and the authors present a direct approach for the assessment of drug distribution and localization. In the same paper, insights into hexamerization of IgG1 mutant assemblies (the mechanism by which IgG mAbs facilitate complementary activation) were reported. Their tandem MS/MS workflow enabled insight into the stoichiometry of antigen binding, even for protein assemblies with molecular weights over 1 million Da, and concluded that IgG hexamerization does not significantly affect bivalent antigen binding. These results were expected to an extent as IgG hexamer interactions have previously been shown to be confined to the Fc region [155].

Charge reducing agents have been used to reduce spectral complexity. Recently, Pacholarz et al. [156] have used the charge reducing agent triethylammonium acetate (TEAA) to help preserve intact mAb structure during

ADC analysis and DAR calculation (paper submitted). Marcoux et al. [157] also used native MS in combination with the charge state reducing reagent imidazole to reduce the charge states observed for the lysine-linked ADC trastuzumab emtansine (T-DM1) to below 20+. With the addition of imidazole and use of a high-resolution Orbitrap® mass spectrometer, the authors were able to avoid the overlapping of peaks corresponding to the D0 and D8 species of sequential charge states (zero and eight drug moieties linked, respectively), even without deglycosylation. It was emphasized also that DAR calculations should be performed from raw data to avoid the introduction of bias upon the application of deconvolution parameters. A solution to this type of bias was recently published by Firth et al. [57] whereby use of enzymatic digestions meant that default deconvolution parameters could be applied to the raw data. MaxEnt1 (UNIFI, Waters) parameters, applied to data for intact or fragmented proteins greater than 40 kDa, can require significant manipulation, which often leads to skewed results that would invariably affect DAR calculations. However, the addition of middle-up enzymatic cleavage steps using IdeS and DTT reduction meant that three different fragments (all ~25 kDa) for thiol-linked ADCs were generated and then analyzed using their in-house LC-MS method with minimal chromatographic separation. Analysis of LC, Fc, and Fd fragments meant that default parameters could be used, which offered mass accuracy within less than 20 ppm (<1 Da).

In terms of MS developments for ADC analysis, the most recently published method was proposed by Gautier et al. [158] who demonstrated the use of a tandem mass tag (TMT) to identify "hotspot" lysine residues on IgG1 mAbs. TMT labeling relies upon the same chemistry as drug conjugation in ADCs, and therefore this approach enables straightforward identification of mAbs suitable for lysine-linked ADC development. Attachment of the TMT label was confirmed using LC-MS/MS, and the comparison of glycosylated and deglycosylated IgG1s concluded that glycan presence can sterically protect lysine residues from TMT labeling and hence block cytotoxic drug conjugation. This technique could provide fundamental information with regard to controlling drug load and specificity within lysine-linked ADCs.

10.3.3 Biosimilar Characterization

Comparative studies between authentic batches of Herceptin®, manufactured in different geographical locations (the EU and the United States), and a possible biosimilar candidate of trastuzumab have been carried out by Firth et al. (paper submitted) using LC-MS and a series of enzymatic digestions, as described in Section 3.1, to quantify the glycosylation profiles. The relative MS responses, presented as a function of percent, are shown in Table 10.1. The trastuzumab biosimilar sample clearly contained significantly less afucosylated glycan. With the addition of surface plasmon resonance binding analysis and

Table 10.1 Averaged relative MS responses for the quantification of nonglycosylated, afucosylated, and fucosylated Fc/2 subunits of EU and US Herceptin® and a trastuzumab sample following digestion with IdeS, EndoS, and EndoS2.

	Mean relative MS response (%)		
Sample	Fc/2	Fc/2 + GlcNAc	Fc/2 + GlcNAc + Fuc
US Herceptin®	0.957	10.2	88.9
EU Herceptin®	0.951	14.7	84.3
Trastuzumab	0.476	4.35	95.2

ADCC potency studies, the trastuzumab sample was demonstrated to have significantly lower binding affinities toward the FcɣIIIa receptor and reduced potency of approximately 50% compared with the authentic Herceptin® batches. It was observed that there were also differences between the two batches of Herceptin®, with the EU lot containing a higher level of afucosylation compared with the US lot. With afucosylation playing a role in the secondary mechanism of action, it was interesting that different lots of the approved, authentic Herceptin® contained different proportions of this important attribute, but with little effect on potency. The findings illustrated the importance of evaluating multiple lots of innovator product in order to understand the structure and the effect of variability in attributes such as glycosylation toward defining acceptance criteria for biosimilarity.

Work published by Damen et al. also reports differences in the glycosylation profiles between different batches of the originator trastuzumab [20]. Their analysis was carried out at the intact level using LC-ESI-MS, and deconvoluted masses were closely matched to the calculated masses for the main glycoforms. Although qualitatively the different batches were the same, quantitatively they were not. Comparison of a candidate biosimilar with the equivalent innovator mAb characterized by LC-MS highlighted how MS can be used to identify mass differences and therefore differences in amino acid sequence that ultimately decide whether a proposed biosimilar is likely to meet acceptance criteria even at the intact level [68]. The authors report observing a difference of 64 Da between the two samples. Following reduction with DTT, the location of the mass difference could be attributed to the HC subunits (32 Da on each). Additional nanoLC-MS/MS identified variation of two amino acids within the HC, which accounted for the change in mass. From this information, it could be concluded that the biosimilar mAb was derived from a different allotype altogether. The ability to identify key quality attributes, such as differences in amino acid sequence, in a quick and simple experiment, is essential for mitigating risk from a biosimilar campaign within the early stages. At the intact level, however, the mass measurement accuracy is not capable of differentiating small mass differences or minor modifications such as deamidation or oxidation. Therefore bottom-up MS approaches have been studied by Chen et al. [159]. A three-step strategy was employed including peptide sequencing via collision-induced dissociation/electron transfer dissociation (CID/ETD), alongside nontargeted comparisons of the tryptic map and targeted comparisons of minor modifications for three samples: originator trastuzumab and two biosimilar products, one without known variants and one with two amino acid variants. Varying levels of modifications were detected across all three samples.

10.4 Concluding Remarks

MS as a characterization technique for large, complex structures such as mAbs has advanced greatly in recent years. The flexible nature of the technique has enabled mass spectrometers to be coupled to a series of other alternative analytical techniques such as liquid chromatography systems and ion mobility separation devices, delivering an additional separation and/or characterization dimension to an analysis. With the relatively recent introduction of biosimilar and ADC mAb-based therapeutics into the market, the need for a robust characterization technique has never been stronger, and MS is viewed as the most promising contender. To realize the true potential of MS toward biosimilar candidate characterization, however, comprehensive analysis of originator mAbs needs to be performed, with acceptance criteria implemented, which account for the mammalian cell-induced variability that is inherently present. PTMs such as glycosylation have been shown to be essential for potency and efficacy of many mAb-based therapeutics due to their critical role in effector function cell recruitment. MS currently plays a key role in the identification and quantification of glycosylation, and with the inclusion of HDX and IM platforms, MS is elevated to a position where it is capable of understanding the effects of different glycosylation profiles upon conformation and receptor binding as well as the dynamics associated with antigen binding. Due to the recent development of mAb-based therapeutics, such as ADCs, the characterization techniques reported are still in their infancy despite significant advances being made. Undoubtedly MS will continue to offer insight into these complex protein systems, and it is hard to envisage another technique that will rival the high-throughput, high-precision and high-accuracy, and cost-effective advantages that MS has to offer.

References

1 Pucca MB, Bertolini TB, Barbosa JE, Galina SVR, Porto GS. Therapeutic monoclonal antibodies: ScFv patents as a marker of a new class of potential biopharmaceuticals. *Braz J Pharm Sci* 2011;**47**(1):31–39.

2 Beck A, Reichert JM. Therapeutic Fc-fusion proteins and peptides as successful alternatives to antibodies. *MAbs* 2011;**3**(5):415–416.

3 Wu B, Sun Y-N, Wu B, Sun Y-N. Pharmacokinetics of peptide-Fc fusion proteins. *J Pharm Sci* 2013;**103**:53–64. J Pharm Sci 2014;103(6):1583–1928.

4 Nelson AL. Antibody fragments: Hope and hype. *MAbs* 2010;**2**(1):77–83.

5 May C, Sapra P, Gerber HP. Advances in bispecific biotherapeutics for the treatment of cancer. *Biochem Pharmacol* 2012;**84**(9):1105–1112.

6 Kontermann RE. Dual targeting strategies with bispecific antibodies. *MAbs* 2012;**4**(2):182–197.

7 Lianos GD, Vlachos K, Zoras O, Katsios C, Cho WC, Roukos DH. Potential of antibody–drug conjugates and novel therapeutics in breast cancer management. *Onco Targets Ther* 2014;**7**:491–500.

8 Chari RVJ, Miller ML, Widdison WC. Antibody–drug conjugates: An emerging concept in cancer therapy. *Angew Chem Int Ed* 2014;**53**(15):3796–3827.

9 Steiner M, Neri D. Antibody–radionuclide conjugates for cancer therapy: Historical considerations and new trends. *Clin Cancer Res* 2011;**17**(20):6406–6416.

10 Hess C, Venetz D, Neri D. Emerging classes of armed antibody therapeutics against cancer. *MedChemComm* 2014;**5**:408.

11 Moran N. Biotech innovators jump on biosimilars bandwagon. *Nat Biotechnol* 2012;**30**(4):297–299.

12 Agostini C, Canonica GW, Maggi E. European Medicines Agency guideline for biological medicinal products: A further step for a safe use of biosimilars. *Clin Mol Allergy* 2015;**13**(3):15–16.

13 Schellekens H. The first biosimilar epoetin: But how similar is it? *Clin J Am Soc Nephrol* 2008;**3**(1):174–178.

14 Mellstedt H. Implications of the development of biosimilars for cancer treatment. *Future Oncol* 2010;**6**(7):1065–1067.

15 ICH. ICH Harmonised Tripartite Guideline. Comparability of Biotechnological/Biological Products Subject to Changes in Their Manufacturing Process Q5E. 2005.

16 Chames P, Van Regenmortel M, Weiss E, Baty D. Therapeutic antibodies: Successes, limitations and hopes for the future. *Br J Pharmacol* 2009;**157**(2):220–233.

17 Gomollón F. Biosimilars: Are they bioequivalent? *Dig Dis* 2014;**32**:82–87.

18 Nellore R. Regulatory considerations for biosimilars. *Perspect Clin Res* 2010;**1**(1):11–14.

19 Berkowitz SA, Engen JR, Mazzeo JR, Jones GB. Analytical tools for characterizing biopharmaceuticals and the implications for biosimilars. *Nat Rev Drug Discov* 2012;**11**(7):527–540.

20 Damen CWN, Chen W, Chakraborty AB, van Oosterhout M, Mazzeo JR, Gebler JC, et al. Electrospray ionization quadrupole ion-mobility time-of-

flight mass spectrometry as a tool to distinguish the lot-to-lot heterogeneity in N-glycosylation profile of the therapeutic monoclonal antibody trastuzumab. *J Am Soc Mass Spectrom* 2009;**20**(11):2021–2033.

21 Beck A, Wagner-Rousset E, Ayoub D, Van Dorsselaer A, Sanglier-cianfe S. Characterization of therapeutic antibodies and related products. *Anal Chem* 2013;**85**:715–736.

22 Rojas R, Apodaca G. Immunoglobulin transport across polarized epithelial cells. *Nat Rev Mol Cell Biol* 2002;**3**(12):944–955.

23 Glennie MJ, van de Winkel JGJ. Renaissance of cancer therapeutic antibodies. *Drug Discov Today* 2003;**8**(11):503–510.

24 Junqueira LC, Carneira J, Kelley RO. *Basic Histology*. McGraw-Hill Publishing Co.; 1998.

25 Poljak RJ, Amzel LM, Avey HP, Chen BL, Phizackerley RP, Saul F. Three-dimensional structure of the Fab' fragment of a human immunoglobulin at 2,8-A resolution. *Proc Natl Acad Sci U S A* 1973;**70**(12):3305–3310.

26 Janeway CAJ, Travers P, Walport M, Shlomchik MJ. *Immunobiology: The immune system in health and disease*. 5th ed. Garland Science: New York; 2001.

27 Wang W, Singh S, Zeng DL, King K, Nema S. Antibody structure, instability, and formulation. *J Pharm Sci* 2007;**96**(1):1–26.

28 Pink JR, Milstein C. Inter heavy-light chain disulphide bridge in immune globulins. *Nature* 1967;**214**:92–94.

29 Milstein C. The heterogeneity of human immunoglobulins. *Biochem J* 1968;**110**(3):26–27.

30 Liu H, May K. Disulfide bond structures of IgG molecules: Structural variations, chemical modifications and possible impacts to stability and biological function. *MAbs* 2012;**4**(1):17–23.

31 Milstein C, Frangione B. Disulphide bridges of the heavy chain of human immunoglobulin G2. *Biochem J* 1971;**121**(2):217–225.

32 Melmer M, Stangler T, Schiefermeier M, Brunner W, Toll H, Rupprechter A, et al. HILIC analysis of fluorescence-labeled N-glycans from recombinant biopharmaceuticals. *Anal Bioanal Chem* 2010;**398**(2):905–914.

33 Wright A, Tao M, Kabat EA, Morrison SL. Antibody variable region glycosylation : Position effects on antigen binding and carbohydrate structure. *Eur Mol Biol Organ* 1991;**10**(10):2717–2723.

34 Wang X, Kumar S, Buck PM, Singh SK. Impact of deglycosylation and thermal stress on conformational stability of a full length murine igG2a monoclonal antibody: Observations from molecular dynamics simulations. *Proteins Struct Funct Bioinform* 2013;**81**(3):443–460.

35 Kayser K, Lin N, Allison D, Donahue L, Caple M. Cell line engineering methods for improving productivity. *Bioprocess Int* 2006;**4**:6–13.

36 Noh SM, Sathyamurthy M, Lee GM. Development of recombinant Chinese hamster ovary cell lines for therapeutic protein production. *Curr Opin Chem Eng* 2013;**2**(4):391–397.

37 Walsh G. Biopharmaceutical benchmarks 2010. *Nat Biotechnol* 2010;**28**(9):1–10.

38 Beck A, Sanglier-cianfe S, Van Dorsselaer A. Biosimilar, biobetter and next generation antibody characterization by mass spectrometry. *Anal Chem* 2012;**84**:4637–4646.

39 Sliwkowski MX, Mellman I. Antibody therapeutics in cancer. *Science* 2013;**341**:1192–1198.

40 An Y, Zhang Y, Mueller H, Shameem M, Chen X. A new tool for monoclonal antibody analysis – application of IdeS proteolysis in IgG domain-specific characterization. *MAbs* 2014;**6**(4):1–15.

41 Leymarie N, Zaia J. Effective use of mass spectrometry for glycan and glycopeptide structural analysis. *Anal Chem* 2012;**84**(7):3040–3048.

42 Bharti A, Ma PC, Salgia R. Biomarker discovery in lung cancer – promises and challenges of clinical proteomics. *Mass Spectrom Rev* 2009;**26**(3):451–466.

43 Yamane-Ohnuki N, Satoh M. Production of therapeutic antibodies with controlled fucosylation. *MAbs* 2009;**1**(3):230–236.

44 Jefferis R. Glycosylation of recombinant antibody therapeutics. *Biotechnol Prog* 2005;**21**(1):11–16.

45 Fernandes D. Demonstrating comparability of antibody glycosylation during biomanufacturing. *Eur Biopharm Rev* 2005:106–110.

46 Roger SD. Biosimilars: How similar or dissimilar are they? *Nephrology (Carlton)* 2006;**11**(4):341–346.

47 Ayoub D, Jabs W, Resemann A, Evers W. Correct primary structure assessment and extensive glyco-profiling of cetuximab by a combination of intact, middle-up, middle-down and bottom-up ESI and MALDI mass spectrometry techniques. *MAbs* 2013;**10**:699–710.

48 Mizushima T, Yagi H, Takemoto E, Shibata-Koyama M, Isoda Y, Iida S, et al. Structural basis for improved efficacy of therapeutic antibodies on defucosylation of their Fc glycans. *Genes Cells* 2011;**16**:1071–1080.

49 Kanda Y, Yamada T, Mori K, Okazaki A, Inoue M, Kitajima-Miyama K, et al. Comparison of biological activity among nonfucosylated therapeutic IgG1 antibodies with three different N-linked Fc oligosaccharides: The high-mannose, hybrid, and complex types. *Glycobiology* 2006;**17**(1):104–118.

50 Okazaki A, Shoji-Hosaka E, Nakamura K, Wakitani M, Uchida K, Kakita S, et al. Fucose depletion from human IgG1 oligosaccharide enhances binding enthalpy and association rate between IgG1 and FcγRIIIa. *J Mol Biol* 2004;**336**:1239–1249.

51 Shields RL, Lai J, Keck R, O'Connell LY, Hong K, Gloria Meng Y, et al. Lack of fucose on human IgG1 N-linked oligosaccharide improves binding to human FcGammaRIII and antibody-dependent cellular toxicity. *J Biol Chem* 2002;**277**(30):26733–26740.

52 Hudis CA. Trastuzumab — mechanism of action and use in clinical practice. *N Engl J Med* 2007;**357**(1):39–51.

53 Shen Y, Liu H. Methods to determine the level of afucosylation in recombinant monoclonal antibodies. *Anal Chem* 2010;**82**(23):9871–9877.

54 Presta LG. Molecular engineering and design of therapeutic antibodies. *Curr Opin Immunol* 2008;**20**(4):460–470.

55 Debaene F, Amandine B, Wagner-Rousset E, Colas O, Ayoub D, Corva N, et al. Innovative native MS methodologies for antibody drug conjugate characterization: High resolution native MS and IM-MS for average DAR and DAR distribution assessment. *Anal Chem* 2014;**86**:10674–10683.

56 Chen J, Yin S, Wu Y, Ouyang J. Development of a native nanoelectrospray mass spectrometry method for determination of the drug-to-antibody ratio of antibody–drug conjugates. *Anal Chem* 2013;**85**(3):1699–1704.

57 Firth D, Bell L, Squires M, Estdale S, McKee C. A rapid approach for characterization of thiol-conjugated ADCs and calculation of drug-antibody ratio by LC-MS. *Anal Biochem* 2015;**485**:34–42.

58 Hamann PR, Hinman LM, Hollander I, Beyer CF, Lindh D, Holcomb R, et al. Gemtuzumab ozogamicin, a potent and selective anti-CD33 antibody–calicheamicin conjugate for treatment of acute myeloid leukemia. *Bioconjug Chem* 2002;**13**(1):47–58.

59 Junutula JR, Raab H, Clark S, Bhakta S, Leipold DD, Weir S, et al. Site-specific conjugation of a cytotoxic drug to an antibody improves the therapeutic index. *Nat Biotechnol* 2008;**26**(8):925–932.

60 Trail PA, Willner D, Lasch SJ, Henderson AJ, Hofstead S, Casazza AM, et al. Cure of xenografted human carcinomas by BR96-doxorubicin immunoconjugates. *Science* 1993;**261**(5118):212–215.

61 Teicher BA, Chari RVJ. Antibody conjugate therapeutics: Challenges and potential. *Clin Cancer Res* 2011;**17**(20):6389–6397.

62 Hamblett KJ, Senter PD, Chace DF, Sun MMC, Lenox J, Cerveny CG, et al. Effects of drug loading on the antitumor activity of a monoclonal antibody drug conjugate. *Clin Cancer Res* 2004;**10**(425):7063–7070.

63 Wakankar A, Chen Y, Gokarn Y, Jacobson FS. Analytical methods for physicochemical characterization of antibody drug conjugates. *MAbs* 2011;**3**(2):161–172.

64 Wang J, Chow SC. On the regulatory approval pathway of biosimilar products. *Pharmaceuticals* 2012;**5**:353–368.

65 Beck A, Debaene F, Diemer H, Wagner-Rousset E, Colas O, Van Dorsselaer A, et al. Cutting-edge mass spectrometry characterization of originator, biosimilar and biobetter antibodies. *J Mass Spectrom* 2015;**50**:285–297.

66 Staub A, Guillarme D, Schappler J, Veuthey J-L, Rudaz S. Intact protein analysis in the biopharmaceutical field. *J Pharm Biomed Anal* 2011;**55**(4):810–822.

67 Schneider CK, Kalinke U. Toward biosimilar monoclonal antibodies. *Nat Biotechnol* 2008;**26**(9):985–990.

68 Xie H, Chakraborty A, Ahn J, Yu YQ, Dakshinamoorthy DP, Gilar M, et al. Rapid comparison of a candidate biosimilar to an innovator monoclonal antibody with advanced liquid chromatography and mass spectrometry technologies. *MAbs* 2010;**2**(4):379–394.

69 Strand V, Cronstein B. Biosimilars: How similar? *Intern Med J* 2014;**44**(3):218–223.

70 Walsh G. Biopharmaceutical benchmarks 2014. *Nat Biotechnol* 2014;**32**(10):992–1000.

71 Shah B, Jiang XG, Chen L, Zhang Z. LC-MS/MS peptide mapping with automated data processing for routine profiling of N-glycans in Immunoglobulins. *J Am Soc Mass Spectrom* 2014;**25**(6):999–1011.

72 Ahn J, Bones J, Yu YQ, Rudd PM, Gilar M. Separation of 2-aminobenzamide labeled glycans using hydrophilic interaction chromatography columns packed with 1.7 um sorbent. *J Chromatogr B Anal Technol Biomed Life Sci* 2010;**878**:403–408.

73 Higel F, Demelbauer U, Seidl A, Friess W, Sörgel F. Reversed-phase liquid-chromatographic mass spectrometric N-glycan analysis of biopharmaceuticals. *Anal Bioanal Chem* 2013;**405**(8):2481–2493.

74 Prater BD, Connelly HM, Qin Q, Cockrill SL. High-throughput immunoglobulin G N-glycan characterization using rapid resolution reverse-phase chromatography tandem mass spectrometry. *Anal Biochem* 2009;**385**(1):69–79.

75 Nielsen RG, Rickard EC, Santa PF, Sharknas DA, Sittampalam GS. Separation of antibody–antigen complexes by capillary zone electrophoresis, isoelectric-focusing and high-performance size-exclusion chromatography. *J Chromatogr* 1991;**539**(1):177–185.

76 Jayo RG, Thaysen-Andersen M, Lindenburg PW, Haselberg R, Hankemeier T, Ramautar R, et al. Simple capillary electrophoresis-mass spectrometry method for complex glycan analysis using a flow-through microvial interface. *Anal Chem* 2014;**86**(13):6479–6486.

77 Diepold K, Bomans K, Wiedmann M, Zimmermann B, Petzold A, Schlothauer T, et al. Simultaneous assessment of Asp isomerization and Asn deamidation in recombinant antibodies by LC-MS following incubation at elevated temperatures. *PLoS One* 2012;**7**(1):1–11.

78 Damen CWN, Rosing H, Schellens JHM, Beijnen JH. Quantitative aspects of the analysis of the monoclonal antibody trastuzumab using high-performance liquid chromatography coupled with electrospray mass spectrometry. *J Pharm Biomed Anal* 2008;**46**(3):449–455.

79 Gilar M, Yu Y-Q, Ahn J, Xie H, Han H, Ying W, et al. Characterization of glycoprotein digests with hydrophilic interaction chromatography and mass spectrometry. *Anal Biochem* 2011;**417**(1):80–88.

80 Wuhrer M, de Boer AR, Deelder AM. Structural glycomics using hydrophilic interaction chromatography (HILIC) with mass spectrometry. *Mass Spectrom Rev* 2009;**28**:192–206.

81 Pitt JJ. Principles and applications of liquid chromatography-mass spectrometry in clinical biochemistry. *Clin Biochem Rev* 2009;**30**:19–34.

82 Lauber MA, Yu Y-Q, Brousmiche DW, Hua Z, Koza SM, Magnelli P, et al. Rapid preparation of released N-glycans for HILIC analysis using a labeling reagent that facilitates sensitive fluorescence and ESI-MS detection. *Anal Chem* 2015;**87**:5401–5409.

83 Tarentino AL, Gómez CM, Plummer TH Jr. Deglycosylation of asparagine-linked glycans by peptide:N-glycosidase F. *Biochemistry* 1985;**24**(17):4665–4671.

84 Thompson NJ, Rosati S, Heck AJR. Performing native mass spectrometry analysis on therapeutic antibodies. *Methods* 2014;**65**(1):11–17.

85 Taron CH, Duke R. N-glycan composition profiling for quality testing of biotherapeutics. *BioPharm Int* 2015;**28**(12):59–64.

86 Goetze AM, Zhang Z, Liu L, Jacobsen FW, Flynn GC. Rapid LC-MS screening for IgG Fc modifications and allelic variants in blood. *Mol Immunol* 2011;**49**(1–2):338–352.

87 Stacey M, Nashabeh W. Carbohydrate analysis of a chimeric recombinant monoclonal antibody by capillary electrophoresis with laser-induced fluorescence detection. *Anal Chem* 1999;**71**(22):5185–5192.

88 Gennaro LA, Salas-Solano O. On-line CE-LIF-MS technology for the direct characterization of N-linked glycans from therapeutic antibodies. *Anal Chem* 2008;**80**(10):5185–5192.

89 Alvarez M, Tremintin G, Wang J, Eng M, Kao YH, Jeong J, et al. On-line characterization of monoclonal antibody variants by liquid chromatography-mass spectrometry operating in a two-dimensional format. *Anal Biochem* 2011;**419**:17–25.

90 Wei H, Mo J, Tao L, Russell RJ, Tymiak AA, Chen G, et al. Hydrogen/deuterium exchange mass spectrometry for probing higher order structure of protein therapeutics: Methodology and applications. *Drug Discov Today* 2014;**19**(1):95–102.

91 Englander SW, Sosnick TR, Englander JJ, Mayne L. Mechanisms and uses of hydrogen exchange. *Curr Opin Struct Biol* 1996;**6**(1):18–23.

92 Katta V, Chait BT. Hydrogen/deuterium exchange electrospray ionization mass spectrometry : A method for probing protein conformational changes in solution. *J Am Chem Soc* 1993;**115**:6317–6321.

93 Englander SW. Hydrogen exchange and mass spectrometry: A historical perspective. *J Am Soc Mass Spectrom* 2006;**17**(11):1481–1489.

94 Katta V, Chait BT. Conformational changes in proteins probed by hydrogen-exchange electrospray-ionization mass spectrometry. *Rapid Commun Mass Spectrom* 1991;**5**(4):214–217.

95 Thompson NJ, Rosati S, Rose RJ, Heck AJR. The impact of mass spectrometry on the study of intact antibodies: from post-translational modifications to structural analysis. *Chem Commun* 2013;**49**(6):538–548.

96 Zhang A, Hu P, Macgregor P, Xue Y, Fan H, Suchecki P, et al. Understanding the conformational impact of chemical modifications on monoclonal antibodies with diverse sequence variation using hydrogen/deuterium exchange mass spectrometry and structural modeling. *Anal Chem* 2014;**86**(7):3468–3475.

97 Mo J, Tymiak AA. Chen G. Structural mass spectrometry in biologics discovery: Advances and future trends. *Drug Discov Today* 2012;**17**(23–24):1323–1330.

98 Pan LY, Salas-Solano O, Valliere-Douglass JF. Conformation and dynamics of interchain cysteine-linked antibody–drug conjugates as revealed by hydrogen/deuterium exchange mass spectrometry. *Anal Chem* 2014;**86**(5):2657–2664.

99 Fitzgerald MC, West GM. Painting proteins with covalent labels: What's in the picture? *J Am Soc Mass Spectrom* 2009;**20**(6):1193–1206.

100 Zhang Q. Epitope mapping of a 95 kDa antigen in complex with antibody by solution-phase amide backbone hydrogen/deuterium exchange monitored by Fourier transform ion cyclotron resonance mass spectrometry. *Anal Chem* 2011;**83**:7129–7136.

101 Baerga-ortiz A, Hughes CA, Mandell JG, Komives EA. Epitope mapping of a monoclonal antibody against human thrombin by H/D-exchange mass spectrometry reveals selection of a diverse sequence in a highly conserved protein. *Protein Sci* 2002;**11**:1300–1308.

102 Houde D, Arndt J, Domeier W, Berkowitz S, Engen JR. Characterization of IgG1 conformation and conformational dynamics by hydrogen/deuterium exchange mass spectrometry. *Anal Chem* 2009;**81**(7):2644–2651.

103 Rose RJ, Van Berkel PHC, Van Den Bremer ETJ, Labrijn AF, Vink T, Schuurman J, et al. Mutation of Y407 in the CH3 domain dramatically alters glycosylation and structure of human IgG. *MAbs* 2013;**5**(2):219–228.

104 Zhang A, Singh SK, Shirts MR, Kumar S, Fernandez EJ. Distinct aggregation mechanisms of monoclonal antibody under thermal and freeze–thaw stresses revealed by hydrogen exchange. *Pharm Res* 2012;**29**(1):236–250.

105 Zhang Z, Zhang A, Xiao G. Improved protein hydrogen/deuterium exchange mass spectrometry platform with fully automated data processing. *Anal Chem* 2012;**84**(11):4942–4949.

106 Iacob RE, Bou-Assaf GM, Makowski L, Engen JR, Berkowitz SA, Houde D. Investigating monoclonal antibody aggregation using a combination of H/DX-MS and other biophysical measurements. *J Pharm Sci* 2013;**102**(12):4315–4329.

107 Edgeworth MJ, Phillips JJ, Lowe DC, Kippen AD, Higazi DR, Scrivens JH. Global and local conformation of human IgG antibody variants rationalizes loss of thermodynamic stability. *Angew Chem Int Ed* 2015;**54**:15156–15159.

108 Visser J, Feuerstein I, Stangler T, Schmiederer T, Fritsch C, Schiestl M. Physicochemical and functional comparability between the proposed

biosimilar rituximab GP2013 and originator rituximab. *BioDrugs* 2013;**27**(5):495–507.

109 Jensen PF, Larraillet V, Schlothauer T, Kettenberger H, Hilger M, Rand KD. Investigating the interaction between the neonatal Fc receptor and monoclonal antibody variants by hydrogen/deuterium exchange mass spectrometry. *Mol Cell Proteomics* 2015;**14**(1):148–161.

110 Wang X, Li Q, Davies M. Development of antibody arrays for monoclonal antibody higher order structure analysis. *Front Pharmacol* 2013;**4**:1–8.

111 Tito MA, Miller J, Walker N, Griffin KF, Williamson ED, Despeyroux-Hill D, et al. Probing molecular interactions in intact antibody: antigen complexes, an electrospray time-of-flight mass spectrometry approach. *Biophys J* 2001;**81**(6):3503–3509.

112 Banerjee S, Mazumdar S. Electrospray ionization mass spectrometry: A technique to access the information beyond the molecular weight of the analyte. *Int J Anal Chem* 2012;**2012**:1–40.

113 Wilm M, Mann M. Analytical properties of the nanoelectrospray ion source. *Anal Chem* 1996;**68**:1–8.

114 Rosati S, Yang Y, Barendregt A, Heck AJR. Detailed mass analysis of structural heterogeneity in monoclonal antibodies using native mass spectrometry. *Nat Protoc* 2014;**9**(4):967–976.

115 Pacholarz KJ, Porrini M, Garlish RA, Burnley RJ, Taylor RJ, Henry AJ, et al. Dynamics of intact immunoglobulin G explored by drift-tube ion-mobility mass spectrometry and molecular modeling. *Angew Chem Int Ed* 2014;**53**(30):7765–7769.

116 Debaene F, Wagner-Rousset E, Colas O, Ayoub D, Corvaïa N, Van Dorsselaer A, et al. Time resolved native ion-mobility mass spectrometry to monitor dynamics of igg4 fab arm exchange and "bispecific" monoclonal antibody formation. *Anal Chem* 2013;**85**:9785–9792.

117 Shelimov KB, Clemmer DE, Hudgins RR, Jarrold MF. Protein structure *in vacuo*: Gas-phase conformations of BPTI and cytochrome *c*. *J Am Chem Soc* 1997;**119**:2240–2248.

118 Clemmer DE, Hudgins RR, Jarrold MF. Naked protein conformations: Cytochrome *c* in the gas phase. *J Am Chem Soc* 1995;**117**:10141–10142.

119 Pringle SD, Giles K, Wildgoose JL, Williams JP, Slade SE, Thalassinos K, et al. An investigation of the mobility separation of some peptide and protein ions using a new hybrid quadrupole/travelling wave IMS/oa-ToF instrument. *Int J Mass Spectrom* 2007;**261**(1):1–12.

120 Giles K, Pringle SD, Worthington KR, Little D, Wildgoose JL, Bateman RH. Applications of a travelling wave-based radio-frequency-only stacked ring ion guide. *Rapid Commun Mass Spectrom* 2004;**18**(20):2401–2414.

121 Olivova P, Chen W, Chakraborty AB, Gebler JC. Determination of N-glycosylation sites and site heterogeneity in a monoclonal antibody by electrospray quadrupole ion-mobility time-of-flight mass spectrometry. *Rapid Commun Mass Spectrom* 2008;**22**:29–40.

122 Bagal D, Valliere-Douglass JF, Balland A, Schnier PD. Resolving disulfide structural isoforms of IgG2 monoclonal antibodies by ion mobility mass spectrometry. *Anal Chem* 2010;**82**(16):6751–6755.

123 Pacholarz K, Peters SJ, Garlish RA, Henry AJ, Taylor RJ, Humphreys DP, et al. Molecular insights to the thermal stability of mAbs with variable temperature Ion mobility mass spectrometry. *ChemBioChem* 2016;**17**:46–51.

124 Knapman TW, Berryman JT, Campuzano I, Harris SA, Ashcroft AE. Considerations in experimental and theoretical collision cross-section measurements of small molecules using travelling wave ion mobility spectrometry-mass spectrometry. *Int J Mass Spectrom* 2010;**298**(1–3):17–23.

125 Thalassinos K, Slade SE, Jennings KR, Scrivens JH, Giles K, Wildgoose J, et al. Ion mobility mass spectrometry of proteins in a modified commercial mass spectrometer. *Int J Mass Spectrom* 2004;**236**:55–63.

126 Kim YJ, Doyle ML. Structural mass spectrometry in protein therapeutics discovery. *Anal Chem* 2010;**82**(17):7083–7089.

127 Pritchard C, Connor GO, Ashcroft AE. The role of Ion mobility spectrometry – mass spectrometry in the analysis of protein reference standards. *Anal Chem* 2013;**85**:7205–7212.

128 Jones LM, Zhang H, Cui W, Kumar S, Sperry JB, Carroll JA, et al. Complementary MS methods assist conformational characterization of antibodies with altered S-S bonding networks. *J Am Soc Mass Spectrom* 2013;**24**(6):835–845.

129 Tian Y, Han L, Buckner AC, Ruotolo BT. Collision induced unfolding of intact antibodies: Rapid characterization of disulfide bonding patterns, glycosylation, and structures. *Anal Chem* 2015;**87**:11509–11515.

130 Dillon TM, Ricci MS, Vezina C, Flynn GC, Liu YD, Rehder DS, et al. Structural and functional characterization of disulfide isoforms of the human IgG2 subclass. *J Biol Chem* 2008;**283**(23):16206–16215.

131 Wypych J, Li M, Guo A, Zhang Z, Martinez T, Allen MJ, et al. Human IgG2 antibodies display disulfide-mediated structural isoforms. *J Biol Chem* 2008;**283**(23):16194–16205.

132 Martinez T, Guo A, Allen MJ, Han M, Pace D, Jones J, et al. Disulfide connectivity of human immunoglobulin G2 structural isoforms. *Biochemistry* 2008;**47**(28):7496–7508.

133 Liu YD, Wang T, Chou R, Chen L, Kannan G, Stevenson R, et al. IgG2 disulfide isoform conversion kinetics. *Mol Immunol* 2013;**54**(2):217–226.

134 Liu YD, Chen X, Enk JZ, Plant M, Dillon TM, Flynn GC. Human IgG2 antibody disulfide rearrangement *in vivo*. *J Biol Chem* 2008;**283**(43):29266–29272.

135 Liu YD, Chou RY-T, Dillon TM, Poppe L, Spahr C, Shi SDH, et al. Protected hinge in the immunoglobulin G2-A2 disulfide isoform. *Protein Sci* 2014;**23**(12):1753–1764.

136 Zhang A, Fang J, Chou RY-T, Bondarenko PV, Zhang Z. Conformational difference in human IgG2 disulfide isoforms revealed by hydrogen/ deuterium exchange mass spectrometry. *Biochemistry* 2015;**54**:1956–1962.

137 Peters SJ, Smales CM, Henry AJ, Stephens PE, West S, Humphreys DP. Engineering an improved IgG4 molecule with reduced disulfide bond heterogeneity and increased Fab domain thermal stability. *J Biol Chem* 2012;**287**(29):24525–24533.

138 Mao Y, Ratner MA, Jarrold MF. Molecular dynamics simulations of the charge-induced unfolding and refolding of unsolvated cytochrome *c*. *J Phys Chem B* 1999;**103**(45):10017–10021.

139 Dickinson ER, Jurneczko E, Parcholarz K, Clarke DJ, Reeves M, Ball KL, et al. Insights to the conformations of three structurally diverse proteins: Cytochrome *c*, p53 and MDM2, provided by variable temperature ion mobility mass spectrometry. *Anal Chem* 2015;**87**:3231–3238.

140 Pacholarz KJ, Barran PE. Distinguishing loss of structure from subunit dissociation for protein complexes with variable temperature Ion mobility mass spectrometry. *Anal Chem* 2015;**87**(12):6271–6279.

141 Berezovskaya Y, Porrini M, Barran PE. The effect of salt on the conformations of three model proteins is revealed by variable temperature ion mobility mass spectrometry. *Int J Mass Spectrom* 2013;**345–347**:8–18.

142 Du Y, May K, Xu W, Liu H. Detection and quantitation of afucosylated n-linked oligosaccharides in recombinant monoclonal antibodies using enzymatic digestion and LC-MS. *J Am Soc Mass Spectrom* 2012;**23**(May):1241–1249.

143 Wenig K, Chatwell L, von Pawel-Rammingen U, Björck L, Huber R, Sondermann P. Structure of the streptococcal endopeptidase IdeS, a cysteine proteinase with strict specificity for IgG. *Proc Natl Acad Sci U S A* 2004;**101**(50):17371–17376.

144 von Pawel-Rammingen U, Johansson BP, Björck L. IdeS, a novel streptococcal cysteine proteinase with unique specificity for immunoglobulin G. *EMBO J* 2002;**21**(7):1607–1615.

145 Firth D, Upton R, Bell L, Guy C, Caldwell P, Estdale S, Barran PE. Orthogonal assessment of biotherapeutic glycosylation: A case study correlating N-glycan core afucosylation with mechanism of (Submitted).

146 Ouyang J. Drug-to-antibody ratio (DAR) and drug load distribution by hydrophobic interaction chromatography and reversed phase high-performance liquid chromatography. *Methods Mol Biol* 2013;**1045**:275–283.

147 Jackson D, Atkinson J, Guevara CI, Zhang C, Kery V, Moon S-J, et al. *In vitro* and *in vivo* evaluation of cysteine and site specific conjugated herceptin antibody–drug conjugates. *PLoS One* 2014;**9**(1):e83865.

148 McDonagh CF, Turcott E, Westendorf L, Webster JB, Alley SC, Kim K, et al. Engineered antibody–drug conjugates with defined sites and stoichiometries of drug attachment. *Protein Eng Des Sel* 2006;**19**(7):299–307.

149 Beck A, Wurch T, Bailly C, Corvaia N. Strategies and challenges for the next generation of therapeutic antibodies. *Nat Rev Immunol* 2010;**10**(5):345–352.

150 Wagner-Rousset E, Janin-Bussat M-C, Colas O, Excoffier M, Ayoub D, Haeuw J-F, et al. Antibody–drug conjugate model fast characterization by LC-MS following IdeS proteolytic digestion. *MAbs* 2014;**6**(1):273–285.

151 Lazar AC, Wang L, Blättler WA, Amphlett G, Lambert JM, Zhang W. Analysis of the composition of immunoconjugates using size-exclusion chromatography coupled to mass spectrometry. *Rapid Commun Mass Spectrom* 2005;**19**(13):1806–1814.

152 Valliere-Douglass JF, Hengel SM, Pan LY. Approaches to interchain cysteine-linked ADC characterization by mass spectrometry. *Mol Pharm* 2015;**12**:1774–1783.

153 Valliere-Douglass JF, McFee WA, Salas-Solano O. Native intact mass determination of antibodies conjugated with monomethyl Auristatin e and F at interchain cysteine residues. *Anal Chem* 2012;**84**(6):2843–2849.

154 Hengel SM, Sanderson R, Valliere-Douglass J, Nicholas N, Leiske C, Alley SC. Measurement of *in vivo* drug load distribution of cysteine-linked antibody–drug conjugates using microscale liquid chromatography mass spectrometry. *Anal Chem* 2014;**86**(7):3420–3425.

155 Diebolder CA, Beurskens FJ, de Jong RN, Koning RI, Strumane K, Lindorfer MA, et al. Complement is activated by IgG hexamers assembled at the cell surface. *Science* 2014;**343**(March):1260–1263.

156 Pacholarz KJ, Barran PE. Use of charge reducing agent to enable intact mass analysis of cysteine-linked antibody-drug-conjugates by native mass spectrometry. *EuPA Open Proteomics* 2016;**11**:23–27.

157 Marcoux J, Champion T, Colas O, Wagner-Rousset E, Corvaïa N, Van Dorsselaer A, et al. Native mass spectrometry and ion mobility characterization of trastuzumab emtansine, a lysine-linked antibody drug conjugate. *Protein Sci* 2015;**24**(8):1210–1223.

158 Gautier V, Boumeester AJ, Lössl P, Heck AJR. Lysine conjugation properties in human IgGs studied by integrating high-resolution native mass spectrometry and bottom-up proteomics. *Proteomics* 2015;**15**(16):2756–2765.

159 Chen S-L, Wu S-L, Huang L-J, Huang J-B, Chen S-H. A global comparability approach for biosimilar monoclonal antibodies using LC–tandem MS based proteomics. *J Pharm Biomed Anal* 2013;**80**:126–135.

Index

Analysis of Protein Post-Translational Modifications by Mass Spectrometry,
First Edition. Edited by John R. Griffiths and Richard D. Unwin.
© 2017 John Wiley & Sons, Inc. Published 2017 by John Wiley & Sons, Inc.

WILEY SERIES ON MASS SPECTROMETRY

Series Editors

Dominic M. Desiderio

Departments of Neurology and Biochemistry University of Tennessee Health Science Center

Joseph A. Loo

Department of Chemistry and Biochemistry UCLA

Founding Editors

Nico M. M. Nibbering (1938–2014)
Dominic M. Desiderio

Nobuhiro Takahashi and Toshiaki Isobe · *Proteomic Biology Using LC-MS: Large Scale Analysis of Cellular Dynamics and Function*

Agnieszka Kraj and Jerzy Silberring (Editors) · *Proteomics: Introduction to Methods and Applications*

Ganesh Kumar Agrawal and Randeep Rakwal (Editors) · *Plant Proteomics: Technologies, Strategies, and Applications*

Rolf Ekman, Jerzy Silberring, Ann M. Westman-Brinkmalm, and Agnieszka Kraj (Editors) · *Mass Spectrometry: Instrumentation, Interpretation, and Applications*

Christoph A. Schalley and Andreas Springer · *Mass Spectrometry and Gas-Phase Chemistry of Non-Covalent Complexes*

Riccardo Flamini and Pietro Traldi · *Mass Spectrometry in Grape and Wine Chemistry*

Mario Thevis · *Mass Spectrometry in Sports Drug Testing: Characterization of Prohibited Substances and Doping Control Analytical Assays*

Sara Castiglioni, Ettore Zuccato, and Roberto Fanelli · *Illicit Drugs in the Environment: Occurrence, Analysis, and Fate Using Mass Spectrometry*

Ángel Garciá and Yotis A. Senis (Editors) · *Platelet Proteomics: Principles, Analysis, and Applications*

Luigi Mondello · *Comprehensive Chromatography in Combination with Mass Spectrometry*

Jian Wang, James MacNeil, and Jack F. Kay (Editors) · *Chemical Analysis of Antibiotic Residues in Food*

Walter A. Korfmacher (Editor) · *Mass Spectrometry for Drug Discovery and Drug Development*

Toshimitsu Niwa (Editor) · *Uremic Toxins*

Igor A. Kaltashov, Stephen J. Eyles · *Mass Spectrometry in Structural Biology and Biophysics: Architecture, Dynamics, and Interaction of Biomolecules, 2nd Edition*

Alejandro Cifuentes (Editor) · *Foodomics: Advanced Mass Spectrometry in Modern Food Science and Nutrition*

Christine M. Mahoney (Editor) · *Cluster Secondary Ion Mass Spectrometry: Principles and Applications*

Despina Tsipi, Helen Botitsi, and Anastasios Economou (Editors) · *Mass Spectrometry for the Analysis of Pesticide Residues and their Metabolites*

Xianlin Han · *Lipidomics: Comprehensive Mass Spectrometry of Lipids*

Jack F. Kay, James D. MacNeil, and Jian Wang (Editors) · *Chemical Analysis of Non-antimicrobial Veterinary Drug Residues in Food*

Wilfried M. A. Niessen and Ricardo A. Correa C. · *Interpretation of MS-MS Mass Spectra of Drugs and Pesticides*

John R. Griffiths and Richard D. Unwin (Editors) · *Analysis of Protein Post-Translational Modifications by Mass Spectrometry*

Henk Schierbeek · *Mass Spectrometry and Stable Isotopes in Nutritional and Pediatric Research*